Understanding Spin Dynamics

Understanding Spin Dynamics

Danuta Kruk

Published by

Pan Stanford Publishing Pte. Ltd.
Penthouse Level, Suntec Tower 3
8 Temasek Boulevard
Singapore 038988

Email: editorial@panstanford.com
Web: www.panstanford.com

British Library Cataloguing-in-Publication Data
A catalogue record for this book is available from the British Library.

Understanding Spin Dynamics

ISBN 978-981-4463-49-2 (Hardcover)
ISBN 978-981-4463-50-8 (eBook)

Printed in the USA

Contents

Preface

Nuclear magnetic resonance (NMR) and electron spin resonance (ESR) experimental techniques are broadly used and highly appreciated in 'molecular science' as they are a powerful tool for studying dynamical processes in condensed matter. Nevertheless (one could say unfortunately) in most cases this knowledge is not easily accessible. The difficulties lie in the fact that spin resonance is a quantum phenomenon and to understand the obtained results one has to firstly understand the quantum-mechanical principles of the underlying processes. Actually this is well described by the term 'spin dynamics'; spin resonances are about dynamics: spin dynamics (i.e., quantum mechanics) and molecular dynamics (i.e., classical motion). This combination is not a disadvantage but a challenge and 'spin dynamics' itself is a part of fascinating science.

Quite often I get the question: *Which equation should I use to interpret my data*? What should I say? Maybe that there are no closed-form recipes except of a few simple cases and that one should firstly carefully consider the quantum-mechanical properties of the molecular system ...? This answer is correct, but not much helpful. Thus, I do not offer it. Instead of that I decided to write this book. I very much hope that it offers understandable and useful answers to a variety of questions appearing in connection to spin dynamics and spin resonance phenomena. I also believe that it will help to understand the principles, which are illustrated in this book by various examples. In consequence, the readers will be able to develop their own approaches and modify the existing descriptions depending on the system upon consideration, because as per Mark Twain, "Get your facts first, then you can distort them as you please."

My great debt of thanks goes to my colleagues (the list would be very long) for their encouragement and help. I am very grateful to my husband, Robert, and my children, Sabina, Przemek, and Karolina, who stood by me all this time. I also thank Ms. Shivani Sharma, Pan Stanford Publishing, for her assistance.

This work was partially supported by funds for science in years 2009–2013 as research project no. NN202105936 (Polish Ministry of Science and Education).

Danuta Kruk
Summer 2015

Chapter 1

Classical Description of Spin Resonance

This introductory chapter is devoted to a classical description of spin resonance phenomena. As far as quantum mechanics is concerned, one only needs to accept the fact that protons and electrons are characterized by a quantity referred to as spin quantum number (spin). The phenomenological Bloch equations describing precession and relaxation of magnetisation are introduced and discussed.

1.1 Larmor Precession and Bloch Equation

Nuclei and electrons possess magnetic moments (μ_I and μ_S, respectively) which are determined by their spins (I and S, respectively) [1, 2, 4, 5]:

$$\mu_I = g_I \frac{e\hbar}{2m_p} \sqrt{I\,(I+1)} \tag{1.1a}$$

$$\mu_S = g_e \frac{e\hbar}{2m_e} \sqrt{S\,(S+1)} \tag{1.1b}$$

where e is the elementary charge of the proton, $\hbar$ denotes Planck constant, m_p and m_e are the proton and electron masses, $g_e = 2.0023$, and g_I (Lande's factor) depends on the considered

Understanding Spin Dynamics
Danuta Kruk

ISBN 978-981-4463-49-2 (Hardcover), 978-981-4463-50-8 (eBook)
www.panstanford.com

nucleus. The quantities $\frac{e\hbar}{2m_\mathrm{p}}$ and $\frac{e\hbar}{2m_\mathrm{e}}$ are referred to as the nuclear magneton, μ_N, and electron Bohr magneton, μ_B, respectively. The ratio of the magnetic moments for electron and proton, $\mu_\mathrm{S}/\mu_\mathrm{I} = 658.2$, results from the ratio between their masses. It is useful to introduce nuclear and electron gyromagnetic factors, γ_I and γ_S, respectively:

$$\gamma_\mathrm{I} = g_\mathrm{I}\frac{e}{2m_\mathrm{p}} \tag{1.2a}$$

$$\gamma_\mathrm{S} = g_\mathrm{e}\frac{e}{2m_\mathrm{e}} \tag{1.2b}$$

Proton gyromagnetic factor, $\gamma_\mathrm{I} = \gamma_\mathrm{H} = 2.68 \times 10^8\ \mathrm{T^{-1}s^{-1}}$ is the largest nuclear gyromagnetic factor, while $\gamma_\mathrm{S} = 658.2\gamma_\mathrm{H}$.

When a macroscopic ensemble of nuclei (electrons) is placed in an external magnetic field $\vec{B}_0$ one can detect a macroscopic magnetization, $\vec{M}_\mathrm{I}$ for nuclei and $\vec{M}_\mathrm{S}$ for electrons, oriented along the direction of $\vec{B}_0$. The magnetization values per unit volume are given as [1, 5]:

$$M_\mathrm{I} = \frac{N\gamma_\mathrm{I}^2\hbar^2 B_0}{3k_\mathrm{B}T} I\,(I+1) \tag{1.3a}$$

$$M_\mathrm{S} = \frac{N\gamma_\mathrm{S}^2\hbar^2 B_0}{3k_\mathrm{B}T} S\,(S+1) \tag{1.3b}$$

where N is the number of nuclei (electrons) per unit volume; k_B, the Boltzmann constant, and T the temperature.

In the magnetic field, $\vec{B}_0$, the macroscopic magnetization $\vec{M}$ undergoes precession (referred to as Larmor precession) described by the Bloch equation [1–5]:

$$\frac{d\vec{M}\,(t)}{dt} = \gamma\vec{M}\,(t) \times \vec{B}_0 \tag{1.4}$$

where γ denotes nuclear or electron gyromagnetic factor, $\gamma = \gamma_\mathrm{I}$ or $\gamma = \gamma_\mathrm{S}$, respectively. This equation can be rewritten for individual components of the magnetization $\vec{M} = \left[M_x, M_y, M_z\right]$:

$$\frac{dM_x\,(t)}{dt} = \gamma\left[\vec{M}\,(t) \times \vec{B}_0\right]_x \tag{1.5a}$$

$$\frac{dM_y\,(t)}{dt} = \gamma\left[\vec{M}\,(t) \times \vec{B}_0\right]_y \tag{1.5b}$$

$$\frac{dM_z(t)}{dt} = \gamma\left[\vec{M}(t) \times \vec{B}_0\right]_z \quad (1.5c)$$

Applying definition of vector product, one can write:

$$\frac{dM_x(t)}{dt} = \gamma\left[M_y(t)\,B_z - M_z(t)\,B_y\right] \quad (1.6a)$$

$$\frac{dM_y(t)}{dt} = \gamma\left[M_z(t)\,B_x - M_x(t)\,B_z\right] \quad (1.6b)$$

$$\frac{dM_z(t)}{dt} = \gamma\left[M_x(t)\,B_y - M_y(t)\,B_x\right] \quad (1.6c)$$

This set of equations can be considerably simplified when $B_x = B_y = 0$. In such a case one obtains:

$$\frac{dM_\perp(t)}{dt} = -i\gamma B_0 M_\perp(t) \quad (1.7a)$$

$$\frac{dM_z(t)}{dt} = 0 \quad (1.7b)$$

where $M_\perp = M_x + iM_y$. The solution of Eq. 1.7 yields:

$$M_x(t) = \sqrt{M_x^2(0) + M_y^2(0)}\cos(\gamma B_0 t) \quad (1.8a)$$

$$M_y(t) = -\sqrt{M_x^2(0) + M_y^2(0)}\sin(\gamma B_0 t) \quad (1.8b)$$

$$M_z(t) = M_z(0) \quad (1.8c)$$

Equation 1.8 means that the transversal magnetization, $M_\perp$, rotates around the z-axis (the direction of the external magnetic field, $\vec{B}_0$) with angular frequency, $\omega_0 = \gamma B_0$ (referred to as Larmor frequency), while the longitudinal magnetization, M_z remains unchanged, the initial magnetization is denoted as $M_z(0)$.

Discussing the time evolution of the magnetization one has to take into account the relaxation processes. Relaxation is a reinstatement of the nuclear or electron magnetization to its equilibrium configuration after it has been perturbed. The nature of the perturbation and the relaxation mechanisms are discussed later in Chapter 2. The time needed for the longitudinal component of the magnetization (parallel to $\vec{B}_0$) to recover to the equilibrium is referred to as a longitudinal relaxation time T_1, called a spin-lattice relaxation time ($R_1 = T_1^{-1}$ is referred to as spin-lattice relaxation rate), while the recovery of the transverse magnetization (perpendicular to magnetic field $\vec{B}_0$) is characterized by a spin-spin relaxation time, T_2 ($R_2 = T_2^{-1}$ is referred to as spin-spin relaxation time). One of the most common relaxation mechanisms is associated with dipole–dipole interactions between elementary magnetic moments in the sample, which fluctuate in time due to molecular dynamics (rotation, translation, vibrations, etc.). The spin–spin relaxation time cannot be longer than the spin-lattice relaxation time, that is, $T_2 \leq T_1$. To include the relaxation effects, Eq. 1.6 has to be generalized to the form:

$$\frac{dM_x(t)}{dt} = \gamma\left[M_y(t)\,B_z - M_z(t)\,B_y\right] - \frac{M_x(t)}{T_2} \tag{1.9a}$$

$$\frac{dM_y(t)}{dt} = \gamma\left[M_z(t)\,B_x - M_x(t)\,B_z\right] - \frac{M_y(t)}{T_2} \tag{1.9b}$$

$$\frac{dM_z(t)}{dt} = \gamma\left[M_x(t)\,B_y - M_y(t)\,B_x\right] - \frac{M_z(t) - M_0}{T_1} \tag{1.9c}$$

Equation 1.9 implies that the evolution of the magnetization components can be described by single relaxation times, that is, the relaxation process is single exponential. This holds only for simple systems as it will be shown in the forthcoming chapters. Equation 1.9 can be easily solved under the assumption that $B_x = B_y = 0$. The solution yields:

$$M_x(t) = \left[M_x(0)\cos(\omega_0 t) - M_y(0)\sin(\omega_0 t)\right]\exp(-t/T_2) \tag{1.10a}$$

$$M_y(t) = \left[-M_x(0)\sin(\omega_0 t) + M_y(0)\cos(\omega_0 t)\right]\exp(-t/T_2) \tag{1.10b}$$

$$M_z(t) = M_{eq} + \left(M_0 - M_{eq}\right)\exp(-t/T_1) \tag{1.10c}$$

where $M_x(0)$ and $M_y(0)$ denote the initial values of the M_x and M_y components of the magnetization, while M_{eq} denotes the magnetization value at equilibrium (determined by the Boltzmann distribution).

As anticipated, relaxation processes are caused by magnetic fields fluctuating in time. This situation can be mimicked by introducing a time dependent magnetic field, $\vec{B}_1$, oscillating in the

xy plane (i.e., perpendicular to $\vec{B}_0$) with angular frequency ω. Then the set of Eq. 1.6 takes the form:

$$\frac{dM_x(t)}{dt} = \gamma\left[M_y(t)\,B_0 + M_z(t)\,B_1\sin(\omega_0 t)\right] - \frac{M_x(t)}{T_2} \quad (1.11a)$$

$$\frac{dM_y(t)}{dt} = \gamma\left[M_z(t)\,B_1\cos(\omega_0 t) - M_x(t)\,B_0\right] - \frac{M_y(t)}{T_2} \quad (1.11b)$$

$$\frac{dM_z(t)}{dt} = -\gamma\left[M_x(t)\,B_1\sin(\omega_0 t) + M_y(t)\,B_1\cos(\omega_0 t)\right] - \frac{M_z(t) - M_0}{T_1} \quad (1.11c)$$

In the steady state, $\frac{dM_x(t)}{dt} = \frac{dM_y(t)}{dt} = \frac{dM_z(t)}{dt} = 0$, the solution of Eq. 1.11 in a frame rotating with frequency ω is as follows [1, 4, 5]:

$$M_x = \frac{\omega_1 T_2^2(\omega_0 - \omega)}{1 + \omega_1^2 T_1 T_2 + [T_2(\omega_0 - \omega)]^2} M_0 \quad (1.12a)$$

$$M_y = \frac{\omega_1 T_2}{1 + \omega_1^2 T_1 T_2 + [T_2(\omega_0 - \omega)]^2} M_0 \quad (1.12b)$$

$$M_z = \frac{1 + T_2^2(\omega_0 - \omega)^2}{1 + \omega_1^2 T_1 T_2 + [T_2(\omega_0 - \omega)]^2} M_0 \quad (1.12c)$$

where $\omega_1 = \gamma B_1$. The presented classical description applies only to the macroscopic magnetization. The dynamics of individual magnetic moments (spins) has to be described in terms of quantum mechanics.

References

1. Abragam, A. (1961). *The Principles of Nuclear Magnetism* (New York: Oxford University Press).
2. Bargmann, V., Michel, L. and Telegdi, V. L. (1959). Precession of the polarization of particles moving in a homogeneous electromagnetic field. *Phys. Rev. Lett.*, **2**, pp. 435–436.
3. Bloch, F. (1946). Nuclear induction. *Phys. Rev.*, **70**, pp. 460–473.
4. Levitt, M. H. (2001). *Spin Dynamics* (Wiley: Chichester).
5. Slichter, C. P. (1990). *Principles of Magnetic Resonance* (Berlin: Springer-Verlag).

Chapter 2

Introduction to Spin Relaxation

This chapter is devoted to the mechanisms of spin relaxation. It is explained how the combination of quantum-mechanical properties with classical Brownian dynamics determines the relaxation behavior of spin systems. Relaxation processes are described as a result of transitions between spin states (energy levels) induced by local magnetic fields stochastically fluctuating in time due to rotational and translational dynamics of molecules. The derivations are first conducted on the level of "kinetic equations" describing changes in populations of spin states. Then progressively, the quantum mechanical formalism of spin relaxation phenomena is introduced.

2.1 The Nature of Relaxation Processes

Let us consider an ensemble of molecules containing protons (^1H), that is, nuclei of spin quantum number $I = 1/2$. In an external magnetic field, $\vec{B}_0$, there are two energy levels available for spin 1/2 nuclei (Fig. 2.1). They correspond to the magnetic quantum numbers $m_I = -1/2$ and $m_I = 1/2$, respectively. The energy level splitting occurs due to Zeeman interaction of the spin with the magnetic field, $\vec{B}_0$ and the interaction is described by the Hamiltonian [1, 17, 27, 28, 37]:

Understanding Spin Dynamics
Danuta Kruk

ISBN 978-981-4463-49-2 (Hardcover), 978-981-4463-50-8 (eBook)
www.panstanford.com

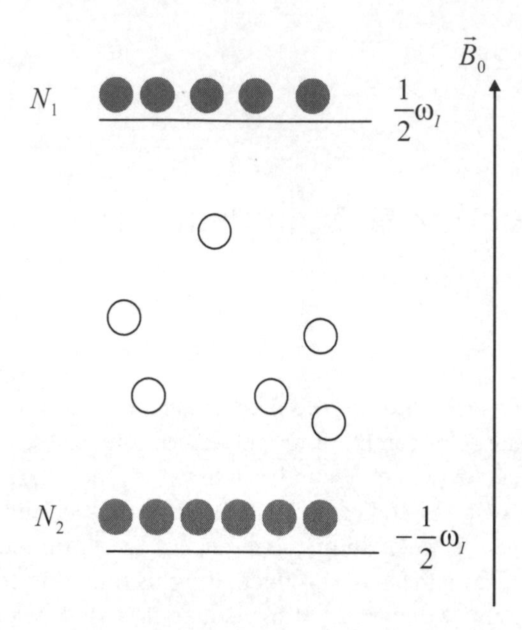

Figure 2.1 Schematic picture of the energy level structure of spin $I = 1/2$ in an external magnetic field $\vec{B}_0$. The energy levels are populated according to Boltzmann distribution. The spins can change the state (energy level) due to relaxation processes.

$$H_Z(I) = \gamma_I B_0 I_z = \omega_I I_z \tag{2.1}$$

where γ_I denotes gyromagnetic factor of the considered nucleus (^{1}H in this case, i.e., $\gamma_I = \gamma_H$) while $\omega_I = \gamma_I B_0$ is the (angular) frequency of spin precession in the external magnetic field referred to as Larmor frequency. The angular frequency, ω_I (in rad/s), is related to the frequency ν_I (in Hz), as $\omega_I = 2\pi\nu_I$. From now on we shall not explicitly use the term "angular" when referring to ω_I. In the quantum-mechanical picture of the magnetic resonance phenomenon the frequency, ω_I, determines the energy level splitting, $\Delta E = E_{1/2} - E_{-1/2} = \omega_I$, which is proportional to the applied magnetic field, $\vec{B}_0$. The energy levels are populated according to the Boltzmann distribution [1, 4, 34, 36]:

$$\frac{N_1}{N_2} = \exp\left(-\frac{\Delta E}{k_B T}\right) = \exp\left(-\frac{\omega_I}{k_B T}\right) \tag{2.2}$$

where N_1 and N_2 denote populations of the upper and lower energy levels, while k_B is the Boltzmann constant, and T denotes temperature in K. At room temperature, for $B_0 = 7.05$ T (that corresponds to Larmor frequency of 300 MHz for ^{1}H) the population ratio is $N_1/N_2 = 0.99995$. This result might be somewhat surprising; one understands that to get a macroscopic, measurable magnetization many spins (nuclei) have to be involved.

When the equilibrium distribution is disturbed, the system is able to return to the equilibrium, that is, to relax. This is, in fact, the idea of relaxation experiments. One polarizes nuclei (spins) in a strong magnetic field, and then the field is reduced to a lower value at which the populations of the energy levels readjust, reaching a new equilibrium distribution. This "return to equilibrium" takes some time which is referred to as spin–lattice relaxation time, T_1, as explained in Chapter 1. The energy which is needed for the spin transitions is taken from a "lattice". By lattice one understands here all degrees of freedom available to the system which can be used as a source of energy. This somewhat abstractive way of thinking becomes clear when we come to examples. Let us consider water molecules. Magnetic moments associated with the proton spins are coupled by magnetic dipole–dipole interactions. Before we come to the Hamiltonian form of the dipole–dipole coupling, at this stage it is enough to say that due to molecular reorientation the interaction fluctuates in time with respect to the direction of the external magnetic field $\vec{B}_0$, referred to as the laboratory axis (L). As a result of the fluctuations, the dipole–dipole coupling serves as a source of energy for the spin transitions.

To establish a link between this qualitative picture and a quantum-mechanical description of the relaxation processes, let us treat the water molecule as a quantum system consisting of two equivalent nuclei, I_1 and I_2 of spin quantum numbers $I = 1/2$ (two protons). The states of the spin system are described by functions labeled as $\{|i\rangle = |m_1, m_2\rangle\}$, where m_1 and m_2 are magnetic quantum numbers for the spins I_1 and I_2, respectively. Thus, the system is characterized by four states, as indicated in Fig. 2.2 [27]. As the spins are identical (equivalent) the energy values associated with the states $|2\rangle = |1/2, -1/2\rangle$ and $|3\rangle = |-1/2, 1/2\rangle$ are equal (the energy levels are degenerated), that is, $E_2 = E_3$.

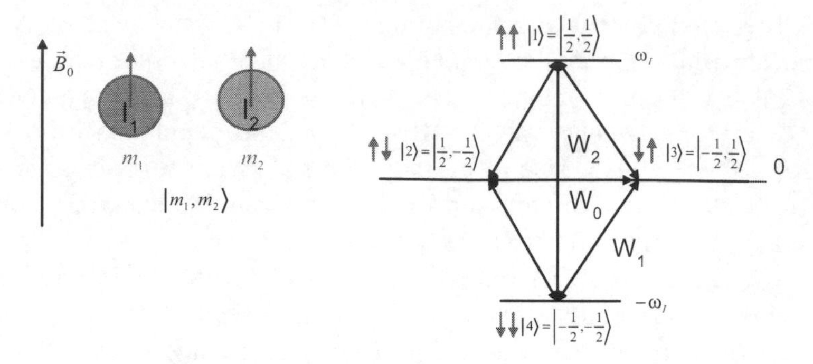

Figure 2.2 A system of two identical (equivalent) spins, $I = 1/2$, energy level structure and transition probabilities; W_0, W_1, and W_2 denote probabilities of zero-, single-, and double-quantum transitions, respectively (explanation in the text).

It is natural to ask how the populations of the spin states evolve in time. At this stage we do not need to think about the mechanism of the spin transitions; let us focus on the kinetic equations, which can be written as follows:

$$\frac{dN_1(t)}{dt} = -(W_{12} + W_{13} + W_{14})\,N_1(t) + W_{21}N_2(t) + W_{31}N_3(t) + W_{41}N_4(t) \tag{2.3a}$$

$$\frac{dN_4(t)}{dt} = W_{14}N_1(t) + W_{24}N_2(t) + W_{34}N_3(t) - (W_{41} + W_{42} + W_{43})\,N_4(t) \tag{2.3b}$$

where N_i denotes the population of the state $|i\rangle$ with the energy E_i.

Figure 2.3 is a graphical representation of the spin transitions which contribute to changes in the populations of the states $|1\rangle = |1/2, 1/2\rangle$ and $|4\rangle = |-1/2, -1/2\rangle$. The probabilities, W_{ij}, correspond to the transitions between the states $|i\rangle$ and $|j\rangle$. Since the number of spins (nuclei) remains unchanged, one gets $W_{ii} = -\sum_{j\neq i} W_{ij}$, we also assume that $W_{ij} = W_{ji}$. The magnetization, M_{Iz} (longitudinal component of I spin magnetization) is proportional to the difference between the populations of the states $|1\rangle = |1/2, 1/2\rangle$ and $|4\rangle = |-1/2, -1/2\rangle$; therefore, only the two states are considered in Eq. 2.3. Before proceeding further one should make an observation. Let us define the total magnetic quantum number of the system as $m_I = m_1 + m_2$. For the transition $|2\rangle \rightarrow |3\rangle$ one gets

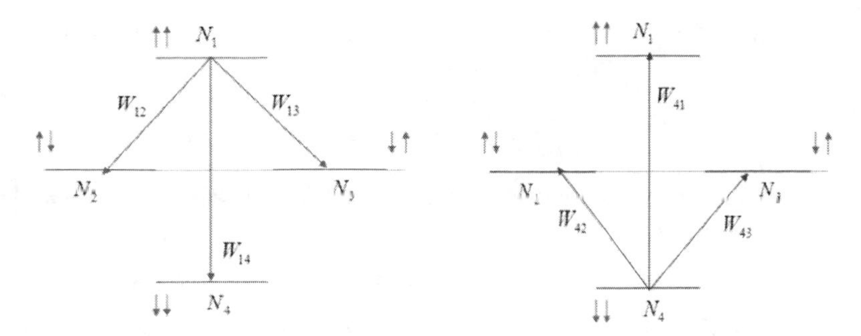

Figure 2.3 Spin transitions leading to magnetization changes in a system of two identical spins 1/2 (Eq. 2.3).

$\Delta m_{\text{I}} = 0$, and therefore, it is called a zero-quantum transition; the corresponding transition probability is denoted as W_0. In analogy, one gets for the transition $|1\rangle \rightarrow |4\rangle$, $|\Delta m_{\text{I}}| = 2$ (double-quantum transition), and the corresponding probability is denoted as W_2. Other transitions are single-quantum with probability W_1 (see Fig. 2.2).

After simple algebra one gets for the population difference:

$$\frac{d}{dt}(N_4 - N_1)(t) = -(2W_1 + 2W_2)(N_4 - N_1)(t) \tag{2.4}$$

This equation has a single exponential solution:

$$\begin{aligned} \Delta N(t) &= (N_4 - N_1)(t) = \Delta N(0) \exp[-2(W_1 + W_2)t] \\ &= \Delta N(0) \exp(-R_1 t) \end{aligned} \tag{2.5}$$

where R_1 denotes the spin–lattice relaxation rate, $R_1 = 1/T_1$.

The fact that the relaxation is single exponential is rather an exception than a rule. It happened because in Eq. 2.3 the terms related to the populations N_2 and N_3 cancel when the difference $(N_4 - N_1)$ is taken. We shall discuss about exponentiality of relaxation later.

The next task is to calculate the transition probabilities W_1 and W_2. One should notice here that the zero-quantum transition (associated with W_0) does not appear in Eq. (2.4). This is not surprising because this transition does not affect the magnetization.

As already stated, the reason for the spin transitions (leading to relaxation) is the dipole–dipole interaction between the spins I_1 and

I_2. The classical expression for the dipole–dipole interaction energy, E_{DD}, is [1, 17, 27, 36]:

$$E_{\mathrm{DD}} = \frac{\mu_0}{4\pi}\frac{1}{r^3}\left(\vec{\mu}_1 \cdot \vec{\mu}_2 - 3\vec{\mu}_1 \cdot \frac{\vec{r}\cdot\vec{r}}{r^3} \cdot \vec{\mu}_2\right) \tag{2.6}$$

where $\vec{\mu}_1$ and $\vec{\mu}_2$ are magnetic dipoles associated with the spins I_1 and I_2, respectively, according to the relationship:

$$\vec{\mu}_{\mathrm{I}} = \gamma_{\mathrm{I}}\hbar I \tag{2.7}$$

while $\vec{r}$ is the inter-spin vector ($\vec{\mu}_{\mathrm{I}}$ denotes magnetic moment associated with spin I). For simplicity we use the same symbol I for the operator and for the spin quantum number, but the meaning is clear from the context. This classical expression has its quantum-mechanical counterpart; it gives the Hamiltonian form of the dipole-dipole interaction [1, 17, 24, 27, 28, 36–39]:

$$H_{\mathrm{DD}}(I_1, I_2) = \sqrt{6}\frac{\mu_0}{4\pi}\frac{\gamma_{\mathrm{I}}^2\hbar}{r^3}\left\{\frac{1}{\sqrt{6}}\left[2I_{1z}I_{2z} - \frac{1}{2}(I_{1+}I_{2-} + I_{1-}I_{2+})\right]\right\} \tag{2.8}$$

where I_+, I_- are raising and lowering operators defined as $I_\pm = I_x \pm i I_y$ [13, 16, 43]. This expression can be written in terms of tensor operators:

$$H_{\mathrm{DD}}(I_1, I_2) = a_{\mathrm{DD}}\, T_0^2(I_1, I_2) \tag{2.9}$$

Comparing Eqs. 2.8 and 2.9 one gets for the dipole–dipole coupling constant:

$$a_{\mathrm{DD}} = \sqrt{6}\frac{\mu_0}{4\pi}\frac{\gamma_{\mathrm{I}}^2\hbar^2}{r^3} \tag{2.10}$$

while the tensor operator T_0^2 is defined as [1, 17, 24, 27, 28, 36–39]:

$$T_0^2(I_1, I_2) = \frac{1}{\sqrt{6}}\left[2I_{1z}I_{2z} - \frac{1}{2}(I_{1+}I_{2-} + I_{1-}I_{2+})\right] \tag{2.11}$$

Equation 2.9 is a special case of a much more general expression [1, 17, 24, 27, 28, 36–39]:

$$H_{\mathrm{DD}}(I_1, I_2) = a_{\mathrm{DD}} \sum_{m=-2}^{2} (-1)^m F_{-m}^2 T_m^2(I_1, I_2) \tag{2.12}$$

where $F_0^2 = 1$, and $F_{\pm 1}^2 = F_{\pm 2}^2 = 0$. This corresponds to the molecular frame representation of the dipole–dipole interaction.

In other words, the Hamiltonian of Eq. 2.12 describes the dipolar coupling in a molecule-fixed frame which axis coincides with the $\vec{r}$ direction (the dipole–dipole axis). In the molecular frame representation the dipolar Hamiltonian contains only the term with $m = 0$. Looking at the explicit form of $T_0^2\,(I_1, I_2)$ (Eq. 2.11) one can easily conclude that interactions described by such a Hamiltonian cannot change the magnetic quantum number of the system, and this is, in fact reflected by the index "0" in the operator $T_0^2\,(I_1, I_2)$, that is, $\Delta m_I = \Delta\,(m_1 + m_2) = m = 0$. This implies that the operator $T_0^2\,(I_1, I_2)$ is associated with the probability W_0, which, as we already know, does not contribute to the relaxation. On the other hand, dipole–dipole interactions are the most common spin relaxation mechanisms. To solve this mystery how they induce relaxation, one should realize that we are not interested in the form of the dipole–dipole interaction in a molecule-fixed frame. The relevant question is how the dipole–dipole coupling looks like in the frame determined by the direction of the external magnetic field, that is, the laboratory (L) frame. This statement can be generalized. There is always an interaction (referred to as "main") which determines the energy level structure of the spin system (it can be given as a superposition of many interactions as we shall see in the forthcoming chapters). Then the interaction, which is supposed to cause relaxation (transitions between the energy levels), has to be considered in the principal axis system of the main interaction.

According to the angular momentum theory the transformation of the Hamiltonian of Eq. 2.8 to the laboratory frame yields [13, 16, 27, 28, 43]:

$$F_{-m}^{2(\mathrm{L})}(t) = \sum_{k=-2}^{2} F_k^{2(\mathrm{DD})} D_{k,-m}^2\,[\Omega\,(t)] = D_{0,-m}^2\,[0, \beta\,(t), \gamma\,(t)] \quad (2.13)$$

The indices (DD) and (L) have been introduced to explicitly point out to which frame we refer; $\Omega\,(t)$ is a set of Euler angles $[\Omega = (\alpha, \beta, \gamma)]$ describing the orientation of $\vec{r}$ (i.e., the dipole–dipole axis, DD) with respect to the laboratory axis, (L), $D_{m,k}^{l}$ denotes Wigner rotation matrices [13, 16, 43]. The last equality in Eq. 2.13 stems from the fact that $F_0^{2(\mathrm{DD})} = 1$ and $F_{\pm 1}^{2(\mathrm{DD})} = F_{\pm 2}^{2(\mathrm{DD})} = 0$, as already explained. Due to the transformation the Hamiltonian of

Eq. 2.9 takes the general form of Eq. 2.12 which now can be explicitly rewritten with the index (L) to refer to the laboratory frame:

$$H_{\mathrm{DD}}^{(\mathrm{L})}(I_1, I_2)(t) = a_{\mathrm{DD}} \sum_{m=-2}^{2} (-1)^m F_{-m}^{2(\mathrm{L})}(t)\, T_m^2(I_1, I_2) \qquad (2.14)$$

where all $F_{-m}^{2(\mathrm{L})}$ terms differ from zero and are given by Eq. 2.13. The conclusion from Eq. 2.14 is that due to the transformation the Hamiltonian becomes time dependent; there is no time-dependence in Eq. 2.9, it is only introduced by Eq. 2.13. This reflects the fact that the dipole–dipole coupling is fixed in the molecule. As due to rotational dynamics the molecular frame stochastically changes its orientation with respect to the direction of the external magnetic field (the laboratory frame), this makes the dipole–dipole coupling time dependent (from the perspective of the laboratory frame). One should note here that rotational motion is not the only source of modulations of spin interactions; the fluctuations can be caused by translational diffusion, internal dynamics of the molecule, or vibrational motion. We shall discuss this later.

As the transformation of Hamiltonian is crucial for a proper description of the quantum-mechanical behavior of spin systems let us consider a molecule containing three spins, I_1, I_2, and I_3 as shown in Fig. 2.4. Now there are three dipole–dipole couplings I_1–I_2, I_2–I_3, and I_1–I_3 described by the Hamiltonians $H_{\mathrm{DD}}(I_1, I_2)$, $H_{\mathrm{DD}}(I_2, I_3)$, and $H_{\mathrm{DD}}(I_1, I_3)$, respectively. All interactions fluctuate in time with respect to the laboratory axis due the same dynamics, that is, the overall molecular tumbling. Nevertheless, while the I_1–I_2 dipole–dipole axis is oriented under the angle $\Omega(t)$ with respect to the (L) frame (so one can directly applied the transformation of Eq. 2.13 to the Hamiltonian $H_{\mathrm{DD}}(I_1, I_2)$), the I_2–I_3 and I_1–I_3 axes are oriented under different angles. To express the Hamiltonian $H_{\mathrm{DD}}(I_1, I_3)$ in the laboratory frame (L) one has to apply two-step transformation. First, the Hamiltonian has to be transferred to the frame determined by the I_1–I_2 axis (which has been chosen as a reference and denoted as the molecular frame (M)) and then, in the second step, from the I_1–I_2 frame to the (L) frame. For this purpose the general transformation rule between two frames, P_1 and P_2 [13, 16, 27, 28, 43]:

$$F_{-m}^{l(\mathrm{P}_2)} = \sum_{k=-l}^{l} F_k^{l(\mathrm{P}_1)} D_{k,-m}^{l}(\Omega_{12}) \qquad (2.15)$$

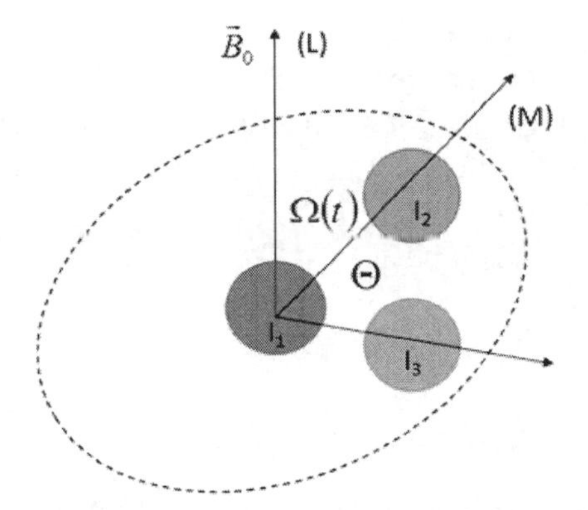

Figure 2.4 Transformation of dipole–dipole Hamiltonians between different reference frames; (M) denotes a molecule-fixed frame.

has to be applied; Ω_{12} is the (time-independent) angle between the frames P_1 and P_2. Thus, as a result of the first transformation applied to $H_{\mathrm{DD}}\,(I_1, I_3)$ one obtains:

$$\begin{aligned} H_{\mathrm{DD}}^{(\mathrm{M})}\,(I_1, I_3)\,(\Theta) &= a_{\mathrm{DD}} \sum_{n=-2}^{2} (-1)^{\mathrm{n}}\, F_{-n}^{2(\mathrm{M})}\,(\Theta)\, T_n^2\,(I_1, I_3) \\ &= \sum_{n=-2}^{2} (-1)^n\, D_{0,-n}^2\,(\Theta)\, T_n^2\,(I_1, I_3) \end{aligned} \tag{2.16}$$

where Θ is the angle between the I_1–I_3 and I_1–I_2 (M) axes. One should note that the Hamiltonian changes its form ($F_{\pm 1}^{2(\mathrm{M})} \neq 0$ and $F_{\pm 2}^{2(\mathrm{M})} \neq 0$), but as the ($M$) axis is also fixed in the molecule, it remains time independent. Due to the second transformation, the Hamiltonian yields the form:

$$\begin{aligned} H_{\mathrm{DD}}^{(\mathrm{L})}\,(I_1, I_3)\,(t) &= a_{\mathrm{DD}} \sum_{m=-2}^{2} (-1)^m\, F_{-m}^{2(\mathrm{L})}\,(t)\, T_m^2\,(I_1, I_3) \\ &= a_{\mathrm{DD}} \sum_{m=-2}^{2} \sum_{n=-2}^{2} (-1)^{m+n}\, D_{0,-n}^2\,(\Omega_{23}) \\ &\quad D_{-n,-m}^2\,(t)\, T_m^2\,(I_1, I_3) \end{aligned} \tag{2.17}$$

Thus, eventually the Hamiltonian becomes time dependent and moreover it reflects the molecular structure (via the angle Θ).

From the form of the dipole–dipole interaction in the laboratory frame (L) one can conclude that this coupling is indeed able to cause single- and double-quantum transitions leading to the relaxation, because now it contains the tensor operators $T^2_{\pm 1}(I_1, I_2)$ and $T^2_{\pm 2}(I_1, I_2)$ (associated with the non-zero functions $F^{2(L)}_{\pm 1}$ and $F^{2(L)}_{\pm 2}$). The operators are defined as [1, 17, 24, 27, 28, 36–39]:

$$T^2_{\pm 1}(I_1, I_2) = \mp \frac{1}{2}\left[I_{1z}I_{2\pm} + I_{1\pm}I_{2z}\right] \tag{2.18}$$

$$T^2_{\pm 2}(I_1, I_2) = \frac{1}{2} I_{1\pm}I_{2\pm} \tag{2.19}$$

The explicit forms of the Wigner rotation matrices are [13, 16, 43]:

$$D^2_{0,0}(\Omega) = P_2(\beta) = \frac{3\cos^2\beta - 1}{2} \tag{2.20}$$

where $P_2(\beta)$ is Legendre polynomial, while

$$D^2_{0,\pm 1}(\Omega) = \mp\sqrt{\frac{3}{2}}\sin\beta\cos\beta\exp(\pm i\gamma) \tag{2.21}$$

$$D^2_{0,\pm 2}(\Omega) = \frac{1}{2}\sqrt{\frac{3}{2}}\sin^2\beta\exp(\pm 2i\gamma) \tag{2.22}$$

where $\Omega = (\alpha, \beta, \gamma)$.

The part of the dipole–dipole Hamiltonian including the $T^2_0(I_1, I_2)$ tensor component:

$$\begin{aligned} H^{\mathrm{res}}_{\mathrm{DD}}(I_1, I_2) &= \sqrt{6}\frac{\mu_0}{4\pi}\frac{\gamma_{\mathrm{I}}^2\hbar}{r^3} \cdot \left[\frac{1}{2}\left(3\cos^2\beta - 1\right)\right] \\ &\quad \times \left\{\frac{1}{\sqrt{6}}\left[2I_{1z}I_{2z} - \frac{1}{2}\left(I_{1+}I_{2-} + I_{1-}I_{2+}\right)\right]\right\} \\ &= a_{\mathrm{DD}} P_2(\beta)\, T^2_0(I_1, I_2) \end{aligned} \tag{2.23}$$

is often called a residual (res) dipole–dipole coupling; we shall see in the forthcoming chapters that this terminology is somewhat misleading.

The transitions probabilities W_1 and W_2 are related to the corresponding matrix elements of the dipolar Hamiltonian (more precisely their squares); for W_2 it is $\langle 1|\, H_{\mathrm{DD}}\, |4\rangle$, while for W_1 one can take for instance the matrix element $\langle 1|\, H_{DD}\, |2\rangle$. These elements can be calculated as follows:

$$\langle 1|\, H_{DD}\, |4\rangle = a_{DD} \left\langle \frac{1}{2}, \frac{1}{2} \right| \sum_{m=-2}^{2} (-1)^m F_{-m}^{2(L)} T_m^2 (I_1, I_2) \left| -\frac{1}{2}, -\frac{1}{2} \right\rangle$$
$$= \frac{1}{2} a_{DD} \left\langle \frac{1}{2}, \frac{1}{2} \right| F_{-2}^{2(L)} I_{1+} I_{2+} \left| -\frac{1}{2}, -\frac{1}{2} \right\rangle = \frac{1}{2} a_{DD} F_{-2}^{2(L)} \tag{2.24}$$

and analogously:

$$\langle 1|\, H_{DD}\, |2\rangle = a_{DD} \left\langle \frac{1}{2}, \frac{1}{2} \right| \sum_{m=-2}^{2} (-1)^m F_{-m}^{2(L)} T_m^2 (I_1, I_2) \left| \frac{1}{2}, -\frac{1}{2} \right\rangle$$
$$= \frac{1}{2} a_{DD} \left\langle \frac{1}{2}, \frac{1}{2} \right| F_{-1}^{2(L)} I_{1+} I_{2z} \left| -\frac{1}{2}, -\frac{1}{2} \right\rangle = \frac{1}{4} a_{DD} F_{-1}^{2(L)} \tag{2.25}$$

where the rules [13, 16, 27, 28, 43]:

$$I_+ \,|I, m_I\rangle = \sqrt{I(I+1) - m_I (m_I + 1)}\, |I, m_I + 1\rangle \tag{2.26}$$

$$I_z \,|I, m_I\rangle = m_I \,|I, m_I\rangle \tag{2.27}$$

have been applied. Orthogonality properties of the functions $\{|i\rangle = |m_1, m_2\rangle\}$: $\langle i|\, i\rangle = 1$ and $\langle i|\, j\rangle = 0$ for $i \neq j$ are essential for the calculations. Performing such calculations one should note that it is enough to take from the total Hamiltonian only the term with the index m fulfilling the condition $m = \Delta m_I$.

Knowing the Hamiltonian matrix elements we can come back to Eq. (2.4) from which one sees that $R_1 = 2(W_1 + W_2)$, and write down an expression for R_1 which is, at this stage, not quite obvious, but it will become clear soon. The expression is as follows:

$$R_1 = 2(W_1 + W_2) = 2(a_{DD})^2$$
$$\times \left[\frac{1}{16} \int_{-\infty}^{+\infty} \left\langle F_{-1}^{2(L)}(t)\, F_{-1}^{2(L)}(0) \right\rangle \exp(-i\omega_I t)\, dt \right.$$
$$\left. + \frac{1}{4} \int_{-\infty}^{+\infty} \left\langle F_{-2}^{2(L)}(t)\, F_{-2}^{2(L)}(0) \right\rangle \exp(-2i\omega_I t)\, dt \right] \tag{2.28}$$

The terms 1/16 and 1/4 come from Eqs. 2.24 and 2.25, as $(1/4)^2$ and $(1/2)^2$, respectively. As already explained, $F_{-m}^{2(\mathrm{L})}(t)$ are stochastic functions of time. Thus, calculating the probabilities, W_1 and W_2 as squares of the Hamiltonian matrix elements one faces the problem at which time, in fact, the $F_{-m}^{2(\mathrm{L})}(t)$ quantities should be taken. This question brings next doubts; one can hardly consider time-dependent transition probabilities. This dilemma can be solved by introducing the concept of correlation function.

The quantity

$$C_m(t) = \left\langle F_m^{2(\mathrm{L})}(t)\, F_m^{2(\mathrm{L})}(0)\right\rangle \tag{2.29}$$

is referred to as a time correlation function and represents a relation between the initial orientation of the molecule and its orientation after time t. The correlation function will be explicitly defined in the next section; including the meaning of the brackets "<>". The quantities entering Eq. 2.28, are Fourier transforms of the correlation functions taken at the corresponding transition frequencies. As $F_{\pm 1}^{2(\mathrm{L})}(t)$ is associated with the operator $T_1^2(I_1, I_2)$ causing single-quantum transitions, the Fourier transform of the correlation function $C_{\pm 1}(t)$ is taken at the frequency ω_{I}, and analogously the Fourier transform for the double-quantum transition $(C_{\pm 2}^2(t))$ is taken at $2\omega_{\mathrm{I}}$.

2.2 Correlation Functions and Spectral Densities

Let us consider a function $A(x)$ of a stochastic variable, $x \equiv x(t)$. The mathematical definition of the correlation function $C(t) = \langle A(x(t))\, A(x(0))\rangle$ yields [1, 15, 17, 20, 27, 28, 36, 42]:

$$C(t) = \langle A(t)\, A(0)\rangle = \int\int A^*(x)\, A(x_0)\, P(x, x_0, t)\, P_{\mathrm{eq}}(x_0)\, dx_0 dx \tag{2.30}$$

where "*" denotes complex conjugation. $P(x, x_0, t)$ is the conditional probability density that the system is at the position x at time t if it has been at the position x_0 at time zero, P_{eq} is the equilibrium distribution; thus, $P_{\mathrm{eq}}(x_0)$ describes the probability of finding the particle at position x_0 when the system is in equilibrium. As we deal with molecular reorientation the correlation function of interest is:

$$\left\langle D^2_{0,-m}(t)\, D^2_{0,-m}(0)\right\rangle = \int\int D^{2*}_{0,-m}(\Omega)\, D^2_{0,-m}(\Omega_0)\, P(\Omega, \Omega_0, t)\, P_{eq}(\Omega_0)\, d\Omega_0 d\Omega \quad (2.31)$$

The definition describes the meaning of the brackets “$<>$”; they denote averaging over the space or, in other words, averaging over all available initial and final (after time t) orientations of the molecule. For isotropic motion, the equilibrium distribution of orientations is given as $P_{eq}(\Omega_0) = 1/4\pi$. A much more complicated problem is the conditional probability, $P(\Omega, \Omega_0, t)$. It depends on the model of motion encoded in the applied diffusion equation. For isotropic molecular tumbling, the diffusion equation has the form [1, 15, 27, 28, 42]:

$$\frac{\partial P(\Omega, \Omega_0, t)}{\partial t} = -D_{rot} \nabla^2 P(\Omega, \Omega_0, t) \quad (2.32)$$

where D_{rot} denotes rotational diffusion coefficient. The solution of this equation is well known and it is given as the following series of Wigner rotation matrices [1, 15, 27, 28, 42]:

$$P(\Omega, \Omega_0, t) = \sum_{l,m} \frac{2l+1}{4\pi} D^{l*}_{0,m}(\Omega_0)\, D^{l}_{0,m}(\Omega) \exp\left(-\frac{t}{\tau_{rot}}\right) \quad (2.33)$$

with the rotational correlation time, $\tau_{rot} = (6D_{rot})^{-1}$. The correlation time describes the time scale at which the molecule loses memory of its initial position. At this stage, one can see why it is convenient to use Wigner rotation matrices; one can profit from their orthogonality properties [13, 16, 43]:

$$\int D^{l*}_{k,m}(\Omega)\, D^{l'}_{k',m'}(\Omega)\, d\Omega = \frac{4\pi}{2l+1} \delta_{ll'} \delta_{kk'} \delta_{mm'} \quad (2.34)$$

ending up with a very simple form of the rotational correlation function [1, 15, 17, 21, 27, 28, 36, 44]:

$$C_{rot}(t) = \left\langle D^2_{0,-m}(t)\, D^{2*}_{0,-m}(0)\right\rangle = \frac{1}{5} \exp\left(-\frac{t}{\tau_{rot}}\right) \quad (2.35)$$

where the factor 1/5 stems from $1/(2l+1)$ for $l = 2$; therefore, τ_{rot} is called rank-two correlation time. As already explained, Fourier transform of a correlation function $C(t)$ is referred to as a spectral density, $J(\omega)$. As Fourier transform of an exponent is a Lorentzian function, thus the spectral density for isotropic rotation is m-independent and given as [1, 17, 21, 27, 28, 36, 44]:

$$J_{\mathrm{rot}}(\omega) = \int_0^\infty C_{\mathrm{rot}}(t) \exp(-i\omega t) = \frac{1}{5}\frac{\tau_{\mathrm{rot}}}{1+(\omega\tau_{\mathrm{rot}})^2} \tag{2.36}$$

The function of Eq. 2.36 reaches its maximum for $\omega\tau_c = 1$, where τ_c denotes a correlation time of a motional processes leading to an exponential correlation function (in the case of rotation $\tau_c = \tau_{\mathrm{rot}}$). The function can be treated as a measure of the probability that a spin transition characterized by a frequency ω can be induced by stochastic fluctuations of local magnetic fields described by a correlation time τ_c. In other words, it shows that dynamical process of the time scale of $\tau_c \cong \omega^{-1}$ are most efficient as a source of spin relaxation at the resonance frequency, ω.

2.3 The Simplest Relaxation Formula

By substituting the expressions for a spectral density derived in the previous section into Eq. 2.28 one gets:

$$\begin{aligned} R_1(\omega_I) &= \frac{1}{4}\left(\sqrt{6}\frac{\mu_0}{4\pi}\frac{\gamma_I^2\hbar}{r^3}\right)^2 [J(\omega_I) + 4J(2\omega_I)] \\ &= \frac{3}{2}\left(\frac{\mu_0}{4\pi}\frac{\gamma_I^2\hbar}{r^3}\right)^2 [J(\omega_I) + 4J(2\omega_I)] \end{aligned} \tag{2.37}$$

In Eq. 2.37 we refer to a general form of spectral density $J(\omega)$. As the formula is not restricted to rotational dynamics, the spectral density can be attributed to any motional process and it does not need to be of Lorentzian form (i.e., the corresponding correlation function does not need to be exponential). This is the simplest possible relaxation formula describing the spin–lattice relaxation rate of two equivalent (identical like protons in water molecule) spins $I_1 = I_2 = 1/2$ [1, 9, 14, 17, 27, 28, 36, 40, 41]. The expression gives the relaxation rate, $R_1(\omega_I)$, as a combination of two spectral densities $J(\omega_I)$ and $J(2\omega_I)$. Their form depends on the applied motional model. However, often Eq. 2.37 is simplified by introducing Lorentzian spectral densities and compressing the interaction parameters to a constant, that is:

$$R_1(\omega_I) = K\left[\frac{\tau_c}{1+\omega_I^2\tau_c^2} + \frac{4\tau_c}{1+4\omega_I^2\tau_c^2}\right] \tag{2.38}$$

where, for $I = 1/2$:

$$K = \frac{3}{10}\left(\frac{\mu_0}{4\pi}\frac{\gamma_I^2\hbar}{r^3}\right)^2 \tag{2.39}$$

Equation 2.38 is often referred to as Bloembergen–Purcell–Pound (BPP) formula [9].

This pragmatic attitude to the description of relaxation processes has disadvantages; it is very easy to apply it without thinking on its origin and the underlying motional mechanism. Independently of the level of simplification one should always remember that the expression of Eq. 2.37 is valid only if the energy level structure of the participating spins is determined by the Zeeman coupling, while the relaxation mechanism is provided by dipole–dipole interactions.

Classical relaxation experiments are performed at one magnetic field versus temperature. For simple molecular systems the temperature dependence of the correlation time often follows Arrhenius law [4, 34]:

$$\tau_c = \tau_0 \exp\left(\frac{E_A}{k_B T}\right) \tag{2.40}$$

where E_A is an activation energy, while τ_0 is the high temperature limit of the correlation time. Much more informative are relaxation experiments performed versus frequency (magnetic field) and temperature. Due to commercial availability of fast field cycling (FFC) spectrometers [25], frequency dependent relaxation studies become a very important, well-established source of information about dynamics in condensed matter. Such spectrometers typically operate in the frequency range of 10 kHz–20 MHz for ^{1}H. The frequency dependence of the spin–lattice relaxation rate (time) is often referred to as nuclear magnetic relaxation dispersion (NMRD) profile. In Fig. 2.5 relaxation dispersion profiles simulated for different correlation times τ_c assuming Lorentzian spectral densities (Eq. 2.38) are shown.

The low frequency plateau gives the value $5\,K\tau_c$. The relaxation becomes slower for higher frequencies. For short correlation times

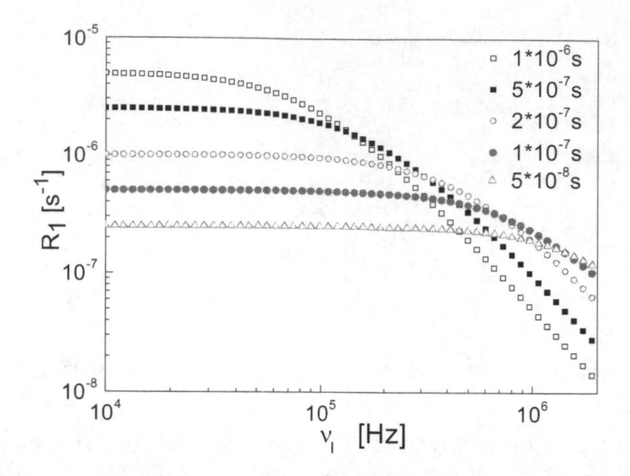

Figure 2.5 Relaxation dispersion profiles for different correlation times, τ_c (it has been set $K = 1\ \mathrm{Hz}^2$ for normalization purposes).

one does not observe pronounced dispersion effects because the relationship $\omega_{\mathrm{I}}\tau_{\mathrm{c}} << 1$ holds. This relation is referred to as the extreme narrowing condition [1, 14, 17, 27, 28, 36].

An alternative way to present the relaxation dispersion results is to use susceptibility representation [31]. The susceptibility function, $\chi(\omega)$, has a simple relation to the relaxation rate:

$$\chi = \omega_{\mathrm{I}} R_1 (\omega_{\mathrm{I}}) \tag{2.41}$$

In fact, using the susceptibility representation means multiplying Eq. 2.39 by ω_{I}. This simple operation has deep implications. The equation reads:

$$\chi = K\left[\frac{\omega_{\mathrm{I}}\tau_{\mathrm{c}}}{1+\omega_{\mathrm{I}}^2\tau_{\mathrm{c}}^2} + \frac{4\omega_{\mathrm{I}}\tau_{\mathrm{c}}}{1+4\omega_{\mathrm{I}}^2\tau_{\mathrm{c}}^2}\right] = K\left[\frac{x}{1+x^2} + \frac{4x}{1+4x^2}\right] \tag{2.42}$$

where $x = \omega_{\mathrm{I}}\tau_{\mathrm{c}}$. This means that one cannot talk any more about the correlation time and frequency separately, but only about their product. One cannot inquire to what correlation time and what frequency a given value of χ corresponds, one can only talk about its x dependence, $\chi \equiv \chi(x)$. In Fig. 2.6 the relaxation dispersion profiles of Fig. 2.5 are shown after multiplying by the frequency.

The first observation from Fig. 2.6 is that actually the shapes of the relaxation curves (in the susceptibility representation) are

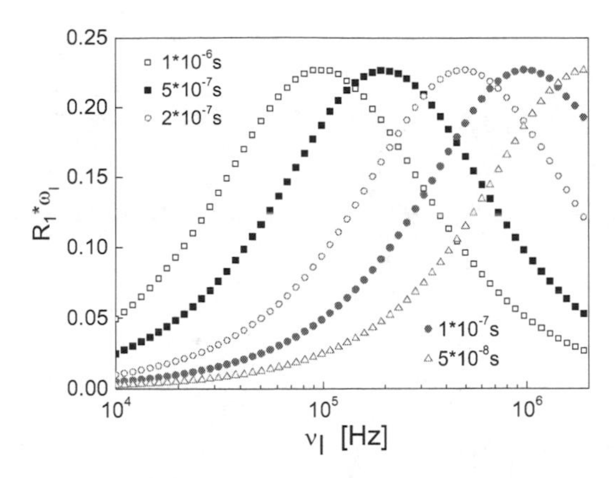

Figure 2.6 Relaxation dispersion results shown in Fig. 2.5 after multiplying by frequency, ω_I.

similar. One can horizontally shift the curves until they overlap, as shown in Fig. 2.7a.

This feature is reflected by Eq. 2.42. All curves represent the same dynamical process. If for a given correlation time τ_c, certain relaxation features are observed at a frequency ω_I, analogous effects will be seen for 0.1 τ_c at the frequency of 10 ω_I. The same conclusion can be drawn from Fig. 2.5, but then a more effort is needed. First, one has to shift the relaxation profiles vertically (along the y-axis) and when they reached the same level they can be shifted horizontally (Fig. 2.7b). One should be, however, aware than the procedure illustrated in Fig. 2.7a,b applies only to very simple cases—from the dynamical viewpoint, the character of motion has to remain unchanged with temperature (only the correlation time τ_c changes)—from the spin (quantum mechanical) viewpoint, the relaxation process must be described by a very simple formula. This last condition is not well defined at this stage, but in the forthcoming chapters, several examples which clarify this point will be shown.

As already said, Eq. 2.37 gives the relationship between the relaxation rates and the spectral densities, but the form of the spectral density function depends on the applied model of motion. Let us consider, for instance, translation diffusion. Then one has to discuss stochastic changes in time of not only the orientation

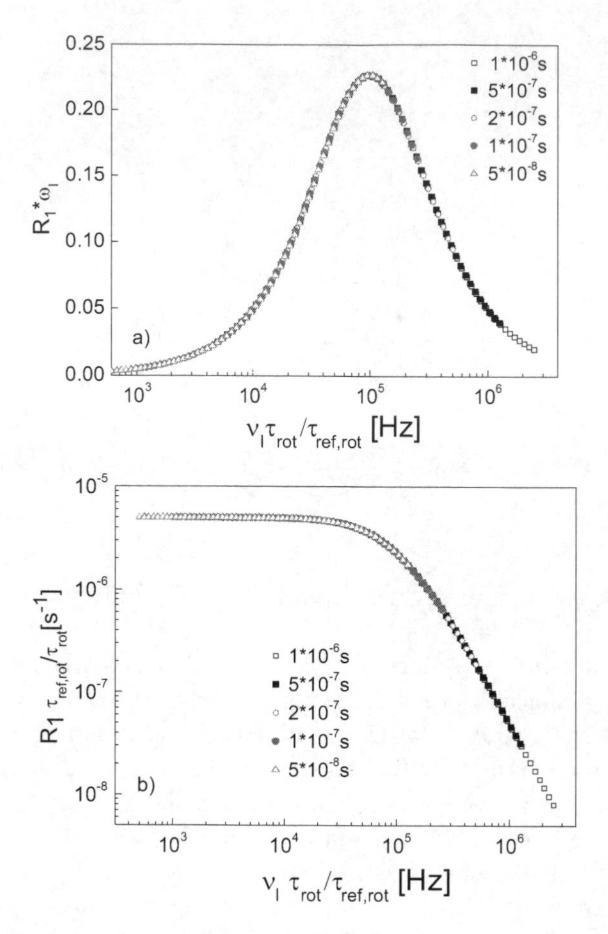

Figure 2.7 Relaxation dispersion results at different temperatures in (a) susceptibility representation and (b) spectral density representation merged to one relaxation curve.

of the dipole–dipole axis, but also of the inter-spin distance, $r(t)$; generally one can say that for translational motion one considers fluctuations of the inter-spin vector, $\vec{r}(t)$. Then, the correlation function of interest has a more complex form [1–3, 5–7, 19, 20, 23, 27, 28, 36]:

$$C_{\text{trans}}(t) = \left\langle \frac{D_{0,m}^2(\Omega(t))}{r^3(t)} \frac{D_{0,m}^2(\Omega(0))}{r^3(0)} \right\rangle \tag{2.43}$$

This function can be calculated from its definition (Eq. 2.31) by solving the appropriate diffusion equation. The simplest model of translational motion is referred to as force-free hard-sphere model [5, 23]. According to the model there are no interactions between molecules in the sense of, for instance, Van der Waals forces (force free) and the molecules are treated as hard sphere carrying spins placed in their center. Assuming reflecting wall boundary conditions, the correlation function is given as [5, 23]:

$$C_{\text{trans}}(t) = \frac{72}{5}\frac{1}{d^3}N\int_0^\infty \frac{u^2}{81+9u^2-2u^4+u^6}\exp\left(-\frac{D_{12}}{d^2}u^2\tau\right)du \tag{2.44}$$

where d is the distance of closest approach ($d = 2r$, where r is the molecular radius), N is the number of spins (nuclei of interest, for instance protons) per unit volume, and D_{12} is the relative translational diffusion coefficient equal to the sum of the self-diffusion coefficients of the participating molecules, when they are identical one gets $D_{12} = 2D$. Then, the corresponding expression for the spectral density $J_{\text{trans}}(\omega)$ yields [23, 28]:

$$\begin{aligned} J_{\text{trans}}(\omega) &= \frac{72}{5}\frac{1}{d^3}N\left[\int_0^\infty \frac{u^2}{81+9u^2-2u^4+u^6}\exp\right. \\ &\quad \times \left.\left(-\frac{u^2t}{\tau_{\text{trans}}}\right)du\right]\exp(-i\omega t)\,dt \\ &= \frac{72}{5}\frac{1}{d^3}N\int_0^\infty \frac{u^2}{81+9u^2-2u^4+u^6}\frac{u^2\tau_{\text{trans}}}{u^4+(\omega\tau_{\text{trans}})^2}du \end{aligned} \tag{2.45}$$

where the correlation time τ_{trans} is defined as $\tau_{\text{trans}} = d^2/D_{12}$. Substituting this expression into Eq. 2.37 one gets:

$$R_1(\omega_{\text{I}}) = \frac{3}{2}\left(\frac{\mu_0}{4\pi}\gamma_{\text{I}}^2\hbar\right)^2\left[J_{\text{trans}}(\omega_{\text{I}}) + 4J_{\text{trans}}(2\omega_{\text{I}})\right] \tag{2.46}$$

where the factor r^{-6} is omitted because now it is included into the correlation function and, in consequence, into the spectral density of Eq. 2.45 (by the distance of closest approach, d, and the concentration of spins, N).

Both, the rotational as well as the translational correlation functions are based on well-defined models of motion encoded in the diffusion equation. Very often motional processes are not so simple and one cannot treat them as force-free; an interaction potential should then be included into the diffusion equation, which in the general case reads [15, 28]:

$$\frac{\partial P(\vec{r},\vec{r}_0,t)}{\partial t} = -\left\{\vec{K}\cdot\hat{D}_{\text{trans}}\cdot\left[\vec{K}+KV(\vec{r})\right] + \vec{J}\cdot\hat{D}_{\text{rot}}\left[\vec{J}+\vec{J}V(\vec{r})\right]\right\}P(\vec{r},\vec{r}_0,\tau) \quad (2.47)$$

where $\vec{K} = -i\nabla, \vec{J} = -i\vec{r}\times\nabla$, while $\hat{D}_{\text{trans}}$ and $\hat{D}_{\text{rot}}$ are translational and rotational diffusion tensors, respectively. $V(\vec{r})$ is effective interaction potential in thermal energy units. Such an equation leads to non-exponential correlation functions, but its solution is a very difficult task. Therefore, often to account for the observed non-exponentiality of correlation functions, a phenomenological approach is applied. A good example is Cole–Davidson spectral density for the rotational dynamics [12, 31]:

$$J_{\text{rot}}(\omega) = \frac{1}{5}\frac{\sin\left[\beta\arctan\left(\omega\tau_{\text{rot}}\right)\right]}{\omega\left[1+\left(\omega\tau_{\text{rot}}\right)^2\right]^{\beta/2}} \quad (2.48)$$

where the parameter β is a measure of the non-exponentiality (for $\beta = 1$, the Cole–Davidson spectral density becomes Lorentzian). One should be aware that this frequently applied function does not have a well-defined physical meaning as there is no explicit form of the interaction potential which, when put into Eq. 2.47 leads to the Cole–Davidson spectral density. Nevertheless, the Cole–Davidson expression can be associated with heterogeneous dynamics, described by the following distribution of correlation times [12, 33]:

$$g_{\text{CD}}(\tau_{\text{rot}}) \equiv g_{\text{CD}}(\tau_{\text{rot}},\tau_0,\beta) \propto \sum_{k=0}^{\infty}(-1)^k\frac{\sin(\pi\beta k)\,\Gamma(\beta k+1)}{k!}\left(\frac{\tau_{\text{rot}}}{\tau_0}\right)^{\beta k} \quad (2.49)$$

which reaches maximum for $\tau_{\text{rot}} = \beta\tau_0$. Here, β determines the width of the distribution; some examples are shown in Fig. 2.8. The distribution is asymmetric; it has a long "tail" for shorter correlation times.

On the other hand, the Cole–Davidson distribution is associated with stretched exponential correlation function [25, 33]:

$$C(t) \propto \exp\left[-\left(\frac{t}{\tau_{\text{rot}}}\right)^{\beta}\right] \quad (2.50)$$

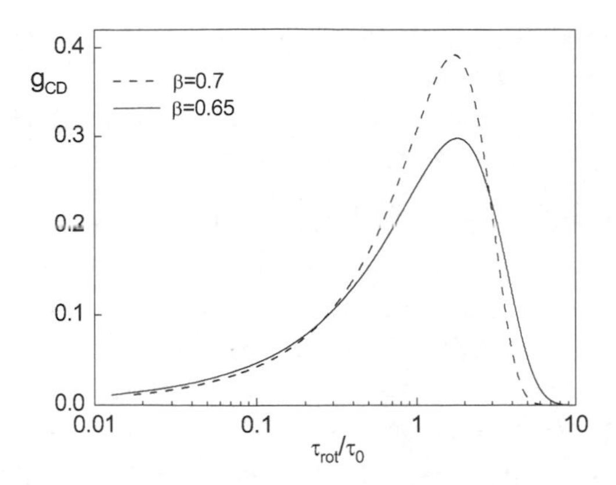

Figure 2.8 Examples of Cole–Davidson distributions of correlation times.

This relation is, however, a kind of approximation. Fourier transform of Eq. 2.50 does not give exactly the spectral density of Eq. 2.48 but a shape which can be well approximated by a Cole–Davidson function, but with β' somewhat different from the original value of β [33]. The stretched exponential function of Eq. 2.43 is a phenomenological description of correlation functions which show deviations from exponentiality for whatever reasons. It does not give any information about the nature of the dynamical process except that it cannot be described by an exponential correlation function. One might say that $\beta < 1$ indicates a kind of cooperativity of molecular motion, that is, this is not a force-free, isotropic motion unaffected by the presence of neighboring molecules.

^{1}H relaxation processes in liquids predominantly stem from dipole–dipole interactions which can be of intra-molecular and inter-molecular origin. Thus, the overall relaxation rate is given as a sum of two contributions:

$$R_1(\omega_I) = R_{1,\text{intra}}(\omega_I) + R_{1,\text{inter}}(\omega_I) \tag{2.51}$$

where the intra-molecular part, $R_{1,\text{intra}}(\omega_I)$ is described as (see Eq. 2.37):

$$R_{1,\text{intra}}(\omega_I) = \frac{3}{2} K_{\text{intra}} \left[J_{\text{rot}}(\omega_I) + 4J_{\text{rot}}(2\omega_I) \right] \tag{2.52}$$

while the inter-molecular relaxation rate, $R_{1,\text{inter}}(\omega_I)$, is given by the formula of Eq. 2.46.

The intra-molecular dipole–dipole interactions fluctuate in time due to rotational dynamics (mediated by some kinds of internal motion, but we neglect this effect at this stage). The coupling constant K_{intra} results from interactions between all pairs of spins in the molecule; its upper limit is $K_{\mathrm{intra}} \leq \sum_{i<j} \left(\frac{\mu_0}{4\pi} \frac{\gamma_I^2 \hbar}{r_{ij}^3} \right)^2$, where the summation goes over pairs of spins in the molecule. The inter-molecular spectral density is associated with translational dynamics and it can be modeled according to Eq. 2.45. In the hydrodynamic limit, that is, when the rotational and translational dynamics is a force-free process, the relationship $\tau_{\mathrm{trans}}/\tau_{\mathrm{rot}} = 9$ holds [1]. It has been shown that for liquids of high viscosity (for instance glycerol, propylene glycol, etc.) the ratio is larger; it yields about 20–50 [29, 31]. This implies, in fact, a time scale separation of the translational and rotational dynamics. In consequence, one can relatively easily separate the intra-molecular and inter-molecular contributions to the overall relaxation [32]. This is shown in Fig. 2.9a.

For illustrative purposes, it has been assumed that the intra-molecular relaxation includes two spins: so the relaxation rate $R_{1,\mathrm{intra}}(\omega_I)$ is given by Eq. 2.37 ($r = 3$ Å has been set) with the spectral densities $J_{\mathrm{rot}}(\omega_I)$ of Eq. 2.36, while the inter-molecular part, $R_{1,\mathrm{inter}}(\omega_I)$, is described by Eq. 2.46 with $J_{\mathrm{trans}}(\omega_I)$ defined by Eq. 2.45 in which $d = 3$ Å and $N = 1/\text{Å}^3$ has been set. One can see from Fig. 2.9a that the inter-molecular contribution prevails at lower frequencies. In this range, the intra-molecular (rotational) contribution is frequency independent, while the inter-molecular (translational) part shows a strong dispersion. This is due to the time-scale separation of the two kinds of motion. It is of interest to analyze the ratio $r = R_{1,\mathrm{intra}}(\omega_I)/R_{1,\mathrm{inter}}(\omega_I)$ versus frequency. This is shown in Fig. 2.9b in which $r = d$ and $N = 1/\text{Å}^3$ has been used. Experimental examples of ^{1}H spin–lattice relaxation processes separated into intra-molecular and inter-molecular contributions one can find, for instance in Ref. [29, 31]. Performing relaxation experiments at a single frequency one should be aware how these two contributions can vary with temperature. Examples of that are shown in Fig. 2.10 [30].

The translational spectral density, $J_{\mathrm{trans}}(\omega_I)$ (Eq. 2.45) can be expanded at low frequencies as [7, 18, 19, 22, 23, 32, 35]:

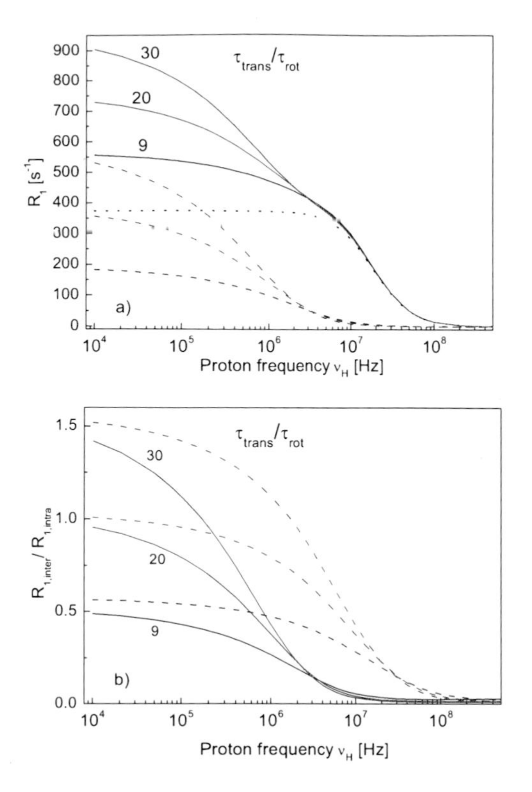

Figure 2.9 (a) Intra-molecular (dotted line), inter-molecular (dashed lines), and total (solid lines) ^{1}H spin–lattice relaxation rates for different ratios τ_{trans}/τ_{rot}, $\tau_{rot} = 1$ ns (other parameters are given in the text) and (b) $R_{1,inter}/R_{1,intra}$ for different τ_{trans}/τ_{rot}, $\tau_{rot} = 1$ ns (solid lines), $\tau_{rot} = 0.1$ ns (dashed lines).

$$J_{trans}(\omega_I) \cong a - b\sqrt{\omega_I} \tag{2.53}$$

where $b = \frac{2^{3/2}\pi}{45 D_{12}^{3/2}}$. This relationship allows determining the relative translational diffusion coefficient D_{12} from the slope of $R_1(\omega_I)$ plotted versus $\sqrt{\omega_I}$ [7, 18, 19, 22, 23, 32, 35]:

$$R_1(\omega_I) \cong R_1(0) - B\sqrt{\omega_I} = R_1(0) - \frac{\sqrt{2}\pi}{15} \times \left(1 + 4\sqrt{2}\right)\left(\frac{\mu_0}{4\pi}\gamma_I^2\hbar\right)^2 N D_{12}^{-3/2}\sqrt{\omega_I} \tag{2.54}$$

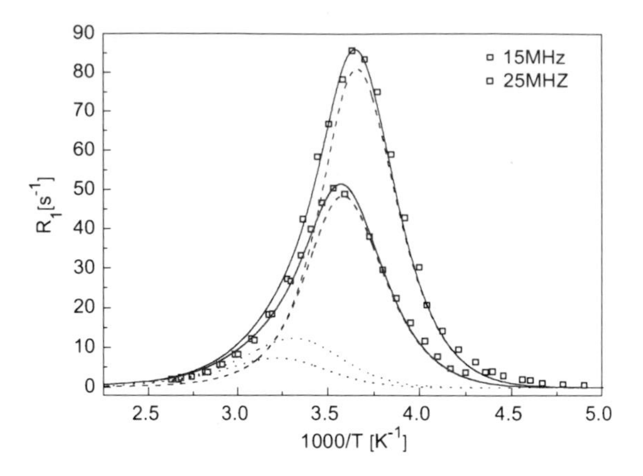

Figure 2.10 ^{1}H spin–lattice relaxation rate for glycerol; solid squares 15 MHz, open squares 25 MHz. Fits of a sum of intra-molecular and inter-molecular contributions (solid lines); intra-molecular parts (dashed lines); and inter-molecular parts (dotted lines). Reprinted with permission from Ref. [30]. Copyright 2012, AIP Publishing LLC.

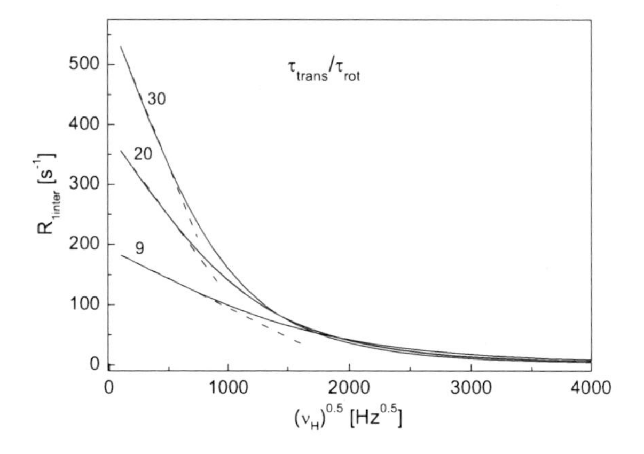

Figure 2.11 Inter-molecular ^{1}H spin–lattice relaxation rate, $R_{1,\text{inter}}$, shown in Fig. 2.9a ($\tau_{\text{rot}} = 1$ ns) versus square root of ν_{H}; the linearity range at low frequencies is indicated (dashed lines).

One can see from Fig. 2.11 that $R_{1,\text{inter}}(\omega_{\text{I}})$ plotted versus $\sqrt{\nu_{\text{I}}}$ shows a linear behavior at low frequencies. Due to the time scale separation (as already explained, the lower bound limit of $\tau_{\text{trans}}/\tau_{\text{rot}}$ is 9) the inter-molecular contribution to the overall relaxation,

$R_{1,\text{intra}}(\omega_I)$, can be included in Eq. 2.54 into the constant $R_1(0)$. This way of determining D_{12} has been used for a variety of liquids [32].

One should, however, be aware that the simple recipe of Eq. 2.54 applies only when the inter-molecular relaxation rate can be described by Eqs. 2.37 and 2.46 and this holds only for identical spins 1/2. Otherwise, Eq. 2.54 has to be adjusted to the appropriate form of the relaxation formula and then it will become much more complex.

2.4 Bi-Exponential Relaxation

Spin relaxation processes are often non-exponential, that is, the magnetization does not evolve in time toward the equilibrium in an exponential way. The explanation of this effect is that the system is dynamically heterogeneous, that is, it consists of several (in the case of bi-exponential relaxation, two) sub-systems which exhibit different dynamics (for instance, dynamically non-equivalent crystal sub-lattices). Such a system would indeed relax bi-exponentially; however, one should be aware that, in fact, there are no systems which consist of completely separated parts. The sub-systems can "communicate" by exchanging atoms (ions); proton exchange is very common. Besides the exchange effects, spins which are relatively close interact (for instance by dipole–dipole couplings), and this is unavoidable. Exchange effects act as an averaging mechanism; the sub-systems are not separated because they exchange atoms. In consequence, instead of a bi-exponential relaxation, one sees a single-exponential, averaged relaxation process. The role of mutual spin interactions, however, is entirely different; they are the reason for bi-exponential (generally, multi-exponential) relaxation, even though the system is dynamically homogenous.

The relaxation process is bi-exponential when the system consists of two types of non-equivalent spins, that is, when the molecule contains two kinds of NMR active nuclei, for instance, ^{1}H and ^{19}F, ^{1}H and ^{13}C, and ^{1}H and ^{2}H (although in the last case the problem is even more complex due to quadrupolar interactions of ^{2}H nuclei). In fact, one does not need different nuclei to observe bi-exponentiality (in general multi-exponentiality) of relaxation; it is enough when the molecule consists of non-equivalent protons, like for instance in glycerol molecules. There are three protons of the hydroxyl groups, OH and five other protons. These two groups have slightly different resonance frequencies due to local magnetic fields (different chemical surrounding); this effect is referred to as chemical shift. It is rather difficult to find a system which is heterogeneous from the quantum-mechanical point of view (i.e., it contains non-equivalent spins), but it is homogenous as far as the dynamics is concerned. Considering, for instance, molecules containing aromatic rings and methyl groups, one can say that protons belonging to these molecular units are not equivalent with respect to the chemical surrounding (specific transition frequencies) as well as with respect to dynamics, methyl groups rotate fast, while the motion of aromatic rings is slower and rather anisotropic. Summarizing, the conclusion of these consideration is that even if the system is dynamically homogenous, the relaxation can be non-exponential due to mutual dipole–dipole interactions between spins which are characterized by different transition (resonance) frequencies.

Let us consider a system containing two spins $I = 1/2$ and $S = 1/2$. In this book, we denote electron spins by S; in this case however, we have just in mind two non-equivalent spins (both of them can be associated with nuclei). Nevertheless, a system containing nuclear and electron spins is a very good example illustrating the issue of bi-exponential relaxation. In Fig. 2.12, a schematic view of I–S

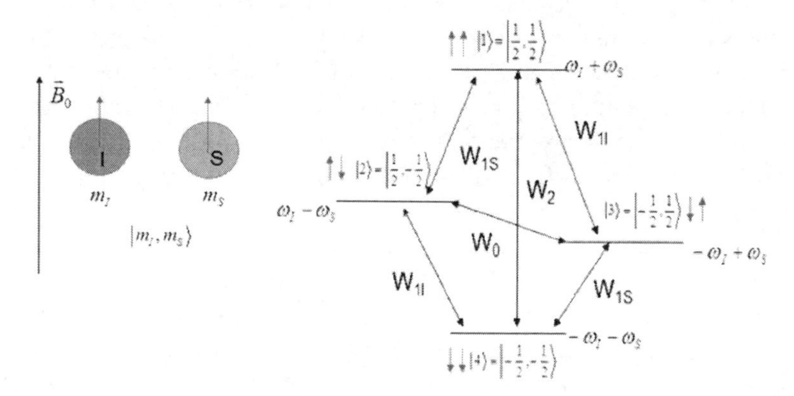

Figure 2.12 A system of two non-equivalent spins $I = 1/2$ and $S = 1/2$, energy level structure and transition probabilities.

spin system is shown with energy levels and transition probabilities labeled in analogy to the system of two equivalent nuclei I_1–I_2 presented in Fig. 2.2.

The system is characterized by four energy levels corresponding to the spin states in the external magnetic field. In analogy to the functions $\{|i\rangle = |m_1, m_2\rangle\}$ for the I_1–I_2 case, the I–S system is described by the functions $\{|i\rangle = |m_{\mathrm{I}}, m_{\mathrm{S}}\rangle\}$ where m_{I} and m_{S} are magnetic quantum numbers of I and S spins, respectively. The energy associated with the quantum state described by the function $|m_{\mathrm{I}}, m_{\mathrm{S}}\rangle$ yields:

$$E\,(m_{\mathrm{I}}, m_{\mathrm{S}}) = m_{\mathrm{I}}\omega_{\mathrm{I}} + m_{\mathrm{S}}\omega_{\mathrm{S}} \tag{2.55}$$

thus, the energy levels are not degenerated. The kinetic equations (in analogy to Eq. 2.3) read:

$$\frac{dN_1\,(t)}{dt} = -\,(W_{1\mathrm{I}}+W_{1\mathrm{S}}+W_2)\,N_1\,(t)+W_{1\mathrm{I}}N_2\,(t)+W_{1\mathrm{S}}N_3\,(t)+W_2N_4\,(t) \tag{2.56a}$$

$$\frac{dN_2\,(t)}{dt} = W_{1\mathrm{S}}N_1\,(t)-(W_{1\mathrm{S}} + W_0 + W_{1\mathrm{I}})\,N_2\,(t)+W_0N_3\,(t)+W_{1\mathrm{I}}N_4\,(t) \tag{2.56b}$$

$$\frac{dN_3\,(t)}{dt} = W_{1\mathrm{I}}N_1\,(t)+W_0N_2\,(t)-(W_{1\mathrm{S}} + W_0 + W_{1\mathrm{I}})\,N_3\,(t)+W_{1\mathrm{I}}N_4\,(t) \tag{2.56c}$$

$$\frac{dN_4\,(t)}{dt} = W_2N_1\,(t)+W_{1\mathrm{I}}N_2\,(t)+W_{1\mathrm{S}}N_3\,(t)-(W_{1\mathrm{I}} + W_{1\mathrm{S}} + W_2)\,N_4\,(t) \tag{2.56d}$$

Comparing Eq. 2.3 with Eq. 2.56 one sees that in the last case it is necessary to distinguish between the single-quantum transition probabilities involving the spins I and S, $W_{1\mathrm{I}}$ and $W_{1\mathrm{S}}$, respectively. One has to separately analyze the time evolution of the I and S spin magnetizations. The magnetizations, M_{Iz} and M_{Sz} (the index "z" denotes longitudinal magnetizations, that is, the components of the magnetization which are parallel to the external magnetic field), are proportional to the differences in the populations of the spin states:

$$M_{\mathrm{Iz}} \propto N_{\mathrm{I}} = N_1 + N_2 - N_3 - N_4 \tag{2.57a}$$

$$M_{\mathrm{Sz}} \propto N_S = N_1 + N_3 - N_2 - N_4 \tag{2.57b}$$

Combining Eq. 2.56 with Eq. 2.57 one gets for the magnetization evolution:

$$\frac{dM_{\mathrm{Iz}}(t)}{dt} = -\left(W_0 + 2W_{1\mathrm{I}} + W_2\right)\left(M_{\mathrm{Iz}}(t) - M_{\mathrm{Iz}}^{\mathrm{eq}}\right) - \left(W_2 - W_0\right)\left(M_{\mathrm{Sz}}(t) - M_{\mathrm{Sz}}^{\mathrm{eq}}\right) \quad (2.58\mathrm{a})$$

$$\frac{dM_{\mathrm{Sz}}(t)}{dt} = -\left(W_2 - W_0\right)\left(M_{\mathrm{Iz}}(t) - M_{\mathrm{Iz}}^{\mathrm{eq}}\right) - \left(W_0 + 2W_{1\mathrm{S}} + W_2\right)\left(M_{\mathrm{Sz}}(t) - M_{\mathrm{Sz}}^{\mathrm{eq}}\right) \quad (2.58\mathrm{b})$$

In Eq. 2.58, we have explicitly introduced the equilibrium magnetizations, $M_{\mathrm{I(S)z}}^{\mathrm{eq}}$, and pointed out that we consider deviations of the magnetizations from their equilibrium values. Equation 2.58 forms a coupled set of linear differential equations which has a bi-exponential solution. Before we provide the solution, let us make a step toward a quantum-mechanical formalism and rewrite the equations as [1, 8, 10, 11, 17, 26–28, 36]:

$$\frac{d\left(\langle I_{\mathrm{z}}\rangle - \langle I_{\mathrm{z}}\rangle_{\mathrm{eq}}\right)}{dt} = -R_{\mathrm{II}}\left(\langle I_{\mathrm{z}}\rangle - \langle I_{\mathrm{z}}\rangle_{\mathrm{eq}}\right) - R_{\mathrm{IS}}\left(\langle S_{\mathrm{z}}\rangle - \langle S_{\mathrm{z}}\rangle_{\mathrm{eq}}\right) \quad (2.59\mathrm{a})$$

$$\frac{d\left(\langle S_{\mathrm{z}}\rangle - \langle S_{\mathrm{z}}\rangle_{\mathrm{eq}}\right)}{dt} = -R_{\mathrm{IS}}\left(\langle I_{\mathrm{z}}\rangle - \langle I_{\mathrm{z}}\rangle_{\mathrm{eq}}\right) - R_{\mathrm{SS}}\left(\langle S_{\mathrm{z}}\rangle - \langle S_{\mathrm{z}}\rangle_{\mathrm{eq}}\right) \quad (2.59\mathrm{b})$$

$\langle I_{\mathrm{z}}\rangle$ and $\langle S_{\mathrm{z}}\rangle$ describe expected values of the operators I_{z} and S_{z} (we shall discuss in detail this notation and its physical meaning in the next chapters; at this stage it is enough to say that the longitudinal magnetization components are proportional to these quantities). The relaxation rates R_{II}, R_{SS}, R_{IS}, and R_{SI} are given as linear combinations of the transition probabilities W_0, $W_{1\mathrm{S}}$, $W_{2\mathrm{S}}$, and W_2. The solution of the set of Eq. 2.59 is bi-exponential, that is, the magnetization evolves in time as [1, 8, 10, 11, 17, 26–28, 36]:

$$M_{\mathrm{I(S)_z}}(t) = C_{\mathrm{I(S)}}^{+}\exp\left(-R_1^{+}t\right) + C_{\mathrm{I(S)}}^{-}\exp\left(-R_1^{-}t\right) \quad (2.60)$$

where

$$R_1^{\pm} = \frac{1}{2}\left[\left(R_{\mathrm{II}} + R_{\mathrm{SS}}\right) \pm \sqrt{\left(R_{\mathrm{II}} - R_{\mathrm{SS}}\right)^2 + 4R_{\mathrm{IS}}R_{\mathrm{SI}}}\right] \quad (2.61)$$

The transition probabilities W_0, W_{1S}, W_{2S}, and W_2 are related to a square of the corresponding matrix elements $\langle 2|\, H_{\mathrm{DD}}\, |3\rangle$, $\langle 1|\, H_{\mathrm{DD}}\, |2\rangle$, $\langle 1|\, H_{\mathrm{DD}}\, |3\rangle$, and $\langle 1|\, H_{\mathrm{DD}}\, |4\rangle$, respectively, in analogy to

Eqs. 2.24 and 2.25. As $\langle 2|\, H_{\mathrm{DD}}\, |3\rangle = \frac{1}{2\sqrt{6}} a_{\mathrm{DD}} F_{0}^{2(\mathrm{L})}$, $\langle 1|\, H_{\mathrm{DD}}\, |2\rangle = \langle 1|\, H_{\mathrm{DD}}\, |3\rangle = \frac{1}{4} a_{\mathrm{DD}} F_{-1}^{2(\mathrm{L})}$, and $\langle 1|\, H_{\mathrm{DD}}\, |4\rangle = \frac{1}{2} a_{\mathrm{DD}} F_{-2}^{2(\mathrm{L})}$, the relaxation rates yield [1, 8, 10, 11, 17, 26–28, 36]:

$$R_{\mathrm{II}} = W_0 + 2W_{1\mathrm{I}} + W_2 = \frac{1}{2}\left(\frac{\mu_0}{4\pi}\frac{\gamma_{\mathrm{I}}\gamma_{\mathrm{S}}\hbar}{r^3}\right)^2 \times \left[J(\omega_{\mathrm{I}} - \omega_{\mathrm{S}}) + 3J(\omega_{\mathrm{I}}) + 6J(\omega_{\mathrm{I}} + \omega_{\mathrm{S}})\right] \quad (2.62)$$

$$R_{\mathrm{SS}} = W_0 + 2W_{1S} + W_2 = \frac{1}{2}\left(\frac{\mu_0}{4\pi}\frac{\gamma_{\mathrm{I}}\gamma_{\mathrm{S}}\hbar}{r^3}\right)^2 \times \left[J(\omega_{\mathrm{I}} - \omega_{\mathrm{S}}) + 3J(\omega_{\mathrm{S}}) + 6J(\omega_{\mathrm{I}} + \omega_{\mathrm{S}})\right] \quad (2.63)$$

$$R_{\mathrm{IS}} = R_{\mathrm{SI}} = W_2 - W_0 = \frac{1}{2}\left(\frac{\mu_0}{4\pi}\frac{\gamma_{\mathrm{I}}\gamma_{\mathrm{S}}\hbar}{r^3}\right)^2 \left[6J(\omega_{\mathrm{I}} + \omega_{\mathrm{S}}) - J(\omega_{\mathrm{S}} - \omega_{\mathrm{I}})\right] \quad (2.64)$$

The relaxation rates, R_{II} and R_{SS}, are referred to as auto-relaxation rates, while R_{IS} and R_{SI} are called cross-relaxation coefficients.

It is very important to realize that even when for a system of non-equivalent spins, I–S, one experimentally observes a single exponential magnetization evolution, this does not mean that the process is "by nature" single exponential. The reason for seeing only one exponent might be that one of the relaxation processes, $R_1^{\pm}$, is too fast (R_1^{+}) or too slow (R_1^{-}), or its contribution is small (the contributions are given by the constants $C_{\mathrm{I(S)}}^{\pm}$). Thus, the fact that experimentally the relaxation seems to be single-exponential does not justify neglecting the cross-relaxation terms and setting, for instance, for the I spin relaxation $R_1 = R_{\mathrm{II}}$. One should be aware that neglecting the cross-relaxation terms does not give $R_1 = R_{\mathrm{II}}$, but $R_1 = \frac{1}{2}(R_{\mathrm{II}} + R_{\mathrm{SS}})$, and the relaxation rates R_{II} and R_{SS} can considerably differ when the transition frequencies ω_{I} and ω_{S} are different (due to the terms $J(\omega_{\mathrm{I}})$ and $J(\omega_{\mathrm{S}})$ in Eq. 2.62 and 2.63, respectively). There is a case when the I spin relaxation is indeed single exponential (for a system of non-equivalent spins). This happens when the S spin relaxation is very fast, that is, $R_{\mathrm{SS}} >> R_{\mathrm{II}}$. Then, the S spin system reaches its equilibrium already after a very short time, what implies that $\langle S_{\mathrm{z}}\rangle = \langle S_{\mathrm{z}}\rangle_{\mathrm{eq}}$. In consequence, the second term in Eq. 2.59a vanishes and one gets for the I spin magnetization a simple equation:

$$\frac{d\left(\langle I_z\rangle - \langle I_z\rangle_{eq}\right)}{dt} = -R_{II}\left(\langle I_z\rangle - \langle I_z\rangle_{eq}\right) \tag{2.65}$$

with an exponential solution including then indeed only the relaxation rate R_{II}, that is, $R_1 = R_{II}$. Looking at Eqs. 2.62–2.64, one should also notice that the cross-relaxation terms can hardly be treated as negligible. For instance, when $\omega_S >> \omega_I$ (this is the case for an electron-nucleus spin system; for protons $\omega_S \cong 658\omega_I$), one gets $R_{IS} = R_{SI} \cong 0.5R_{SS}$.

It is important for the form of the spectral densities entering Eqs. 2.62–2.64, whether both spins involved participate in the same dynamical processes (as already said, the "quantum-mechanical" heterogeneity is often accompanied by dynamical heterogeneity). When the motion of the molecular units including the I and S spins is the same and this is, for instance, isotropic tumbling, the spectral densities are given by Eq. 2.36. Otherwise (i.e., when the spins are involved in different dynamics), an effective correlation time, $\tau_{c,eff}$, for the modulations of the I–S dipole–dipole coupling should be defined [1, 8, 10, 11, 17, 26–28, 36]:

$$\tau_{c,eff}^{-1} = \tau_{cI}^{-1} + \tau_{cS}^{-1} \tag{2.66}$$

where τ_{cI} and τ_{cS} are correlation times characterizing the I and S spin dynamics. Then, when it is justified to assume an exponential correlation function (Lorentzian spectral density), Eq. 2.36 should be modified to the form:

$$J_{rot}(\omega) = \frac{1}{5}\frac{\tau_{c,eff}}{1+\omega^2\tau_{c,eff}^2} \tag{2.67}$$

When the dynamics is of translational origin, D_{12} in Eq. 2.45 is defined as $D_{12} = D_1 + D_2 = D_I + D_S$, where $D_1 = D_I$ and $D_2 = D_S$ are self-diffusion coefficients of the molecules carrying the I and S spins, respectively.

Eventually, it is worth to come back to Eq. 2.60 which describes the general case when the magnetization evolves in time in a bi-exponential way. This equation can be explicitly rewritten as:

$$M_{Iz}(t) = C_1 \exp\left(-R_1^+ t\right) + C_2 \exp\left(-R_1^- t\right) \tag{2.68a}$$

$$M_{Sz}(t) = C_1 \frac{R_1^+ - R_{II}}{R_{IS}} \exp\left(-R_1^+ t\right) + C_2 \frac{R_1^- - R_{SS}}{R_{IS}} \exp\left(-R_1^- t\right) \tag{2.68b}$$

where, the constants C_1 and C_2 can be determined from the initial conditions for the magnetizations $M_{\mathrm{Iz}}(0)$ and $M_{\mathrm{Sz}}(0)$. They yield:

$$C_1 = \frac{R_{\mathrm{IS}} M_{\mathrm{Sz}}(0) + \left(R_{\mathrm{SS}} - R_1^-\right) M_{\mathrm{Iz}}(0)}{R_1^+ - R_1^- + R_{\mathrm{II}} - R_{\mathrm{SS}}} \quad (2.69a)$$

$$C_2 = M_{\mathrm{Iz}}(0) - C_1 \quad (2.69b)$$

In fact, $M_{\mathrm{Iz}}(t)$ and $M_{\mathrm{Sz}}(t)$ in Eq. 2.68 denote the differences between the current magnetizations and their equilibrium values (see Eq. 2.59).

References

1. Abragam, A. (1961). *The Principles of Nuclear Magnetism* (New York: Oxford University Press).
2. Albrand, J. P., Taieb, M. C., Fries, P. H. and Belorizky, E. (1981). Frequency dependence of the effects of eccentricity on nuclear magnetic relaxation in liquids. ^{13}C relaxation of neopentane due to free radicals. *J. Chem. Phys.*, **75**, pp. 2141–2147.
3. Albrand, J. P., Taieb, M. C., Fries, P. H. and Belorizky, E. (1983). NMR study of spectral densities over a large frequency range for intermolecular relaxation in liquids: Pair correlation effects. *J. Chem. Phys.*, **78**, pp. 5809–5816.
4. Atkins, P. and de Paula, J. (2009). *Physical Chemistry* (New York: Oxford University Press).
5. Ayant, Y., Belorizky, E., Alizon, J. and Gallice, J. (1975). Calcul des densités Spectrales résultant d'un mouvement aléatoire de translation en relaxation par interaction dipolaire magnetique dans les liquids. *J. Phys. (Paris)*, **36**, pp. 991–1004.
6. Ayant, Y., Belorizky, E., Fries, P. and Rosset, J. (1977). Time correlation functions of isotropic intermolecular site-site interactions in liquids: Effects of the site eccentricity and of the molecular distribution. *J. Phys. France,* **38**, pp. 325–341.
7. Belorizky E. and Fries, P. H. (1981). Comment on nuclear spin relaxation by translational diffusion in liquids: Effects of the eccentricity of the nuclei. *J. Phys. C,* **14**, pp. 521–522.
8. Bertini, I., Luchinat, C. and Parigi, G. (2001). *Solution NMR of Paramagnetic Molecules* (Amsterdam: Elsevier).

9. Bloembergen, N., Purcell, E. M. and Pound, R. V. (1948). Relaxation effects in nuclear magnetic resonance absorption. *Phys. Rev.*, **73**, pp. 679–712.
10. Bloembergen, N. (1957). Proton relaxation times in paramagnetic solutions. *J. Chem. Phys.*, **27**, pp. 572–573.
11. Bloembergen, N. and Morgan, L. O. (1961). Proton relaxation times in paramagnetic solutions: Effects of electron spin relaxation. *J. Chem. Phys.*, **34**, pp. 842–850.
12. Böttcher, C. J. F. and Bordewijk, P. (1973). *Theory of Electric Polarization*, Vol. 2. (Amsterdam: Elsevier).
13. Brink, D. M. and Satchler, G. R. (1979). *Angular Momentum* (Oxford: Clarendon Press).
14. Canet, D. (2005). Introduction: General theory of nuclear relaxation. *Adv. Inorg. Chem.*, **57**, pp. 3–40.
15. Doi, M. and Edwards, S. F. (1986). *The Theory of Polymer Dynamics* (Oxford: Clarendon Press).
16. Edmunds, A. R. (1974). *Angular Momentum in Quantum Mechanics* (Princeton: Princeton University Press).
17. Ernst, R. R., Bodenhausen, G. and Wokaun, A. (1994). *Principles of Nuclear Magnetic Resonance in One and Two Dimensions* (Oxford: Clarendon Press).
18. Fries, P. H. (1983). Dipolar nuclear spin relaxation in liquids and plane fluids undergoing chemical reactions. *Mol. Phys.*, **48**, pp. 503–526.
19. Fries, P. H. and Belorizky, E. (1983). Monte Carlo calculation of the intermolecular dipolar spin relaxation in a liquid solution. *J. Chem. Phys.*, **79**, 1166–1170.
20. Gillespie, D. T. (1992). *Markov Processes* (Boston: Academic Press).
21. Grant, D. M. and Brown, R. A. (1996). *Encyclopedia of Nuclear Magnetic Resonance, Relaxation of Coupled Spin Systems from Rotational Diffusion* (Chichester: Wiley), pp. 4003–4018.
22. Harmon, J. F. (1970). Low frequency spin–lattice relaxation in glycerol. *Chem. Phys. Lett.*, **7**, pp. 207–208.
23. Hwang, L. P. and Freed, J. H. (1975). Dynamic effects of pair correlation function on spin relaxation by translational diffusion in liquids. *J. Chem. Phys.*, **63**, pp. 4017–4025.
24. Jeener, J. (1982). Superoperators in magnetic resonance. *Adv. Magn. Reson.*, **10**, pp. 1–51.

25. Kimmich, R. and Anoardo, E. (2004). Field-cycling NMR relaxometry. *Prog. Nucl. Magn. Reson. Spectrosc.*, **44**, pp. 257–320.
26. Kowalewski, J., Kruk, D. and Parigi, G. (2005). NMR relaxation in solution of paramagnetic complexes: Recent theoretical progress for $S \geq 1$. *Adv. Inorg. Chem.*, **57**, pp. 41–104.
27. Kowalewski, J. and Mäler, L. (2006). *Nuclear Spin Relaxation in Liquids: Theory, Experiments and Applications* (New York: Taylor & Francis).
28. Kruk, D. (2007). *Theory of Evolution and Relaxation of Multi-Spin Systems* (Bury St Edmunds: Arima).
29. Kruk, D., Meier, R. and Roessler, E. (2011). Translational and rotational diffusion of glycerol by means of field cycling 1H NMR relaxometry. *J. Phys. Chem. B*, **115**, pp. 951–957.
30. Kruk, D., Korpala, A., Rössle, E., Earle, K. A., Medycki, W., Moscicki, J. (2012) 1H NMR relaxation in glycerol solutions of nitroxide radicals: effects of translational and rotational dynamics. *J. Chem. Phys.*, **136**, pp. 114504–114512.
31. Kruk, D., Hermann, A. and Rossler, E. A. (2012). Field-cycling NMR relaxometry of viscous liquids and polymers. *Prog. Nucl. Magn. Reson. Spectrosc.*, **63**, pp. 33–64.
32. Kruk, D., Meier, R. and Rossler, E. A. (2012). *Phys. Rev. E*, **85**, pp. 020201–020205.
33. Lindsey, C. P. and Patterson, G. D. (1980). Detailed comparison of the Williams-Watts and Cole-Davidson functions. *J. Chem. Phys.*, **73**, pp. 3348–3357.
34. Shavit, A. and Gutfinger, C. (2009). *Thermodynamics, From Concepts to Applications* (New York: Taylor & Francis).
35. Sholl, C. A. (1981). Nuclear spin relaxation by translational diffusion in liquids and solids: High-and low-frequency limits. *J. Phys. C*, **14**, pp. 447–452.
36. Slichter, C. P. (1990). *Principles of Magnetic Resonance* (Berlin: Springer-Verlag).
37. Smith, S. A., Palke, W. E. and Gerig, J. T. (1992). The Hamiltonians of NMR. Part I, *Concepts Magn. Reson.*, **4**, pp. 107–144.
38. Smith, S. A., Palke, W. E. and Gerig, J. T. (1992). The Hamiltonians of NMR. Part II, *Concepts Magn. Reson.*, **4**, pp. 181–204.
39. Smith, S. A., Palke, W. E. and Gerig, J. T. (1993). The Hamiltonians of NMR. Part III, *Concepts Magn. Reson.*, **4**, pp. 151–177.

40. Solomon, I. (1955). Relaxation processes in a system of two spins. *Phys. Rev.*, **99**, pp. 559–565.

41. Solomon, I. and Bloembergen, N. (1956). Nuclear magnetic interactions in the HF molecule. *J. Chem. Phys.*, **25**, pp. 261–266.

42. Van Kampen, N. G. (1981). *Stochastic Processes in Physics and Chemistry* (Amsterdam: North Holland).

43. Varshalovich, D. A., Moskalev, A. N. and Khersonkii, V. K. (1988). *Quantum Theory of Angular Momentum* (Singapore: Word Scientific Publishing).

44. Woessner, D. E. (1996). *Encyclopedia of Nuclear Magnetic Resonance, Brownian Motion and Correlation Times* (Chichester: Wiley), pp. 1068–1084.

Chapter 3

Formal Theory of Spin Relaxation

In Chapter 2 the mechanism of spin relaxation phenomena has been explained in a semi-phenomenological, intuitive way. However, for more complex (real) spin systems one can hardly deal with spin dynamics in terms of "kinetic equations." Comprehensive theories of spin relaxation are based on the concept of a density operator, which is introduced in this chapter. The formal theory of spin relaxation is presented in terms of the density operator formalism.

3.1 The Concept of Density Operator

For a spin I interacting with an external magnetic field, $\vec{B}_0$, there are $(2I+1)$ allowed states, corresponding to the magnetic spin quantum numbers m_{I} taking the values, $-I, -I+1, \ldots, I-1, I$. For the simplest case of $I = 1/2$ there are two quantum states corresponding to $m_{\mathrm{I}} = \pm 1/2$, which have been denoted in Chapter 2 as $|1\rangle = |1/2\rangle$ and $|2\rangle = |-1/2\rangle$. In Chapter 2, the concept of spin Hamiltonian (energy operator) was used to describe the interaction of the magnetic moment associated with a spin I with the external magnetic field, that is, the Zeeman interaction (Eq. 2.1). The classical expression for the energy of this interaction, $E = -\vec{\mu} \cdot \vec{B}_0$ (scalar

Understanding Spin Dynamics
Danuta Kruk

ISBN 978-981-4463-49-2 (Hardcover), 978-981-4463-50-8 (eBook)
www.panstanford.com

product of the magnetic moment, $\vec{\mu}$ and the magnetic field, $\vec{B}_0$) has been written in terms of spin operators. Defining the reference frame in such a way that the magnetic field $\vec{B}_0$ is oriented along its z-axis (laboratory frame), the classical expression can be simplified to the form, $E = -\mu_z B_0$. The expression can be straightforwardly transformed to the quantum-mechanical representation (Eq. 2.1 in Chapter 2) in which the Zeeman Hamiltonian, H_Z, represents the energy E, while the longitudinal (parallel to the external magnetic field) component of the magnetic moment, μ_Z, is represented by the operator I_z.

For the "semi-quantum-mechanical" calculations of Chapter 2, it was enough to accept the fact that in an external magnetic field there are discrete spin states (energy levels) associated with the magnetic quantum numbers, m_I. This fact is described by the time-independent Schrödinger equation, which, in this case reads:

$$H_Z |m_I\rangle = \omega_I I_z |m_I\rangle = m_I \omega_I |m_I\rangle = E(m_I)|m_I\rangle \tag{3.1}$$

Equation 3.1 means that the Hamiltonian operator, H_Z, acting on the spin function $|m_I\rangle$ gives the energy corresponding to the m_I state, $E(m_I)$. The $|m_I\rangle$ quantities are referred to as eigenfunctions of the spin system, in "bra-ket" notation which will be often used in the book. One could also denote the eigenfunctions as, for instance, ψ_{m_I} (this is only a matter of convention).

According to the time-independent Schrödinger equation, a function $|i\rangle$ is an eigenfunction of a system described by a time-independent Hamiltonian, H_0, if this Hamiltonian acting on $|i\rangle$ gives the corresponding energy, E_i: $H_0 |i\rangle = E_i |i\rangle$. The functions $|m_I\rangle$ are often referred to as "Zeeman functions" as they are eigenfunctions of spin systems for which the time-independent Hamiltonian, H_0, contains only the Zeeman Hamiltonian, that is, $H_0 = H_Z$. We shall see in the forthcoming chapters that when H_0 contains other terms, the $|m_I\rangle$ functions are not anymore the eigenfunctions of the spin system. Nevertheless, the new eigenfunctions can always be given as linear combinations of the Zeeman functions, $|m_I\rangle$.

Let us now consider the "classical" magnetic moment $\vec{\mu}$. The expectation value of the z-component of the magnetic moment,

$\langle \mu_z \rangle (t)$, can be calculated as:

$$\langle \mu_z \rangle (t) = \int [\mu \sin\theta] \cdot p(\theta(t)) \sin\theta d\theta \tag{3.2}$$

where $\mu_z = \mu \sin\theta$, $p(\theta(t))$ is the probability that the magnetic moment $\vec{\mu}$ is oriented by angle θ with respect to the magnetic field $\vec{B}_0$ at time t, the integration $(\sin\theta d\theta)$ is taken over all orientations of the magnetic moment ($\sin\theta$ stems from transformation to spherical coordinate system). The quantum-mechanical counterpart of Eq. 3.2 yields:

$$\langle I_z \rangle (t) = \int \psi^* (t) I_z \psi (t) d\sigma \tag{3.3}$$

where $\psi(t)$ denotes an eigenfunction of the system, while $\psi^*(t)$ is its complex conjugation. The quantity $\psi^*(t)\psi(t) = |\psi(t)|^2$ corresponds to the probability $p(\theta(t))$, the integration over molecular orientations has been replaced by integration over spin space (the variable σ denotes a "spin coordinate"). The crucial feature of Eq. 3.3 is that the eigenfuntions are treated as time-dependent quantities. The expectation value, $\langle I_z \rangle (t)$, is associated with the longitudinal component of the I spin magnetization, $M_{Iz}(t)$, and it is natural that it evolves in time. This happens, however, not due to a time dependence of the spins states (eigenfunctions) as they are just determined by the magnetic quantum numbers, but due to the fact that the populations of these states evolve in time. This effect is described by the time-dependent Schrödinger equation:

$$\frac{\partial \psi(t)}{\partial t} = iH\psi(t) \tag{3.4}$$

Now, the Hamiltonian H includes time-independent as well as time-dependent terms. Equation 3.4 describes how the spin changes its quantum state in time. One can say that, for instance, initially a proton occupies the energy level of $m_I = 1/2$ and we are interested how its position changes in time.

When the Hamiltonian H is time-independent one gets from Eq. 3.4:

$$\psi_i(t) = \psi_i(0) \exp(-iE_i t) \tag{3.5}$$

In Eq. 3.5 it has been taken into account that the system is described by a set of eigenfunctions $\{\psi_i \equiv \psi_i(0)\}$. Equation 3.5

shows that the time dependence of the eigenfunctions can be shifted to another quantity—in this case the oscillating factor $\exp(-iE_i t)$. This implies that Eq. 3.3 can be rewritten as:

$$\langle I_z \rangle (t) = \int (\psi^* I_z \psi)\, \rho(t)\, d\sigma \tag{3.6}$$

where $\rho(t)$ is a time dependent operator referred to as density operator [1–18].

Now, it is convenient to come back to the "bra-ket" notation in which Eq. 3.6 takes the form:

$$\begin{aligned} \langle I_z \rangle (t) &= \sum_i \sum_j \rho_{ij}(t) \langle j | I_z | i \rangle \\ &= \sum_i \sum_j \langle i | \rho(t) | j \rangle \langle j | I_z | i \rangle = Tr\{\rho(t) I_z\} \end{aligned} \tag{3.7}$$

The last equality in Eq. 3.7 shows that time evolution of the expectation value of an arbitrary operator, $\langle O \rangle (t)$, can be obtained from the matrix operation [1, 3, 5, 7–9, 12, 13, 16]:

$$\langle O \rangle (t) = Tr\{\rho(t)\, O\} \tag{3.8}$$

This relationship will be extensively exploited throughout the book.

One sees that describing time evolution of a spin system means, in fact, evaluating time dependence of the density operator $\rho(t)$. For this purpose one should solve the time-dependent Schrödinger equation (Eq. 3.4) with the Hamiltonian H including time-independent (for instance, Zeeman) and time-dependent (for instance dipole–dipole) interactions. The system changes its state due to time-dependent interactions, while the static couplings only determine the available states.

The density operator can be obtained from the Liouville von Neumann equation [1, 3, 5, 7–18] (which can be derived from the time-dependent Schrödinger equation):

$$\frac{d\rho(t)}{dt} = -i\,[H(t), \rho(t)] \tag{3.9}$$

where the commutator $[P, Q]$ between two operators is defined as $[P, Q] = PQ - QP$. This issue will be explained in the next section. Before that it is useful to get some insight into the physical interpretation of the density operator. In Eq. 3.7 we have introduced the matrix representation of the density operator $\rho_{ij} = \langle i|\, \rho\, |j\rangle$ in a basis constructed from eigenfunctions $\{|i\rangle\}$ of a spin system. As already pointed out, the eigenfunctions $|i\rangle$ can always be expressed as linear combinations of the Zeeman functions $|m_I\rangle$, but in most cases they are not just the Zeeman functions (this happens only when there are no other time-independent interactions besides the Zeeman coupling).

In the forthcoming sections instead of the notation $|i\rangle$ for the eigenfunctions, we denote them as $|\alpha\rangle$, as this is more common in literature on spin dynamics. The diagonal elements of the density operator, $\rho_{\alpha\alpha}(t)$, represent the probabilities that the spin state described by the eigenfunction $|\alpha\rangle$ is occupied at time t. In other words one can say that $\rho_{\alpha\alpha}(t)$ describes the relative population of the state $|\alpha\rangle$ at time t. It is somewhat more difficult to understand the meaning of the off-diagonal elements $\rho_{\alpha\beta} = \langle\alpha|\, \rho\, |\beta\rangle, \alpha \neq \beta$. We shall come to this subject later.

3.2 The Liouville von Neumann Equation and Relaxation Rates

Except of some sophisticated cases, magnetic resonance experiments are performed in the laboratory frame, that is, in a frame in which the z-axis is determined by the direction of the external magnetic field. This statement is of large importance as we need to classify spin interactions as time-dependent or time-independent. For this purpose it is necessary to specify the reference frame with respect to which this classifications should be done; for instance, as explained in Chapter 2, dipole–dipole interactions are static in a molecule-fixed frame (when there is no internal dynamics) and time-dependent in the laboratory frame. The most natural choice is the laboratory axis system. The total Hamiltonian entering the Liouville von Neumann equation can be then decomposed into static (time-independent), H_0, and time-dependent, $H_1(t)$, parts:

$$H = H_0 + H_1(t) \tag{3.10}$$

To remove the explicit influence of H_0 one can make the substitutions:

$$\tilde{\rho}(t) = \exp(iH_0t)\,\rho(t)\exp(-iH_0t) \quad (3.11a)$$
$$\tilde{H}_1(t) = \exp(iH_0t)\,H_1(t)\exp(-iH_0t) \quad (3.11b)$$

In literature this is referred to as the transformation to interaction representation [1, 3, 5, 7–16]. Equation 3.9 simplifies then to the form:

$$\frac{d\tilde{\rho}(t)}{dt} = -i\left[\tilde{H}_1(t), \tilde{\rho}(t)\right] \quad (3.12)$$

The formal solution of Eq. 3.12, obtained by integrating its both sides over time, yields:

$$\tilde{\rho}(t) = \rho(0) + i\int_0^t \left[\tilde{\rho}(t'), \tilde{H}_1(t')\right] dt' \quad (3.13)$$

As the sub-integral function $\tilde{\rho}(t')$ is not known, this expression does not provide, in fact, any solution. Nevertheless, one can apply an iteration procedure, setting:

$$\tilde{\rho}(t') = \rho(0) + i\int_0^{t'} \left[\tilde{\rho}(t''), \tilde{H}_1(t'')\right] dt'' \quad (3.14)$$

and substituting Eq. 3.14 into Eq. 3.13, which after applying, in addition, the variable substitution, $\tau = t - t'$, leads to [1, 3, 5, 7–16]:

$$\frac{d\tilde{\rho}(t)}{dt} = i\left\langle\left[\tilde{\rho}(0), \tilde{H}_1(t)\right]\right\rangle - \int_0^t \left\langle\left[\left[\tilde{\rho}(0), \tilde{H}_1(t+\tau)\right], \tilde{H}_1(t)\right]\right\rangle d\tau \quad (3.15)$$

In Eq. (3.15) we have introduced the symbol '$\langle\rangle$' to explicitly indicate that we are interested in averaged values of the commutators, as we consider an ensemble of spin systems (for instance, many water molecules); $\langle\rangle$ denotes averaging over a molecular ensemble.

When the molecules undergo an isotropic motion, the time-dependent Hamiltonian, $H_1(t)$, averages to zero, that is, $\langle H_1(t)\rangle = 0$. This implies that $\left\langle\left[\tilde{\rho}(0), \tilde{H}_1(t)\right]\right\rangle = 0$, when $\tilde{H}_1(t)$ and $\tilde{\rho}(0)$ are not correlated, and the factorization $\left\langle \tilde{H}_1(t)\,\tilde{\rho}(0)\right\rangle = \left\langle \tilde{H}_1(t)\right\rangle \langle\tilde{\rho}(0)\rangle$

holds. Assuming now that in the sub-integral function (Eq. 3.15) $\tilde{\rho}(0) \cong \tilde{\rho}(t)$ and extending the upper limit of integration to infinity, one obtains [1, 3, 5, 7–16]:

$$\frac{d\tilde{\rho}(t)}{dt} = -\int_{0}^{\infty} \left\langle \left[\left[\tilde{\rho}(t), \tilde{H}_1(t+\tau)\right], \tilde{H}_1(t)\right]\right\rangle d\tau \qquad (3.16)$$

Setting $\tilde{\rho}(0) \cong \tilde{\rho}(t)$ one assumes that during the time t the density operator does not change significantly. The relative change of the density operator $\frac{\tilde{\rho}(t)-\tilde{\rho}(0)}{\tilde{\rho}(0)}$ is determined by the factor $t\omega_1^2\tau_c$, where ω_1 is the amplitude (in angular frequency units) of the interaction described by the Hamiltonian H_1 and τ_c denotes the correlation time characterizing the fluctuations of the H_1 interaction [1, 5, 9, 16]. At the same time the extension of the integration limit to infinity is permitted only when the contribution to the integral for times $\tau > t$ is negligible compared to the contribution of integrating from 0 to t. The correlation function for the Hamiltonian $\left\langle \tilde{H}_1(t+\tau)\, \tilde{H}_1(t)\right\rangle$ only lasts for a short time interval $\tau \leq \tau_c$ (i.e., in a good approximation it vanishes for $\tau > \tau_c$). Thus, when the condition $\omega_1\tau_c << 1$ (referred to as the Redfield condition) is fulfilled, one can set $\tilde{\rho}(0) \cong \tilde{\rho}(t)$ as well as extend the integration to infinity [1, 3, 5, 7–9, 12–18].

For practical purposes it is very convenient to write Eq. 3.16 in a matrix representation in the basis $\{|\alpha\rangle\}$ formed by the eigenfunction of the spin system. Detailed derivations of the matrix form of Eq. 3.16 are given in several books and reviews [1, 3, 5–9, 12–18]. Here we intend to avoid, if possible, long calculations, not directly accompanied by examples. Therefore, we shall not repeat the derivations. Instead of that let us write the final expression for the time evolution of the individual elements of the matrix representation of the density operator $\rho_{\alpha\alpha'}(t) = \langle\alpha|\,\rho(t)\,|\alpha'\rangle$ [1, 3, 5–9, 12–18]:

$$\frac{d\rho_{\alpha\alpha'}(t)}{dt} = -i\omega_{\alpha\alpha'}\rho_{\alpha\alpha'}(t) + \sum_{\substack{\beta\beta' \\ \omega_{\alpha\alpha'}=\omega_{\beta\beta'}}} \Gamma_{\alpha\alpha'\beta\beta'}\rho_{\beta\beta'}(t) \qquad (3.17)$$

One should note that in Eq. 3.17 we came back to the original density operator, $\rho(t)$ instead of its interaction frame representation, $\tilde{\rho}(t)$; Eq. 3.17 requires explanations. First of all, this is a set of coupled differential equations containing $(2I+1)^2 = N^2$ equations. The dimension is determined by the number of the matrix elements, that is, the number of pairs of the eigenfunctions $(|\alpha\rangle, |\beta\rangle)$. The matrix $[\Gamma]$ with elements $\Gamma_{\alpha\alpha'\beta\beta'}$ is called Redfield relaxation-evolution matrix. The elements $\Gamma_{\alpha\alpha'\beta\beta'}$ link the time evolution of the component $\rho_{\alpha\alpha'}(t)$ of the density operator with the component $\rho_{\beta\beta'}(t)$. The elements $\Gamma_{\alpha\alpha'\beta\beta'}$ are complex; their real parts $R_{\alpha\alpha'\beta\beta'}$ represent relaxation rates (relaxation matrix elements), while the imaginary parts $\Lambda_{\alpha\alpha'\beta\beta'}$ are referred to as dynamic frequency shift elements [1, 5, 7, 14–18]:

$$\Gamma_{\alpha\alpha'\beta\beta'} = R_{\alpha\alpha'\beta\beta'} + i\Lambda_{\alpha\alpha'\beta\beta'} \tag{3.18}$$

while $\omega_{\alpha\alpha'} = E_\alpha - E_{\alpha'}$. One should note that the summation in Eq. 3.17 includes only terms for which $\omega_{\alpha\alpha'} = \omega_{\beta\beta'}$; in other words, there is no connection between pairs of states characterized by different transition frequencies.

The relaxation rates $R_{\alpha\alpha'\beta\beta'}$ are given as linear combinations of generalized spectral density functions $\tilde{J}(\omega)$ [1, 3, 5–9, 12–18]:

$$\begin{aligned} R_{\alpha\alpha'\beta\beta'} &= -\tilde{J}_{\alpha\beta\alpha'\beta'}(\omega_{\alpha\beta}) - \tilde{J}_{\alpha\beta\alpha'\beta'}(\omega_{\beta'\alpha'}) \\ &\quad + \delta_{\alpha'\beta'}\sum_{\gamma}\tilde{J}_{\alpha\gamma\beta\gamma}(\omega_{\gamma\beta}) + \delta_{\alpha\beta}\sum_{\gamma}\tilde{J}_{\beta'\gamma\alpha'\gamma}(\omega_{\beta'\gamma}) \end{aligned} \tag{3.19}$$

where the $\tilde{J}_{\alpha\beta\alpha'\beta'}(\omega)$ quantities are defined as:

$$\tilde{J}_{\alpha\beta\alpha'\beta'}(\omega) = \int_0^\infty \left\langle \langle\alpha| H_1(t) |\beta\rangle \langle\alpha'| H_1(0) |\beta'\rangle \right\rangle \exp(-i\omega t) d\tau \tag{3.20}$$

Expressing the Hamiltonian $H_1(t)$ in the general manner as (Eq. 2.14):

$$H_1(t) = \xi \sum_{m=-l}^{l} (-1)^m F^l_{-m}(t)\, T^l_m \tag{3.21}$$

where ξ denotes a characteristic coupling constant of the interaction represented by the Hamiltonian $H_1(t)$ (for instance the dipole-dipole coupling, a_{DD}), while T^l_m is the m-component of a spin tensor operator of rank l, one gets a more explicit expression for the spectral density $\tilde{J}_{\alpha\beta\alpha'\beta'}(\omega)$:

$$\tilde{J}_{\alpha\beta\alpha'\beta'}(\omega) = \xi^2 \sum_{m=1}^{l} \langle\alpha| T_m^l |\beta\rangle \langle\alpha'| T_m^l |\beta'\rangle \times \int_0^\infty \langle F_{-m}^{l*}(t) F_{-m}^l(0)\rangle \exp(-i\omega t)d\tau$$
$$= J_m(\omega)\,\xi^2 \sum_{m=1}^{l} \langle\alpha| T_m^l |\beta\rangle \langle\alpha'| T_m^l |\beta'\rangle \qquad (3.22)$$

One distinguishes rank-one ($l = 1$) and rank-two ($l = 2$) spin interactions, which are expressed in terms of T_m^1 and T_m^2 operators, respectively, $m = -l, \ldots, l$. Dipole–dipole interactions discussed in Chapter 2 are an example of rank-two spin interactions.

Equation 3.22 contains only a single summation (over m) as there is no correlation between the quantities F_m^l for different m values, $\langle F_m^{l*}(t) F_{m'}^l(0)\rangle = 0$ for $m \neq m'$.

One can see a link between Eqs. 3.22 and 2.24 and Eqs. 2.25 and 2.28. The coefficients (1/4 and 1/16) in front of the integrals of Eq. 2.28 stem from the term $\sum_{m=1}^{2} \langle\alpha| T_m^2 |\beta\rangle \langle\alpha'| T_m^2 |\beta'\rangle$ (explicitly calculated in Eqs. 2.24 and 2.25), while $J_m(\omega) = \int_0^\infty C_m(t)\exp(-i\omega t)$, where $C_m(t)$ denotes the correlation function $C_m(t) = \langle F_m^{2(\mathrm{L})}(t) F_m^{2(\mathrm{L})}(0)\rangle$.

Introducing in the previous section the concept of the density operator, $\rho(t)$, it has been said that the diagonal elements, $\rho_{\alpha\alpha}(t)$, represent the relative populations of the quantum states $|\alpha\rangle$. They are linked to the longitudinal component of the I spin magnetization, $M_{\mathrm{Iz}}(t)$ via the quantity $\langle I_z\rangle(t)$. The diagonal elements (populations) are referred to as zero-quantum (0Q) coherences as only the Hamiltonian elements $\langle\alpha| T_0^l |\alpha\rangle$ ($m = 0$) differ from zero. The off-diagonal elements represent single-quantum ($|m| = 1$) and double-quantum ($|m| = 2$) spin coherences. For instance, for a system of two spins 1/2 (as considered in Chapter 2), the states $|\alpha\rangle = |1/2, 1/2\rangle$ and $|\beta\rangle = |1/2, -1/2\rangle$ are connected by the T_1^2 tensor component (Eq. 2.25), which means that $\rho_{\alpha\beta}$ represents a single-quantum (1Q) spin coherence, while the pair of states

$|\alpha\rangle = |1/2, 1/2\rangle$ and $|\gamma\rangle = |-1/2, -1/2\rangle$ linked by the T_2^2 tensor component forms a double-quantum spin coherence as $\langle\alpha| T_2^2 |\gamma\rangle \neq 0$ (Eq. 2.24). Single- and double-quantum coherences are also associated with spin operators. For instance, 1Q coherences are linked to the transversal magnetization component $M_{\mathrm{Ix}}(t)$ via the spectral density $\langle I_x\rangle(t)$.

3.3 Liouville Space and Redfield Kite

After this introduction we are ready for a complete, quantum-mechanical treatment of relaxation processes. Let us again begin with the simple case of two identical spins $I_1 = I_2 = 1/2$. As already explained, there are four Zeeman functions $\{|\alpha\rangle \left|m_{I_1}, m_{I_2}\right\rangle = |m_1, m_2\rangle\}$, for this system, $|1\rangle = |1/2, 1/2\rangle$, $|2\rangle = |1/2, -1/2\rangle$, $|3\rangle = |-1/2, 1/2\rangle$, and $|4\rangle = |-1/2, -1/2\rangle$. We know already that when the only time-independent interaction is the Zeeman coupling described by the Hamiltonian:

$$H_0 = H_z(I_1) + H_z(I_2) = \omega_{\mathrm{I}}(I_{1z} + I_{2z}) \tag{3.23}$$

(as the spins are identical, they are characterized by the same frequency, ω_{I}), the functions $\{|\alpha\rangle\}$ are eigenfunctions of this spin system with the corresponding energies, $E(m_1, m_2) = \omega_{\mathrm{I}}(m_1 + m_2)$. The set of the eigenfunctions $\{|\alpha\rangle\}$ forms a basis which is referred to as the Hilbert space [1, 3, 5–9, 12–18]; its dimension is $N = (2I_1 + 1)(2I_2 + 1)$ (which gives 4 in this case). One can construct from the functions $|\alpha\rangle$ N^2 (16 in this case) pairs of functions $\{|\alpha\rangle, |\alpha'\rangle\} \equiv \{|\alpha\rangle\langle\alpha'|\}$ which also form a basis which is referred to as the Liouville space [1, 3, 5–9, 12–18]. The Liouville space vectors, $|\alpha\rangle\langle\beta|$ are equivalent to the elements of the density operator, $\rho_{\alpha\alpha'}$.

As already said, the diagonal elements $\rho_{\alpha\alpha} \equiv |\alpha\rangle\langle\alpha|$ represent the relative populations of the corresponding spin states, $|\alpha\rangle$. This implies that Eq. 2.3 can be rewritten in the form:

$$\frac{d\rho_{11}(t)}{dt} = -R_{1111}\rho_{11}(t) - R_{1122}\rho_{22}(t) - R_{1133}\rho_{33}(t) - R_{1144}\rho_{44}(t) \tag{3.24a}$$

$$\frac{d\rho_{44}(t)}{dt} = -R_{1144}\rho_{11}(t) - R_{2244}\rho_{22}(t) - R_{3344}\rho_{33}(t) - R_{4444}\rho_{44}(t) \tag{3.24b}$$

where instead of the probabilities, W_0, W_1, and W_2, the relaxation rates $R_{\alpha\alpha\beta\beta}$ linking the changes in the population of the state $|\alpha\rangle$ ($\rho_{\alpha\alpha} \equiv |\alpha\rangle\langle\alpha|$) with the current population of the state $|\beta\rangle$ ($\rho_{\beta\beta} \equiv |\beta\rangle\langle\beta|$) are introduced. One can see now that Eq. 3.24 is a special case of Eq. 3.17. As $\omega_{\alpha\alpha} = 0$, Eq. 3.17 takes the following form for the populations $\rho_{\alpha\alpha}$:

$$\frac{d\rho_{\alpha\alpha}(t)}{dt} = \sum_{\substack{\beta\beta' \\ \omega_{\beta\beta'}=0}} \Gamma_{\alpha\alpha\beta\beta'}\rho_{\beta\beta'}(t) \tag{3.25}$$

Thus, as $\omega_{\beta\beta} = 0$ Eq. 3.25 links the populations of all spin states. As already said, the populations, $\rho_{\alpha\alpha} \equiv |\alpha\rangle\langle\alpha|$, are called 0Q coherences as only the zero-order ($m = 0$) component of the tensor operator leads to a non-zero value of the quantity, $\langle\alpha| T^l_m |\alpha\rangle$. One should notice here the consistency; the 0Q coherences do not imply changes in the spin energy ($\omega_{\alpha\alpha} = m\omega_{\mathrm{I}} = 0$) and therefore, they describe the populations of the quantum states.

Equation 3.24 corresponds to Eq. 3.25 when the summation in Eq. 3.25 includes only terms with $\beta = \beta'$. This happens when the energy levels of the spin system are not degenerated; then the condition $\omega_{\beta\beta'} = 0$ is fulfilled only for $\beta = \beta'$. However, in the case of two equivalent spins, the states $|2\rangle = |1/2, -1/2\rangle$ and $|3\rangle = |-1/2, 1/2\rangle$ are degenerated, that is, $E_2 = E_3 = 0$. This means that the set of Eqs. 3.24a,b should be extended by including the terms ρ_{23} and ρ_{32}:

$$\frac{d\rho_{11}(t)}{dt} = -R_{1111}\rho_{11}(t) - R_{1122}\rho_{22}(t) - R_{1133}\rho_{33}(t) \\ -R_{1144}\rho_{44}(t) - R_{1123}\rho_{23}(t) - R_{1132}\rho_{32}(t) \tag{3.26a}$$

$$\frac{d\rho_{44}(t)}{dt} = -R_{1144}\rho_{11}(t) - R_{2244}\rho_{22}(t) - R_{3344}\rho_{33}(t) \\ -R_{4444}\rho_{44}(t) - R_{2344}\rho_{23}(t) - R_{3244}\rho_{32}(t) \tag{3.26b}$$

These two equations should be complemented by the next two equations for ρ_{22} and ρ_{33}:

$$\frac{d\rho_{22}(t)}{dt} = -R_{1122}\rho_{11}(t) - R_{2222}\rho_{22}(t) - R_{2233}\rho_{33}(t) \\ -R_{2244}\rho_{44}(t) - R_{2223}\rho_{23}(t) - R_{2232}\rho_{32}(t) \tag{3.27a}$$

$$\frac{d\rho_{33}(t)}{dt} = -R_{1122}\rho_{11}(t) - R_{2222}\rho_{22}(t) - R_{3333}\rho_{33}(t) - R_{3344}\rho_{44}(t) - R_{3323}\rho_{23}(t) - R_{3332}\rho_{32}(t) \quad (3.27b)$$

and next two for ρ_{23} and ρ_{32}:

$$\frac{d\rho_{23}(t)}{dt} = -R_{1123}\rho_{11}(t) - R_{2223}\rho_{22}(t) - R_{2333}\rho_{33}(t) - R_{2344}\rho_{44}(t) - R_{2323}\rho_{23}(t) - R_{2332}\rho_{32}(t) \quad (3.28a)$$

$$\frac{d\rho_{32}(t)}{dt} = -R_{1132}\rho_{11}(t) - R_{2232}\rho_{22}(t) - R_{3333}\rho_{33}(t) - R_{3244}\rho_{44}(t) - R_{3223}\rho_{23}(t) - R_{3232}\rho_{32}(t) \quad (3.28b)$$

In this way we can conclude that for a system of two identical spins 1/2 there is six 0Q coherences which evolve in time in a coupled way according to the set of Eqs. 3.26, 3.27, and 3.28.

From the experimental viewpoint we are, however, interested in the evolution of the longitudinal magnetization which is given by the difference $(\rho_{44} - \rho_{11})$ (Eq. 2.4). The results obtained here should agree with those of Chapter 2. Dealing at that time with the populations we have not bothered with the states N_2 (corresponding to ρ_{22}) and N_3 (corresponding to ρ_{33}), although they are present in Eq. 2.3. It was not necessary because of the relationship, $W_{12} = W_{13} = W_{24} = W_{34} = W_1$, due to which the contributions of the populations N_2 and N_3 to the time evolution of the quantity $(N_4 - N_1)$ cancel. Thus, one can expect that now relations between the relaxation coefficients (relaxation rates) should lead to the same situation, which has a physical explanation—for identical spins there is no contribution to the magnetization from the states $|2\rangle = |1/2, -1/2\rangle$ and $|3\rangle = |-1/2, 1/2\rangle$.

The relaxation rates are given by Eq. 3.19 with the spectral densities defined by Eq. 3.20. To calculate them one has to specify the spin interaction which is responsible for the relaxation process. For protons, the dominating relaxation mechanism is provided by dipole–dipole interactions. The dipolar Hamiltonian is given by Eqs. 2.13 and 2.14 which are below combined into one formula:

$$H_{DD}^{(L)}(I_1, I_2)(t) = a_{DD} \sum_{m=-2}^{2} (-1)^m D_{0,-m}^{2(L)}(t)\, T_m^2(I_1, I_2) \quad (3.29)$$

with, $T_0^2(I_1, I_2) = \frac{1}{\sqrt{6}}\left[2I_{1z}I_{2z} - \frac{1}{2}(I_{1+}I_{2-} + I_{1-}I_{2+})\right]$, $T_{\pm1}^2(I_1, I_2) = \mp\frac{1}{2}[I_{1z}I_{2\pm} + I_{1\pm}I_{2z}]$, $T_{\pm2}^2(I_1, I_2) = \frac{1}{2}I_{1\pm}I_{2\pm}$, and $a_{DD} = \sqrt{6}\frac{\mu_0}{4\pi}\frac{\gamma_I^2\hbar^2}{r^3}$

(it is useful to remind here the explicit definitions of all quantities). The matrix representation of the Hamiltonian $H_{\mathrm{DD}}^{(\mathrm{L})}(I_1, I_2)(t)$ in the Hilbert space $\{|\alpha\rangle\}$ is given as (it has been shown in Chapter 2 how to calculate the elements $\langle\alpha| H |\beta\rangle$) [9]:

$$\left[H_{\mathrm{DD}}^{(\mathrm{L})}(I_1, I_2)\right](t)$$
$$= a_{\mathrm{DD}} \begin{bmatrix} \frac{1}{2\sqrt{6}} D_{0,0}^2(t) & -\frac{1}{4} D_{0,-1}^2(t) & -\frac{1}{4} D_{0,-1}^2(t) & -\frac{1}{2} D_{0,-2}^2(t) \\ \frac{1}{4} D_{0,1}^2(t) & -\frac{1}{2\sqrt{6}} D_{0,0}^2(t) & -\frac{1}{2\sqrt{6}} D_{0,0}^2(t) & \frac{1}{4} D_{0,-1}^2(t) \\ \frac{1}{4} D_{0,1}^2(t) & -\frac{1}{2\sqrt{6}} D_{0,0}^2(t) & -\frac{1}{2\sqrt{6}} D_{0,0}^2(t) & \frac{1}{4} D_{0,-1}^2(t) \\ \frac{1}{2} D_{0,2}^2(t) & -\frac{1}{4} D_{0,1}^2(t) & -\frac{1}{4} D_{0,1}^2(t) & \frac{1}{2\sqrt{6}} D_{0,0}^2(t) \end{bmatrix} \tag{3.30}$$

Knowing this matrix representation, we can straightforwardly profit from Eq. 3.19 which for the relaxation coefficients $R_{\alpha\alpha\beta\beta}$ takes a very simple form:

$$R_{\alpha\alpha\beta\beta} = 2\tilde{J}_{\alpha\beta\alpha\beta}\left(\omega_{\alpha\beta}\right) \tag{3.31}$$

Then one sees from the definition of the spectral densities $\tilde{J}(\omega)$ of Eq. 3.22 that, for instance:

$$R_{1122} = 2a_{\mathrm{DD}}^2 J(\omega_{\mathrm{I}}) \sum_{m=1}^{2} \langle 1| T_m^2(I_1, I_2) |2\rangle^2 = \frac{1}{8} a_{\mathrm{DD}}^2 J(\omega_{\mathrm{I}}) \tag{3.32}$$

as $\sum_{m=1}^{2} \langle 1| T_m^2(I_1, I_2) |2\rangle^2 = \langle 1| T_1^2(I_1, I_2) |2\rangle^2 = \frac{1}{16}$ (see the matrix of Eq. 3.30). It has been assumed that the spectral density is m-independent, $J_m(\omega) = J(\omega)$. Proceeding further in this way one gets [9]:

$$R_{1122} = R_{1133} = R_{2244} = R_{3344} = \frac{1}{8} a_{\mathrm{DD}}^2 J(\omega_{\mathrm{I}}) = W_1 \tag{3.33a}$$

$$R_{1144} = \frac{1}{2} a_{\mathrm{DD}}^2 J(2\omega_{\mathrm{I}}) = W_2 \tag{3.33b}$$

$$R_{2233} = \frac{1}{12} a_{\mathrm{DD}}^2 J(0) \tag{3.33c}$$

In Eqs. 3.33a,b, the obtained relaxation rates have been compared with the probabilities introduced in Chapter 2. Furthermore, one gets [9]:

$$R_{1111} = R_{4444} = a_{\mathrm{DD}}^2 \left[\frac{1}{4}J(\omega_\mathrm{I}) + \frac{1}{2}J(2\omega_\mathrm{I})\right] \tag{3.34a}$$

$$R_{2222} = R_{3333} = a_{\mathrm{DD}}^2 \left[\frac{1}{12}J(0) + \frac{1}{4}J(\omega_\mathrm{I})\right] \tag{3.34b}$$

which stems from the relationship, $R_{\alpha\alpha\alpha\alpha} = -\sum_{\beta\beta} R_{\alpha\alpha\beta\beta}$. We shall not evaluate here the other relaxation coefficients. As $R_{1123} = R_{2344}$ and $R_{3211} = R_{3244}$ the ρ_{23} and ρ_{32} terms cancel when the difference $\left(\frac{d\rho_{44}}{dt} - \frac{d\rho_{11}}{dt}\right)$ is taken. For similar reasons (Eq. 3.32a) also the ρ_{22} and ρ_{33} terms vanish. Eventually one gets:

$$\begin{aligned}\frac{d}{dt}(\rho_{44} - \rho_{11})(t) &= -2\left(R_{1122} + R_{1144}\right)(\rho_{44} - \rho_{11})(t)\\ &= -R_1 (\rho_{44} - \rho_{11})(t)\end{aligned} \tag{3.35}$$

with the spin–lattice relaxation rate:

$$\begin{aligned}R_1(\omega_\mathrm{I}) &= a_{\mathrm{DD}}^2 \left(\frac{1}{4}J(\omega_\mathrm{I}) + J(2\omega_\mathrm{I})\right)\\ &= \frac{3}{2}\left(\frac{\mu_0}{4\pi}\frac{\gamma_\mathrm{I}^2\hbar}{r_3}\right)^2 [J(\omega_\mathrm{I}) + 4J(2\omega_\mathrm{I})]\end{aligned} \tag{3.36}$$

in agreement with Eq. 2.37.

In the next step, we shall consider the 1Q coherences, that is, the pairs of states (vectors in the Liouville space) for which $\langle\alpha| T_{\pm 1}^{l} |\beta\rangle \neq 0$. There is eight of them, $|1\rangle\langle 2|,\ |1\rangle\langle 3|,\ |2\rangle\langle 4|,\ |3\rangle\langle 4|$ and correspondingly, $|2\rangle\langle 1|,\ |3\rangle\langle 1|,\ |4\rangle\langle 2|,\ |4\rangle\langle 3|$. For the first group the transition frequency is ω_I, while for the second one, $-\omega_\mathrm{I}$. When the energy levels are not degenerated, 1Q coherences evolve independently, because Eq. 3.17 forbids a coupling between coherences associated with different frequencies (we shall show this soon). Nevertheless, due to the degeneracy of the energy levels, the situation with the 1Q coherences for a system of identical spins 1/2 is special. We have to again consider a set of equations:

$$\frac{d\rho_{12}(t)}{dt} = -i\omega_I \rho_{12}(t) + R_{1212}\rho_{12}(t) + R_{1213}\rho_{13}(t) + R_{1224}\rho_{24}(t) + R_{1234}\rho_{34}(t) \quad (3.37a)$$

$$\frac{d\rho_{13}(t)}{dt} = -i\omega_I \rho_{13}(t) + R_{1213}\rho_{12}(t) + R_{1313}\rho_{13}(t) + R_{1324}\rho_{24}(t) + R_{1334}\rho_{34}(t) \quad (3.37b)$$

$$\frac{d\rho_{24}(t)}{dt} = -i\omega_I \rho_{24}(t) + R_{1224}\rho_{12}(t) + R_{1324}\rho_{13}(t) + R_{2424}\rho_{24}(t) + R_{2434}\rho_{34}(t) \quad (3.37c)$$

$$\frac{d\rho_{34}(t)}{dt} = -i\omega_I \rho_{34}(t) + R_{1234}\rho_{12}(t) + R_{1334}\rho_{13}(t) + R_{2434}\rho_{24}(t) + R_{3434}\rho_{34}(t) \quad (3.37d)$$

According to Eq. 3.19 the relaxation rates $R_{\alpha\beta\alpha\beta}$ are given as:

$$R_{\alpha\beta\alpha\beta} = -2\tilde{J}_{\alpha\alpha\beta\beta}(0) + \tilde{J}_{\alpha\alpha\alpha\alpha}(0) + \tilde{J}_{\beta\beta\beta\beta}(0) + 2\tilde{J}_{\alpha\beta\alpha\beta}(\omega_{\alpha\beta}) + \sum_{\gamma\neq\alpha}\tilde{J}_{\alpha\gamma\alpha\gamma}(\omega_{\alpha\gamma}) + \sum_{\gamma\neq\beta}\tilde{J}_{\beta\gamma\beta\gamma}(\omega_{\beta\gamma}) \quad (3.38)$$

This leads to the relation:

$$R_{1212} = R_{1313} = R_{2424} = R_{3434} = \frac{1}{4}\left(\frac{\mu_0}{4\pi}\frac{\gamma_{\mathrm{I}}^2\hbar}{r^3}\right)^2 [5J(0) + 6J(\omega_{\mathrm{I}}) + 6J(2\omega_{\mathrm{I}})] = a \quad (3.39)$$

Using the general formula of Eq. 3.19 for other relaxation rates, one also obtains:

$$R_{1234} = R_{1324} = R_{1334} = R_{1224} = -\frac{3}{8}\left(\frac{\mu_0}{4\pi}\frac{\gamma_{\mathrm{I}}^2\hbar}{r^3}\right)^2 J(\omega_{\mathrm{I}}) = b \quad (3.40)$$

$$R_{2434} = R_{1213} = \frac{1}{4}\left(\frac{\mu_0}{4\pi}\frac{\gamma_{\mathrm{I}}^2\hbar}{r^3}\right)^2 [2J(0) + 3J(\omega_{\mathrm{I}}) + 6J(2\omega_{\mathrm{I}})] = c \quad (3.41)$$

As anticipated, the longitudinal component of the magnetization, M_{Iz}, is associated with the $\langle I_z \rangle$ quantity. Every operator can be represented as a linear combination of the vectors of the Liouville space constructed from pairs of the eigenfunctions, $\rho_{\alpha\alpha'} = |\alpha\rangle\langle\alpha'|$. The $I_z = I_{1z} + I_{2z}$ operator is expressed as:

$$I_z \propto (\rho_{11} - \rho_{44}) \equiv |1\rangle\langle 1| - |4\rangle\langle 4| \quad (3.42)$$

When one is interested in the evolution of the transversal component of the magnetization, for instance M_{Ix}, one needs to consider the corresponding operator, I_{x}. It is more convenient to use the operator $I_{+} = I_{\mathrm{x}} + i I_{\mathrm{y}}$ (then one can take the real part of the obtained quantity). The operator I_{+} has the following representation in the Liouville basis $\{|\alpha\rangle\langle\alpha'|\}$:

$$I_{+} = I_{1+} + I_{2+} \propto (|1\rangle\langle 2| + |3\rangle\langle 4|) + (|1\rangle\langle 3| + |2\rangle\langle 4|) \tag{3.43}$$

After some algebraic manipulations one obtains from Eqs. 3.39–3.41:

$$\frac{d(\rho_{12} + \rho_{34} + \rho_{13} + \rho_{24})(t)}{dt} = -(a + 2b + c)(\rho_{12} + \rho_{34} + \rho_{13} + \rho_{24})(t) \tag{3.44}$$

This means that the transversal magnetization decays exponentially with a relaxation rate, R_2, referred to as spin–spin relaxation rate, which yields:

$$R_2(\omega_{\mathrm{I}}) = a + 2b + c = \frac{1}{4}\left(\frac{\mu_0}{4\pi}\frac{\gamma_{\mathrm{I}}^2\hbar}{r^3}\right)^2 [7J(0) + 6J(\omega_{\mathrm{I}}) + 12J(2\omega_{\mathrm{I}})] \tag{3.45}$$

Eventually, we can consider the double-quantum coherences, ρ_{14} and ρ_{41}, which are associated with the operators $(I_{1+}I_{2+})$ and $(I_{1-}I_{1-})$, respectively (i.e., with the rank-two components of the tensor operator, $T^2_{\pm 2}$). In this case the set of the evolution equations (Eq. 3.17) consists only of one equation:

$$\frac{d\rho_{14}(t)}{dt} = -i2\omega_{\mathrm{I}}\rho_{14}(t) + R_{1414}\rho_{14}(t) \tag{3.46}$$

which implies a single-exponential evolution of the $\rho_{14}(t)$ coherence with the rate:

$$R_{DQ}(\omega_{\mathrm{I}}) = \frac{3}{2}\left(\frac{\mu_0}{4\pi}\frac{\gamma_{\mathrm{I}}^2\hbar}{r^3}\right)^2 [J(\omega_{\mathrm{I}}) + 2J(2\omega_{\mathrm{I}})] \tag{3.47}$$

Now, it is of interest to inquire into the relaxation dynamics of a system consisting of two non-equivalents spins and figure out how the formal treatment is related to the somewhat intuitive description presented in Chapter 2. The basis (Hilbert space) is formed by the eigenfunctions $\{|\alpha\rangle = |m_{\mathrm{I}}, m_{\mathrm{S}}\rangle\}$; let us remind the labeling $|1\rangle = |1/2, 1/2\rangle$, $|2\rangle = |1/2, -1/2\rangle$, $|3\rangle = |-1/2, 1/2\rangle$, and $|4\rangle = |-1/2, -1/2\rangle$. Thus, the operators I_z and S_z are represented as the following combinations of the Liouville space functions, $|\alpha\rangle\langle\alpha|$:

$$\langle I_z\rangle = |1\rangle\langle 1| + |2\rangle\langle 2| - |3\rangle\langle 3| - |4\rangle\langle 4| = \rho_{11} + \rho_{22} - \rho_{33} - \rho_{44} \tag{3.48a}$$

$$\langle S_z\rangle = |1\rangle\langle 1| + |3\rangle\langle 3| - |2\rangle\langle 2| - |4\rangle\langle 4| = \rho_{11} + \rho_{33} - \rho_{22} - \rho_{44} \tag{3.48b}$$

The set of equations for the 0Q coherences is limited only to the populations, that is, it contains four equations, like Eqs. 3.25a,b and 3.26a,b, as there are no coherences $|\alpha\rangle\langle\alpha'|$ $(\alpha \neq \alpha')$ for which $\omega_{\alpha\alpha'} = 0$. This set of equations can be conveniently written in a matrix form containing the explicit expressions (in terms of the spectral densities, $J(\omega)$) for the relaxation rates [9]:

$$\frac{d}{dt}\begin{bmatrix}\rho_{11}\\ \rho_{22}\\ \rho_{33}\\ \rho_{44}\end{bmatrix}(t) = a_{DD}^2\begin{bmatrix} R_{1111} & \frac{1}{8}J(\omega_S) & \frac{1}{8}J(\omega_I) & \frac{1}{2}J(\omega_I+\omega_S)\\ \frac{1}{8}J(\omega_S) & R_{2222} & \frac{1}{12}J(\omega_I-\omega_S) & \frac{1}{8}J(\omega_I)\\ \frac{1}{8}J(\omega_I) & \frac{1}{12}J(\omega_I-\omega_S) & R_{3333} & \frac{1}{8}J(\omega_S)\\ \frac{1}{2}J(\omega_I+\omega_S) & \frac{1}{8}J(\omega_I) & \frac{1}{8}J(\omega_S) & R_{4444}\end{bmatrix} \times \begin{bmatrix}\rho_{11}\\ \rho_{22}\\ \rho_{33}\\ \rho_{44}\end{bmatrix}(t) \tag{3.49}$$

with

$$R_{1111} = R_{2222} = R_{3333} = R_{4444} = -\left[\frac{1}{8}J(\omega_I) + \frac{1}{8}J(\omega_S) + \frac{1}{2}J(\omega_I+\omega_S)\right] \tag{3.50}$$

The dipole–dipole coupling is now defined as $a_{DD} = \sqrt{6}\frac{\mu_0}{4\pi}\frac{\gamma_I\gamma_S\hbar}{r^3}$. Using the representations of the spin operators of Eq. 3.48 one obtains [9]:

$$\frac{d}{dt}\left(\rho_{11}+\rho_{22}-\rho_{33}-\rho_{44}\right)=-R_{\mathrm{II}}\left(\rho_{11}+\rho_{22}-\rho_{33}-\rho_{44}\right)$$
$$-R_{\mathrm{IS}}\left(\rho_{11}+\rho_{33}-\rho_{22}-\rho_{44}\right) \quad (3.51a)$$

$$\frac{d}{dt}\left(\rho_{11}+\rho_{33}-\rho_{22}-\rho_{44}\right)=-R_{\mathrm{SS}}\left(\rho_{11}+\rho_{33}-\rho_{22}-\rho_{44}\right)$$
$$-R_{\mathrm{SI}}\left(\rho_{11}+\rho_{22}-\rho_{33}-\rho_{44}\right) \quad (3.51b)$$

with the relaxation rates R_{II}, R_{SS}, R_{IS}, and R_{SI} are given by Eqs. 2.62–2.64. Discussing 1Q coherences one has to consider 1Q transitions of the I and S spins. The I spin coherences are represented by the elements ρ_{13} (ρ_{31}) and ρ_{24} (ρ_{42}), while the S spin 1Q coherences are associated with ρ_{12} (ρ_{21}) and ρ_{34} (ρ_{43}). Thus, the representations of the I_+ and S_+ operators are:

$$\langle I_+\rangle = |1\rangle\langle 3| + |2\rangle\langle 4| \quad (3.52a)$$

$$\langle S_+\rangle = |1\rangle\langle 2| + |3\rangle\langle 4| \quad (3.52b)$$

This implies that the evolution of the I spin coherences is described by the set of equations:

$$\frac{d\rho_{13}(t)}{dt} = -i\omega_{\mathrm{I}}\rho_{13}(t) + R_{1313}\rho_{13}(t) + R_{1324}\rho_{24}(t) \quad (3.53a)$$

$$\frac{d\rho_{24}(t)}{dt} = -i\omega_{\mathrm{I}}\rho_{24}(t) + R_{1324}\rho_{13}(t) + R_{2424}\rho_{24}(t) \quad (3.53b)$$

As $\omega_{\mathrm{I}} \neq \omega_{\mathrm{S}}$ the I and S spin coherences are decoupled. Using Eqs. 3.29 and 3.30 one gets:

$$R_{1313} = R_{2424} = \frac{1}{24}a_{\mathrm{DD}}^2\left[4J(0) + 3J(\omega_{\mathrm{I}}) + 3J(\omega_{\mathrm{S}})\right.$$
$$\left.+6J(\omega_{\mathrm{I}}+\omega_{\mathrm{S}}) + J(\omega_{\mathrm{S}}-\omega_{\mathrm{I}})\right] \quad (3.54)$$

$$R_{1324} = -\frac{1}{24}a_{\mathrm{DD}}^2 J(\omega_{\mathrm{I}}) \quad (3.55)$$

This implies that the perpendicular component of the I spin magnetization M_{Ix} decays exponentially:

$$M_{\mathrm{Ix}}(t) \equiv Re\langle I_+\rangle(t) = M_{\mathrm{Ix}}(0)\exp\left(-T_{2\mathrm{I}}t\right) \quad (3.56)$$

where the spin–spin relaxation rate is equal to:

$$R_{2\mathrm{I}}(\omega) = \frac{1}{4}\left(\frac{\mu_0}{4\pi}\frac{\gamma_{\mathrm{I}}\gamma_{\mathrm{S}}\hbar}{r^3}\right)^2\left[4J(0) + 2J(\omega_{\mathrm{I}}) + 3J(\omega_{\mathrm{S}})\right.$$
$$\left.+6J(\omega_{\mathrm{I}}+\omega_{\mathrm{S}}) + J(\omega_{\mathrm{S}}-\omega_{\mathrm{I}})\right] \quad (3.57)$$

The get an analogous expression for R_{1S} it is enough to interchange the indices "I" and "S". To complete this picture let us consider the double-quantum coherences—they are represented by the element ρ_{14}; both spins participate in the double-quantum process which is associated with the transition frequency $(\omega_I + \omega_S)$. The double-quantum coherence $\rho_{14}(t) \equiv \langle I_+ S_+ \rangle (t)$ decays with the relaxation rate:

$$R_{DQ} = \frac{3}{2}\left(\frac{\mu_0}{4\pi}\frac{\gamma_I^2 \hbar}{r^3}\right)^2 \left[\frac{1}{2}J(\omega_I) + \frac{1}{2}J(\omega_S) + 2J(\omega_I + \omega_S)\right] \quad (3.58)$$

Following the examples one can observe that Eq. 3.17 can be conveniently written in a matrix form as:

$$\frac{d}{dt}[\rho(t)] = -[\Gamma - i\omega 1] \cdot [\rho(t)] \quad (3.59)$$

The matrices are set up in the Liouville space $\{|\alpha\rangle\langle\alpha'|\}$ constricted from the eigenfunctions $|\alpha\rangle$ of the system. The real part of $[\Gamma]$ is referred to as the Redfield relaxation matrix $[R] = \mathrm{Re}\{[\Gamma]\}$, $[i\omega 1]$ is diagonal and contains the transition frequencies $\omega_{\alpha\alpha'}$ [1, 3, 7–9, 12–18]. In Fig. 3.1 we show the structure of the relaxation matrix $[R]$ for a system of two identical spins 1/2. The matrix is

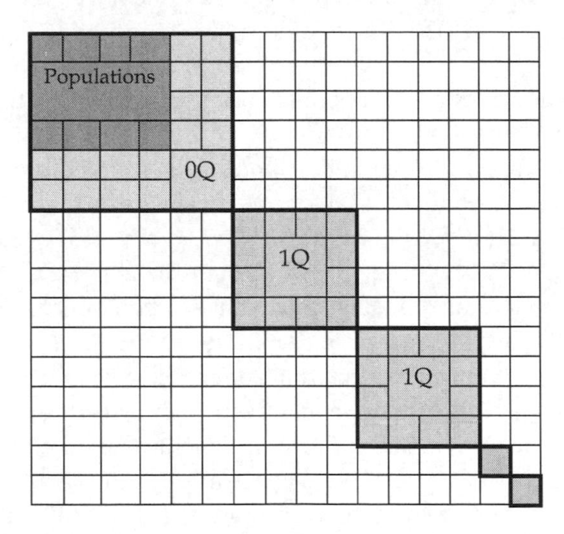

Figure 3.1 The structure of the Redfield relaxation matrix for a system of two identical spins 1/2.

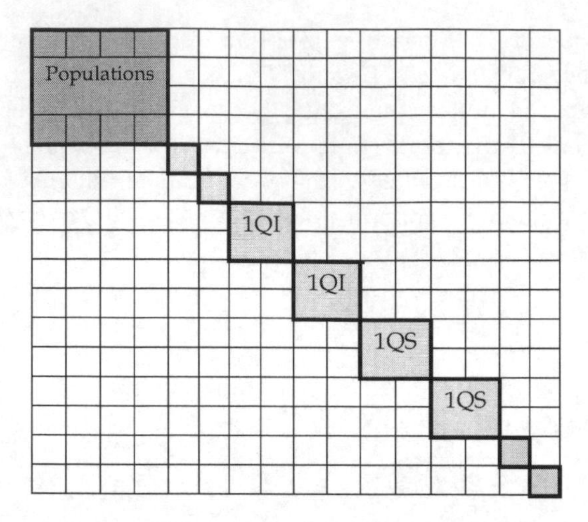

Figure 3.2 The structure of the Redfield relaxation matrix for a system of two non-equivalent spins 1/2.

of the dimension 16 × 16 which is $N^2 = [(2I_1 + 1)(2I_2 + 1)]^2$. The population block $|\alpha\rangle\langle\alpha|$ (4×4) is, in this case coupled to other 0Q coherences, $|2\rangle\langle 3|$ and $|3\rangle\langle 2|$. Then there are the 1Q coherence blocks corresponding to the transition frequency ω_I and $-\omega_I$, and eventually we have the double-quantum coherence (2Q) for $2\omega_I$ and $-2\omega_I$.

In Fig. 3.2 the structure of the Redfield relaxation matrix for two non-equivalent spins 1/2 is shown.

In this case the population block is decoupled from the 0Q coherences, $|2\rangle\langle 3|$ and $|3\rangle\langle 2|$, as they correspond to the transition frequencies $\pm(\omega_I - \omega_S) \neq 0$. Then, we have the 1Q coherences for the I spin, 1QI (Eq. 3.53) characterized by the transition frequency $\pm\omega_I$, and analogously, the 1Q coherences for the S spin, 1QS, corresponding to the frequency $\pm\omega_S$. The matrix is completed by the double-quantum coherences. The reason why the relaxation matrices contain so many zero-elements (the non-colored parts) is encoded in Eq. 3.17 in which the summation includes only the coherences with the same transition frequency. In other words, this equation says that one can neglect couplings between coherences characterized by different frequencies because due to fast oscillations of the terms $\exp\left[-i\left(\omega_{\alpha\alpha'} - \omega_{\beta\beta'}\right)t\right]$ for $\omega_{\alpha\alpha'} \neq$

$\omega_{\beta\beta'}$, the coupling is averaged out; this is referred to as the secular approximation [1, 3, 5, 7–9, 13–18]. Although couplings between different coherences should be carefully considered for individual cases one can say that the relaxation matrix consists of a population block ($N \times N$) and a diagonal (semi-diagonal) part referred to as Redfield kite [1, 3, 5, 7–9, 13–18].

As we perform the calculations in the Liouville space and we shall use it many times throughout the book, it is worth to write the Liouville von Neumann equation (Eq. 3.9) in the Liouville formalism:

$$\frac{d\rho(t)}{dt} = -i L\rho(t) \tag{3.60}$$

where the Liouville operator, L, is defined as $L = -i\,[H, \ldots]$.

3.4 Validity Range of the Perturbation Theory

There are two stages in deriving Eq. 3.17 at which very important assumptions have been made. The perturbation solution (Eq. 3.15) of the Liouville von Neumann equation is limited to the second-order term. This means that the higher order terms are negligible. This holds only when $\omega_1\tau_c << 1$, where ω_1 is the amplitude of the time-dependent Hamiltonian, H_1, in angular frequency units (in our case $\omega_1 = \omega_{DD}$ as it refers to the dipole–dipole coupling), while τ_c is a correlation time of the fluctuations of this interaction. This condition is referred to as the Redfield condition [1, 3, 5, 7–9, 13–18] (see Section 3.2). In Eq. 3.16 $\rho(t)$ has been replaced by $\rho(0)$ and the upper limit of integration has been extended to infinity. This can be done only when the relaxation is much slower than the motion causing it, that is, when $R << \tau_c^{-1}$ (R denotes a relaxation rate). In the low frequency range, when $\omega\tau_c << 1$, the relaxation rate is determined by the factor $\omega_1^2\tau_c$. Thus, it is required that $\omega_1^2\tau_c << \tau_c^{-1}$, that is, $(\omega_1\tau_c)^2 << 1$. Obviously, if this condition is fulfilled also $\omega_1\tau_c << 1$ holds. However, at higher frequencies the relaxation becomes slower. Therefore, it can happen that the relaxation is slower than the motion, $R << \tau_c^{-1}$, but the Redfield condition, $\omega_1\tau_c << 1$, is not fulfilled.

3.5 Spin Relaxation in Time Domain

As discussed in Section 3.4 one can derive expressions for relaxation rates without providing a complete solution of the sets of relaxation equations (for instance Eq. 3.37 or 3.49). This is possible by grouping the equations in such a way that the resulted expressions directly describe the evolution of magnetization components or other measurable quantities referred to as magnetization modes; for instance the double-quantum magnetization mode $\langle I_+S_+\rangle$ associated with R_{DQ} (Eq. 3.47) [3, 5, 7, 8, 13]. This approach leads to closed form expressions for relaxation rates which can (in most cases) be directly accessed in experiments. However, although the description is highly advantageous, it might be difficult to apply it to systems containing many mutually coupled spins (examples are shown in next chapters).

Relaxation equations (like Eq. 3.49 for the populations of a system containing two non-equivalent spins 1/2) can be explicitly solved. Using Eq. 3.49 as an example, the solution takes the form:

$$\begin{bmatrix} \rho_{11} \\ \rho_{22} \\ \rho_{33} \\ \rho_{44} \end{bmatrix}(t) = C_1 \begin{bmatrix} p_{11} \\ p_{12} \\ p_{13} \\ p_{14} \end{bmatrix} \exp(\lambda_1 t) + C_2 \begin{bmatrix} p_{21} \\ p_{22} \\ p_{23} \\ p_{24} \end{bmatrix} \exp(\lambda_2 t) + C_3 \begin{bmatrix} p_{31} \\ p_{32} \\ p_{33} \\ p_{34} \end{bmatrix} \exp(\lambda_3 t) + C_4 \begin{bmatrix} p_{41} \\ p_{42} \\ p_{43} \\ p_{44} \end{bmatrix} \exp(\lambda_4 t) \quad (3.61)$$

where λ_i are eigenvalues of the relaxation matrix R_0 (index "0" denotes 0Q transitions):

$$R_0 = a_{DD}^2 \begin{bmatrix} R_{1111} & \frac{1}{8}J(\omega_S) & \frac{1}{8}J(\omega_I) & \frac{1}{2}J(\omega_I+\omega_S) \\ \frac{1}{8}J(\omega_S) & R_{2222} & \frac{1}{12}J(\omega_I-\omega_S) & \frac{1}{8}J(\omega_I) \\ \frac{1}{8}J(\omega_I) & \frac{1}{12}J(\omega_I-\omega_S) & R_{3333} & \frac{1}{8}J(\omega_S) \\ \frac{1}{2}J(\omega_I+\omega_S) & \frac{1}{8}J(\omega_I) & \frac{1}{8}J(\omega_S) & R_{4444} \end{bmatrix} \quad (3.62)$$

while $[P_i] = [p_{i1}, p_{i2}, p_{i3}, p_{i4}]$ denotes the eigenvector corresponding to λ_i. The constants C_i are determined by the initial populations $\rho_{\alpha\alpha}(0)$. As explained in Chapter 2 even for quite high magnetic fields the populations do not change much at room temperature. Therefore, one can set for all populations, $\rho_{\alpha\alpha}(0) = 1$. This is referred to as high temperature approximation [1, 3, 14–18]. As a result of this treatment one explicitly obtains information about the population of every spin state at time t. Then, knowing how the spin magnetization is related to the individual populations, one can follow its evolution in time.

References

1. Abragam, A. (1961). *The Principles of Nuclear Magnetism* (New York: Oxford University Press).
2. Blum, K. (1989). *Density Matrix Theory and Applications* (New York: Plenum Press).
3. Ernst, R.R., Bodenhausen, G. and Wokaun, A. (1994). *Principles of Nuclear Magnetic Resonance in One and Two Dimensions* (Oxford: Clarendon Press).
4. Fano, U. (1957). Description of states in quantum mechanics by density matrix and operator techniques. *Rev. Mod. Phys.*, **29**, pp. 74–93.
5. Goldman, M. (2001). Formal theory of spin–lattice relaxation. *J. Magn. Reson.*, **149**, pp. 160–187.
6. Hubbard, P.S. (1961). Quantum-mechanical and semiclassical forms of the density operator theory of relaxation. *Rev. Mod. Phys.*, **33**, pp. 249–264.
7. Jeener, J. (1982). Superoperators in magnetic resonance. *Adv. Magn. Reson.*, **10**, pp. 1–51.
8. Kowalewski, J. and Mäler, L. (2006). *Nuclear Spin Relaxation in Liquids: Theory, Experiments and Applications* (New York: Taylor & Francis).
9. Kruk, D. (2007). *Theory of Evolution and Relaxation of Multi-Spin Systems* (Bury St Edmunds: Arima).
10. Kubo, R. (1963). Stochastic Liouville equations. *J. Math. Phys.*, **4**, pp. 174–183.
11. Kubo, R. and Tomita, K. (1954). A general theory of magnetic resonance absorption. *J. Phys. Soc. Jpn.*, **9**, pp. 888–819.

12. Levitt, M.H. (2001). *Spin Dynamics* (Chichester: Wiley).
13. Mayne, C.L. and Smith, S.A. (1996). *Encyclopedia of Nuclear Magnetic Resonance, Relaxation Processes in Coupled-Spin Systems* (Chichester: Wiley), pp. 4053–4071.
14. Redfield, A.G. (1957). On the theory of relaxation processes. *IBM J. Res. Dev.*, **1**, pp. 19–31.
15. Redfield, A.G. (1996). *Encyclopedia of Nuclear Magnetic Resonance, Relaxation Theory: Density Matrix Formulation* (Chichester: Wiley), pp. 4085–4092.
16. Slichter, C.P. (1990). *Principles of Magnetic Resonance* (Berlin: Springer-Verlag).
17. Tomita, K. (1958). A general theory of magnetic resonance saturation. *Progr. Theor. Phys.*, **19**, pp. 541–580.
18. Wangsness, R.K. and Bloch, F. (1953). The dynamic theory of nuclear induction. *Phys. Rev.*, **89**, pp. 728–739.

Chapter 4

Spin Resonance Lineshape Analysis

In Chapter 3 we dealt with spin relaxation processes, which are a result of quantum-mechanical features of spin systems combined with molecular dynamics. Spin relaxation phenomena are a rich source of information on spin interactions and motional mechanisms, provided appropriate theoretical models are available. A complementary way of inquiring into the interplay between quantum-mechanical and dynamical properties of molecular (spin) systems is spin resonance spectroscopy. In this chapter principles of spin resonance lineshape analysis are described.

4.1 The Concept of Spin Resonance Spectrum

To understand the information encoded in nuclear magnetic resonance (NMR) or electron spin resonance (ESR) spectra one should begin with the case of a rigid system, that is, assume that the motion is very slow (temperature is low). Let us consider a single nucleus of the spin quantum number $I = 1/2$. An external magnetic field, B_0, leads, as already explained, to an energy level splitting ($m_I = \pm 1/2$) described by the frequency $\omega_I = \gamma_I B_0$. An analogous situation takes place for an electron spin $S = 1/2$, but then the energy levels ($m_S = \pm 1/2$) are separated by the frequency $\omega_S = \gamma_S$

Understanding Spin Dynamics
Danuta Kruk

ISBN 978-981-4463-49-2 (Hardcover), 978-981-4463-50-8 (eBook)
www.panstanford.com

$B_0 \cong 658\omega_{\mathrm{I}}$ (as $\gamma_{\mathrm{S}} \cong 658\gamma_{\mathrm{I}}$). This means that when plotting the probability of the spin transition between the energy levels versus frequency, $P(\omega)$, one should get delta-Dirac function, $P(\omega) = 1$ for $\omega = \omega_{\mathrm{I(S)}}$ and $P(\omega) = 0$ for $\omega \neq \omega_{\mathrm{I(S)}}$; there are no other transitions except of this one for $\omega = \omega_{\mathrm{I(S)}}$. This example illustrates the main concept of a "rigid" spin resonance spectrum, that is, a spectrum obtained for a system for which motional processes (like rotation, translation, etc.) are slow. In such a case, the spectrum is nothing else but the probability density function, $P(\omega)$, describing the relative fraction of spins for which the transition frequency is given by the ω value; thus $\int P(\omega)\, d\omega = 1$. This is a mathematical property explaining the concept; one should not think that the area below every experimentally recorded spectrum is normalized to unity; however, it is proportional to the number of spins.

The highly idealized case of a single spin transition does not apply to real systems. There are two reasons: first, molecular systems are not absolutely rigid (one does not reach 0 K) and, second, besides the Zeeman interaction leading to the energy level splitting, there are always other spin interactions altering this picture.

Let us focus on the example of ^{2}H nucleus (deuteron) of the spin quantum number $I = 1$, which experiences, besides the Zeeman interaction, a quadrupolar coupling. Quadrupolar interactions are present for nuclei of spin quantum numbers $I \geq 1$ (as such nuclei have a non-spherical charge distribution, which results in an electric quadrupolar moment) placed in a molecular environment creating an electric field gradient at the position of the nucleus.

In analogy to the dipole–dipole Hamiltonian, one can consider the form of the quadrupolar Hamiltonian in its principal axis frame (P) and in the laboratory frame (L). In this case by the (P) frame one understands a principal axis system of the electric field gradient tensor. In the (P) frame the quadrupolar Hamiltonian, $H_{\mathrm{Q}}^{(\mathrm{P})}(I)$, takes the form [1–6]:

$$H_{\mathrm{Q}}^{(\mathrm{P})}(I) = \frac{1}{2}\sqrt{\frac{3}{2}}\frac{a_{\mathrm{Q}}}{I(2I-1)}\left[T_0^2(I) + \frac{\eta}{\sqrt{6}}\left[T_{-2}^2(I) + T_2^2(I)\right]\right]$$
$$= \frac{1}{2}\sqrt{\frac{3}{2}}\frac{a_{\mathrm{Q}}}{I(2I-1)}\sum_{m=-2}^{2}(-1)^m V_{-m}^{2(\mathrm{P})} T_m^2 \qquad (4.1)$$

First of all one should note that the quadrupolar interaction is a single spin interaction, in contrary to the dipole–dipole coupling between two spins (Eq. 2.8). The tensor components, $T_m^2(I)$ are defined as [1–6]:

$$T_0^2(I) = \frac{1}{\sqrt{6}}\left[3I_z^2 - I(I+1)\right] \tag{4.2}$$

$$T_{\pm1}^2(I) = \mp\frac{1}{2}\left[I_z I_\pm + I_\pm I_z\right] \tag{4.3}$$

$$T_{\pm2}^2(I) = \frac{1}{2} I_\pm I_\pm \tag{4.4}$$

(although the component $T_{\pm1}^2(I)$ is not present in Eq. 4.1, it will appear in our considerations soon). Comparing both sides of Eq. 4.1 one gets, $V_0^{2(\mathrm{P})} = 1$, $V_{\pm1}^{2(P)} = 0$, and $V_{\pm2}^{2(\mathrm{P})} = \eta/\sqrt{6}$. The quadrupolar coupling constant is defined as [1–6]:

$$a_\mathrm{Q} = e^2 qQ/\hbar \tag{4.5}$$

where Q is a quadrupolar moment of the nucleus, q is the zz component of the electric field gradient tensor, and η, is the asymmetry parameter:

$$\eta = \left(V_{xx} - V_{yy}\right)/V_{zz} \tag{4.6}$$

where the convention $|V_{zz}| \geq |V_{xx}| \geq \left|V_{yy}\right|$ has been applied. The Cartesian components of the electric field gradient tensor are defined as second derivatives of the electrostatic potential V:

$$V_{\alpha\beta} = \left(\frac{\partial^2 V}{\partial r_\alpha \partial r_\beta}\right)_{\vec{r}=0} \tag{4.7}$$

where the index $\vec{r} = 0$ indicates that the derivative is taken at the position of the nucleus.

The energy level structure of the nucleus is determined by a superposition of the Zeeman and quadrupole interactions:

$$H_0(I) = H_\mathrm{Z}(I) + H_\mathrm{Q}^{(\mathrm{L})}(I) \tag{4.8}$$

In Eq. 4.8 two issues should be noted. First, we have introduced the index "0" to denote the main (time-independent) Hamiltonian which is responsible for the energy level structure of the spin system. This notation will be frequently used in the forthcoming chapters. Second, to properly describe the energy levels, both contributing interactions have to be considered in the same frame, and it is natural to choose the laboratory axes system, determined by the direction of the external magnetic field. To obtain the laboratory representation of the quadrupolar interaction, the transformation of Eq. 2.15 applied to the dipole–dipole interaction has to be adapted to the quadrupolar coupling [3, 4, 7–9]. It yields:

$$V_{-m}^{2(\mathrm{L})} = \sum_{k=-2}^{2} V_k^{2(\mathrm{P})} D_{k,-m}^2 (\Omega_{\mathrm{PL}}) \tag{4.9}$$

As the (P) frame is fixed in a molecule, the energy level structure depends on the molecular orientation described by the Euler angles $\Omega_{\mathrm{PL}} \equiv \Omega$. In analogy to Eq. 2.15, the laboratory form of the quadrupolar Hamiltonian, $H_{\mathrm{Q}}^{(\mathrm{L})}(I)(\Omega(t))$, yields:

$$H_{\mathrm{Q}}^{(\mathrm{L})}(I)[\Omega(t)] = \frac{1}{2}\sqrt{\frac{3}{2}} \frac{a_{\mathrm{Q}}}{I(2I-1)} \sum_{m=-2}^{2} (-1)^m V_{-m}^{2(L)} [\Omega(t)] T_m^2(I) \tag{4.10}$$

Quadrupolar and dipolar interactions are rank-two interactions; this is reflected by the index "2" associated with the tensor operators, T_m^2, and the corresponding functions $V_m^2(t)$ and $F_m^2(t)$ present in the quadrupolar and dipolar Hamiltonians, respectively.

Due to the orientation dependence of the energy level structure, there is a distribution of transition frequencies, that is, for a given frequency ω there is a fraction of molecules oriented with respect to the external magnetic field at such an angle Ω that the transition frequency resulting from the combined effect of the Zeeman and (orientation dependent) quadrupolar couplings is equal to ω. This determines the probability $P(\omega)$, describing the distribution of the transition frequencies. The method of calculating the lineshape function is explained below.

In the absence of the quadrupolar interaction there are three states (energy levels) for $I = 1$, corresponding to $m_{\mathrm{I}} = -1, 0, 1$ resulting from the Zeeman splitting. We label the Zeeman functions (states), $|m_{\mathrm{I}}\rangle$, as $|1\rangle = |1\rangle$, $|2\rangle = |0\rangle$, and $|3\rangle = |-1\rangle$. When a quadrupolar interaction is present, the energy levels contain Zeeman and quadrupolar components and, in consequence, the Zeeman functions $|m_{\mathrm{I}}\rangle$ are not any more the eigenfunctions (eigenvectors) of the system. The idea of eigenvectors is very fundamental in quantum mechanics. A given Hamiltonian H can be represented as a matrix in a basis $\{|O_i\rangle\}$; for instance, the basis $\{|m_i\rangle\}$ for spin quantum number $I = 1$ consists of three vectors, for $m_I = -1,\ 0,\ 1$. A given vector $|\psi_i\rangle$ is an eigenvector (eigenfunction) of the Hamiltonian H, if there is a quantity E_i (eigenvalue) for which:

$$H\,|\psi_i\rangle = E_i\,|\psi_i\rangle \tag{4.11}$$

This holds for $|\psi_i\rangle = |O_i\rangle$ only if the representation of the Hamiltonian H in the basis $\{|O_i\rangle\}$ is diagonal; otherwise $|\psi_i\rangle$ is given as a linear combination of vectors forming the basis $\{|O_i\rangle\}$.

The matrix representation of the entire Hamiltonian, $H_0\,(I)\,(\Omega)$ (Eq. 4.8) in the basis $\{|m_i\rangle\}$ has the form:

$$[H_0\,(I)\,(\Omega)] = \begin{bmatrix} -\omega_I + \frac{1}{4}a_Q V_0^{2(L)}(\Omega) & -\frac{\sqrt{3}}{4}a_Q V_1^{2(L)}(\Omega) & \frac{1}{2}\sqrt{\frac{3}{2}}a_Q V_2^{2(L)}(\Omega) \\ \frac{\sqrt{3}}{4}a_Q V_{-1}^{2(L)}(\Omega) & -\frac{1}{2}a_Q V_0^{2(L)}(\Omega) & -\frac{\sqrt{3}}{4}a_Q V_1^{2(L)}(\Omega) \\ \frac{1}{2}\sqrt{\frac{3}{2}}a_Q V_{-2}^{2(L)}(\Omega) & \frac{\sqrt{3}}{4}a_Q V_{-1}^{2(L)}(\Omega) & \omega_I + \frac{1}{4}a_Q V_0^{2(L)}(\Omega) \end{bmatrix} \tag{4.12}$$

The procedure of setting up a matrix representation of a Hamiltonian will be repeated in this book. In most cases this is the first step of describing a quantum spin system. By diagonalizing the Hamiltonian matrix one gets energy levels (eigenvalues, E_i) of the system and corresponding eigenvectors (eigenfunctions, $|\psi_i\rangle$) which are given as linear combinations of the Zeeman functions. Let us calculate, as an example, the first element of this matrix, $[H_0\,(I)]$, which is defined as:

$$\langle -1|\,H_0\,|-1\rangle = \langle -1|\left(\omega_I I_Z + \frac{1}{2}\sqrt{\frac{3}{2}}\frac{a_Q}{2I\,(I+1)} \times \sum_{m=-2}^{2}(-1)^m\,V_{-m}^{2(L)}T_m^2\,(I)\right)|-1\rangle \tag{4.13}$$

As the matrix element is calculated between the states for which $|\Delta m_I = 0|$ only the $T_0^2\,(I)$ operator leads to a non-zero result; thus:

$$\langle -1|\,H_0\,|-1\rangle = \langle -1|\left(\omega_I I_Z + \frac{a_Q}{4}V_0^{2(L)}\left(3I_z^2 - I\,(I+1)\right)\right)|-1\rangle = -\omega_I + \frac{a_Q}{4}V_0^{2(L)} \tag{4.14}$$

In Eq. 4.14 we have substituted $I = 1$, used Eq. 4.2 for $T_0^2(I)$ and omitted (for simplicity) the orientation dependence, $V_0^{2(\mathrm{L})} \equiv V_0^{2(\mathrm{L})}(\Omega)$. Other elements can be calculated analogously, using the relations: $I_+ |I, m_\mathrm{I}\rangle = \sqrt{I(I+1) - m_\mathrm{I}(m_\mathrm{I}+1)}\,|I, m_\mathrm{I}+1\rangle$ and $I_z |I, m_\mathrm{I}\rangle = m_\mathrm{I} |I, m_\mathrm{I}\rangle$ (Eq. 2.26 and Eq. 2.27). To avoid at this moment matrix diagonalization, let us neglect for a while the off-diagonal elements of the matrix of Eq. 4.12. Then, the system has three orientation-dependent energy levels:

$$E_1(\Omega) = -\omega_\mathrm{I} + \frac{1}{4} a_\mathrm{Q} V_0^{2(\mathrm{L})}(\Omega) \tag{4.15}$$

$$E_2(\Omega) = -\frac{1}{2} a_\mathrm{Q} V_0^{2(\mathrm{L})}(\Omega) \tag{4.16}$$

$$E_3(\Omega) = \omega_\mathrm{I} + \frac{1}{4} a_\mathrm{Q} V_0^{2(\mathrm{L})}(\Omega) \tag{4.17}$$

and the Zeeman functions $|m_\mathrm{I}\rangle$ remain to be the eigenfunctions. One can discuss single- and double-quantum transitions between these energy levels. As already explained by single-quantum transitions, one understands transitions between energy levels for which $|\Delta m_\mathrm{I}| = 1$, that is, in this case the transitions $|1\rangle \to |2\rangle$ and $|2\rangle \to |3\rangle$, with the energy:

$$E_{12}(\Omega) = -E_{23}(\Omega) = -\omega_\mathrm{I} + \frac{3}{4} a_\mathrm{Q} \left\{ D_{0,0}^2(\Omega) + \frac{\eta}{\sqrt{6}} \left[D_{-2,0}^2(\Omega) + D_{2,0}^2(\Omega) \right] \right\} \tag{4.18}$$

where the explicit expression for $V_0^{2(\mathrm{L})}(\Omega)$ (Eq. 4.9), resulted from the transformation rules of tensor operators, has been substituted. Assuming isotropic quadrupolar interaction ($\eta = 0$) one gets for the transition energy (frequency):

$$\omega(\theta) = E_{12}(\theta) = -\omega_\mathrm{I} + \frac{3e^2 qQ}{4\hbar} \left(\frac{3\cos^2\theta - 1}{2} \right) \tag{4.19}$$

For an ensemble of molecules, the observed spectrum can be treated as a collection of weight factors of transition frequencies for different molecular orientations, that is, for a given frequency the weight factor (relative probability of the transition) is measured. This implies that to calculate the lineshape function one has to

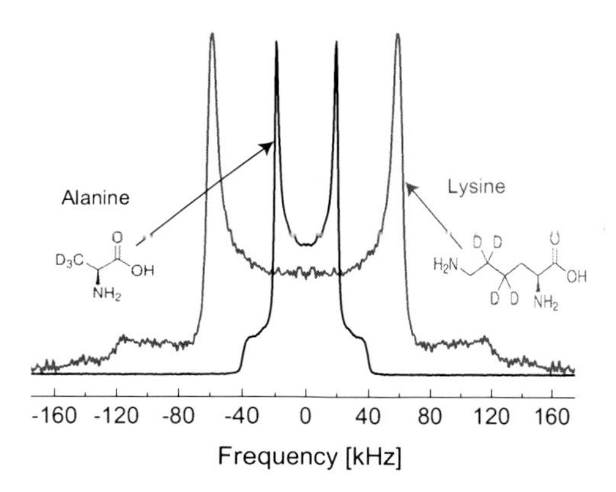

Figure 4.1 Experimental examples of Pake doublets; ^{2}H spectra for partially deuterated alanine and lysine at room temperature.

perform a summation over all molecular orientations, that is, so-called powder averaging (powder is a collection of very small, randomly oriented crystals). Thus, the lineshape function, $L(\omega)$ (one should not mix the notation with Liouville operator), which can be interpreted (when normalized) as a probability of a transition with the frequency ω is given as:

$$L(\omega) = \int_{0}^{\pi} \omega(\theta) \sin^2 \theta d\theta \qquad (4.20)$$

where the term $\sin^2 \theta$ stems from the transformation between Cartesian and spherical coordinate systems and determines the contribution (weigh factor) of the transition with the frequency $\omega(\theta)$. The integration leads to a well-known spectral shape referred to as Pake-spectrum [2, 3, 5, 10]. Experimental examples of such spectra are shown in Fig. 4.1.

Calculating the rigid quadrupole spectrum (Eq. 4.19), we have neglected the off-diagonal terms in Eq. 4.12. The orientation dependent functions, $V^2_{\pm 1}(\Omega)$ and $V^2_{\pm 2}(\Omega)$, have the explicit form (Eq. 4.9):

$$V^2_{\pm 1}(\Omega) = D^2_{0,\pm 1}(\Omega) + \frac{\eta}{\sqrt{6}}\left[D^2_{-2,\pm 1}(\Omega) + D^2_{2,\pm 1}(\Omega)\right] \qquad (4.21)$$

$$V^2_{\pm 2}(\Omega) = D^2_{0,\pm 2}(\Omega) + \frac{\eta}{\sqrt{6}}\left[D^2_{-2,\pm 2}(\Omega) + D^2_{2,\pm 2}(\Omega)\right] \qquad (4.22)$$

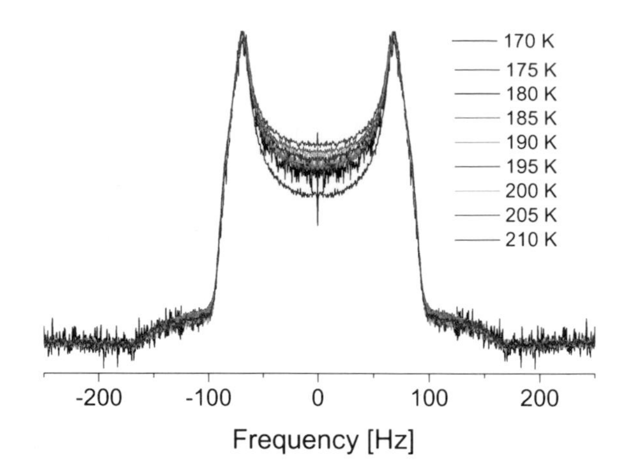

Figure 4.2 ^{2}H spectra for glycerol-d_3 versus temperature.

and they considerably influence the overall lineshape. However, to take them into account one has to perform more advanced calculations which we shall discuss in Section 4.4.

Before proceeding further it is of interest to see that Pake doublets are observed also for much simpler compounds like partially deuterated glycerol; examples are shown in Fig. 4.2. The spectral shapes for glycerol-d_3 slightly evolve with temperature, but the spectrum has the shape of Pake doublet.

4.2 Spin Resonance Spectrum and Motion

Now it is important to understand what molecular motion has to do with spin resonance spectra. For this purpose let us again consider the case of two equivalent nuclear spins $I_1 = I_2 = I = 1/2$. Their energy level structure is determined by the Zeeman interaction. They are mutually coupled by a dipole–dipole interaction which fluctuates in time due to, for instance, tumbling of the molecule to which these spins belong. The $I_1 - I_2$ spin system can undergo single- and double-quantum transitions with the frequencies ω_{I} and $2\omega_{\text{I}}$, respectively. At this stage we are concerned only with single-quantum transitions, that is, the transitions for which $|\Delta m_{\text{I}} = \Delta(m_1 + m_2)| = 1$. As already said, when neglecting the dipole–dipole interaction the spectrum consists of a single line at the frequency ω_{I}. However, when fluctuating dipole–dipole interactions are present, there is a non-zero probability of a spin transition of a frequency ω which somewhat deviates from ω_{I}. The difference between these two frequencies (energies) can be compensated by the frequency (energy) of the stochastic fluctuations of the dipolar interactions. In other words, the missing (or exceeding) part of energy can be taken from (or given back to) the fluctuating magnetic field created by the dipole–dipole coupling. This implies that the spectral line is not infinitely narrow, but it has a certain line-width. When the motion is fast enough to fulfill the condition $\omega_{\mathrm{DD}}\tau_{\mathrm{c}} << 1$ (τ_{c} is a correlation time of the fluctuations, ω_{DD} is the amplitude of the dipole–dipole coupling in angular frequency units), the line is of Lorentzian shape [1, 3, 4, 6, 11, 12].

This feature can be mathematically described in terms of a formal theory of spin resonance lineshape. It is somewhat too early to present at this stage the full lineshape treatment, but we should prepare ourselves for that. Equation 4.20 gives the lineshape function in the limiting case when there is no motion (the system is rigid). In the presence of fast dynamics the lineshape function is given as a matrix product [3, 4, 13–15]:

$$L(\omega) \propto \mathrm{Re}\left\{[I_+]^+ \cdot [M - i\omega 1]^{-1} \cdot [I_+]\right\} \tag{4.23}$$

This requires explanations. Our system is characterized by four states $\left|m_{\mathrm{I}_1}, m_{\mathrm{I}_2}\right\rangle$ ($m_{\mathrm{I}} = \pm 1/2$) which create the basis used for setting up the matrices of Eq. 4.23. Single-quantum transitions are associated with the I_+, I_- operators. For evaluating the lineshape function one can choose one of them (the result will remain unchanged). One should also consider whether one is interested in the spectrum of the I_1 or I_2 spin; in the present case they are identical, $I_1 = I_2 = I$. Saying that the spins are identical (equivalent) does not only mean that the spin quantum number is the same, but also the transition frequencies have to be the same; like for two protons in water molecule. It has been explained in Chapter 3 how to construct from the "ordinary" basis $\{|m_1, m_2\rangle\}$, referred to as Hilbert space, an extended basis referred to as Liouville representation; one has to consider pairs of the Hilbert space vectors (spin functions). Now we are, however, interested only in single-quantum transitions of, let us say, spin I_1. Therefore, we need to consider only two pairs of states $|1/2, -1/2\rangle\langle -1/2, -1/2|$ and $|1/2, 1/2\rangle\langle -1/2, 1/2|$, which differ by $\Delta m_{I_1} = 1$. This implies that we shall deal with 2×2 matrices. Let the matrix $[I_+]$ be a representation of the I_{1+} operator in the sub-basis $\{|1/2, -1/2\rangle\langle -1/2, -1/2|, |1/2, 1/2\rangle\langle -1/2, 1/2|\}$. As

$$I_+ \propto \left|\frac{1}{2}, -\frac{1}{2}\right\rangle\left\langle -\frac{1}{2}, -\frac{1}{2}\right| + \left|\frac{1}{2}, \frac{1}{2}\right\rangle\left\langle -\frac{1}{2}, \frac{1}{2}\right| \quad (4.24)$$

the matrix $[I_+]$ takes the form $[I_+] = [1, 1]$. $[I_+]^+$ is a conjugated form of this matrix, that is $[I_+]^+ = \begin{bmatrix} 1 \\ 1 \end{bmatrix}$. The matrix $[M]$ has already been discussed in Chapter 3, it is referred to as the evolution-relaxation matrix. Denoting the Zeeman (Hilbert) vectors as, $|1\rangle = |1/2, 1/2\rangle$, $|2\rangle = |1/2, -1/2\rangle$, $|3\rangle = |-1/2, 1/2\rangle$, and $|4\rangle = |-1/2, -1/2\rangle$, the lineshape function is given as:

$$L(\omega) \propto \left\{ \begin{bmatrix} 1 \\ 1 \end{bmatrix} \cdot \begin{bmatrix} R_{2424} - i(\omega_{24} - \omega) & R_{1324} \\ R_{1324} & R_{1313} - i(\omega_{13} - \omega) \end{bmatrix}^{-1} \cdot [1, 1] \right\} \quad (4.25)$$

where $\omega_{24} = \omega_{13} = \omega_I$ and R_{2424}, R_{1313}, and R_{1324} denote relaxation matrix elements which can be calculated using the Redfield relaxation theory outlined in Chapter 3 [1, 3, 4, 6, 11, 12]. They are given as the following combinations of spectral densities:

$$R_{2424} = -2\tilde{J}_{2244}(0) + \tilde{J}_{1212}(\omega_{12}) + \tilde{J}_{2323}(\omega_{23}) + 2\tilde{J}_{2424}(\omega_{24}) + \tilde{J}_{1414}(\omega_{14}) + \tilde{J}_{3434}(\omega_{34}) \quad (4.26)$$

$$R_{1313} = -2\tilde{J}_{1133}(0) + \tilde{J}_{1212}(\omega_{12}) + 2\tilde{J}_{1313}(\omega_{13}) + \tilde{J}_{1414}(\omega_{14}) + \tilde{J}_{2323}(\omega_{23}) + \tilde{J}_{3434}(\omega_{34}) \quad (4.27)$$

$$R_{1324} = -\tilde{J}_{1234}(\omega_{13}) - \tilde{J}_{1234}(\omega_{24}) \quad (4.28)$$

Using the matrix representation of the dipole–dipole interaction (Eq. 3.30) and the definition of the spectral densities $\tilde{J}(\omega)$ (Eq. 3.32), one gets for the relaxation rates:

$$R_{2424} = R_{1313} = R_2 = \frac{3}{4}\left(\frac{\mu_0}{4\pi}\frac{\gamma_I^2\hbar}{r^3}\right)^2 [J(0) + 2J(\omega_I) + 2J(2\omega_I)] \quad (4.29)$$

$$R_{1324} = \frac{3}{4}\left(\frac{\mu_0}{4\pi}\frac{\gamma_{\mathrm{I}}^2\hbar}{r^3}\right)^2 J(\omega_{\mathrm{I}}) \tag{4.30}$$

where the functions $J(\omega)$ can be given, for instance, by Eq. 2.36 in the case of simple isotropic molecular tumbling. The relaxation rate $R_2 = R_{2424} = R_{1313}$ is referred to as spin–spin relaxation rate for identical spins 1/2. Neglecting the off-diagonal elements R_{1234} in the matrix $[M - i\omega 1]$ (Eq. 4.25) the lineshape function $L(\omega)$ can be explicitly written as:

$$L(\omega) \propto \frac{T_2}{1 + (\omega - \omega_{\mathrm{I}})^2 T_2^2} \tag{4.31}$$

where $T_2 = R_2^{-1}$ is the spin–spin relaxation time.

As explained in Chapter 3, relaxation rates can be described as linear combinations of spectral densities only when the Redfield condition is fulfilled; in this case it is $\omega_{\mathrm{DD}}\tau_{\mathrm{c}} << 1$ as the dipole-dipole interaction is the relaxation mechanism. This explains our earlier statement that the lineshape is Lorentzian only under this condition.

Finishing this section it is important to note that the off-diagonal relaxation rates R_{1234} stem from the fact that $\omega_{13} = \omega_{24}$ and, in consequence, the coherences ρ_{13} and ρ_{24} are coupled (as discussed in Chapter 3). Therefore, although the simplification of Eq. 4.31 is mathematically attractive, it should be treated with caution.

4.3 Examples of Spin Resonance Spectra

In Section 4.1, we have considered a rigid quadrupolar spectrum for the nuclear spin quantum number $I = 1$. Now, we shall discuss the spectral shape in the presence of motion. The discussion presented below applies also to an electronic spin $S = 1$. The reason for that is the mathematical similarity of quadrupolar interactions for nuclei and zero-field splitting (ZFS) for electrons [4, 14]. Zero-field splitting is present for electronic spins of $S \geq 1$ and it can have different physical origins. For transition metal complexes, it is due to second order effects in the spin–orbit coupling [16–20].

Independently of the origin, the ZFS Hamiltonian has the form (in the laboratory frame):

$$H_{\mathrm{ZFS}}^{(\mathrm{L})}(S) = \sqrt{\frac{2}{3}} D \sum_{m=-2}^{2} (-1)^m U_{-m}^{2(\mathrm{L})} T_m^2(S) \tag{4.32}$$

where the parameter D describes the axial component of the ZFS tensor. In analogy to Eq. 4.9, the spatial tensor components, $U_{-m}^{2(\mathrm{L})}$, represented in the laboratory frame, result from the transformation:

$$U_{-m}^{2(\mathrm{L})} = \sum_{k=-2}^{2} U_k^{2(\mathrm{P})} D_{k,-m}^2 (\Omega_{\mathrm{PL}}) \tag{4.33}$$

with $U_0^{2(\mathrm{P})} = 1$, $U_{\pm 1}^{2(\mathrm{P})} = 0$, and $U_{\pm 2}^{2(\mathrm{P})} = \sqrt{\frac{3}{2}}\frac{E}{D}$, where E is the rhombic component of the ZFS tensor; (P) is a principal frame of the ZFS tensor, the angle $\Omega_{\mathrm{PL}} \equiv \Omega$ describes the orientation of this frame with respect to the laboratory axis, (L). Comparing the expressions for $U_m^{2(\mathrm{P})}$ for ZFS interactions and $V_m^{2(\mathrm{P})}$ for quadrupole interactions ($V_0^{2(\mathrm{P})} = 1$, $V_{\pm 1}^{2(\mathrm{P})} = 0$, and $V_{\pm 2}^{2(\mathrm{P})} = \eta/\sqrt{6}$) one can see that Eq. 4.32 can be directly obtained from Eq. 4.1 by substituting [4, 14]:

$$D \rightarrow \frac{3}{4}\frac{a_{\mathrm{Q}}}{I(2I-1)} \tag{4.34a}$$

$$E \rightarrow \frac{1}{4}\eta\frac{a_{\mathrm{Q}}}{I(2I-1)} \tag{4.34b}$$

To make the example as simple as possible let us assume that the energy level structure of the system is fully determined by the Zeeman interaction, while the ZFS (or quadrupolar coupling) causes relaxation. The origin of the ZFS (quadrupolar coupling) fluctuations can be attributed to molecular tumbling. Thus, the system has three energy levels corresponding to the magnetic quantum numbers $m_{\mathrm{I(S)}} = 1, 0, -1$. In Section 4.1 the Zeeman states associated with the energies (in angular frequency units), $\omega_{\mathrm{I(S)}}$, 0, and $-\omega_{\mathrm{I(S)}}$ ($\omega_{\mathrm{I(S)}} = \gamma_{\mathrm{I(S)}} B_0$) have been labeled as $|1\rangle = |1\rangle$, $|2\rangle = |0\rangle$, and $|3\rangle = |-1\rangle$; we shall keep this notation. Now, the size of the Liouville space is 9, this implies that the matrix $[M]$ (Eq. 4.23) is of dimension 9×9. However, since the spectrum is created only by single-quantum transitions, the relevant part of the matrix is of dimension 6×6 (we do not need to deal with the population block, 3×3, of the matrix), and it can be even further reduced, as we can choose between the operators I_+ and I_-. The operator I_+ for $I = 1$ is represented as:

$$I_+ \propto \{|0\rangle\langle 1| + |-1\rangle\langle 0|\} = \{|2\rangle\langle 1| + |3\rangle\langle 2|\} \tag{4.35}$$

This implies that as long as the matrix of the main Hamiltonian, $H_0(I)$, is diagonal in the Zeeman basis $\{|m_I\rangle\}$ (in this case the main Hamiltonian is actually the Zeeman Hamiltonian $H_0(I) = H_Z(I) = \gamma_I B_0 I_Z = \omega_I I_Z$, so this condition is obviously fulfilled), we can limit ourselves only to the 2×2 block of the matrix $[M]$. Thus, the lineshape function, $L(\omega)$, is described by the matrix product:

$$L(\omega) \propto \mathrm{Re}\left\{[1,1]\cdot\begin{bmatrix} -i\omega_{21} + R_{1212} + i\omega & R_{1223} \\ R_{1223} & -i\omega_{32} + R_{2323} + i\omega \end{bmatrix}^{-1}\cdot\begin{bmatrix}1\\1\end{bmatrix}\right\} \tag{4.36}$$

where $\omega_{\alpha\beta} = \omega_\alpha - \omega_\beta$ that implies $\omega_{21} = \omega_{32} = \omega_I$, while R_{1212}, R_{2323}, and R_{1223} are the relaxation matrix elements. We have omitted the factor $\sqrt{2}$ in the representation of the operator I_+ as the amplitude of the spectrum is given, anyway, in arbitrary units. In analogy to Eq. 4.31, when neglecting the relaxation rate R_{1223}, the lineshape function, $L(\omega)$, is given in terms of Lorentzian functions (as a result of inverting the diagonal matrix $[M]$ incorporated into Eq. 4.36):

$$L(\omega) \propto \mathrm{Re}\left\{\frac{1}{R_{1212} + i(\omega - \omega_I)} + \frac{1}{R_{2323} + i(\omega - \omega_I)}\right\} \tag{4.37}$$

One can expect (this will be confirmed soon) that, in analogy to the case of $I = 1/2$, $R_{1212} = R_{2323} = T_2^{-1}$, where T_2 is the spin–spin relaxation time. Then, Eq. 4.37 simplifies to Eq. 4.31, which means that the spectral lineshape is described by a Lorentzian function with the width determined by the spin–spin relaxation time T_2.

The relaxation rates $R_{\alpha\alpha'\beta\beta'}$ can again be calculated by means of the Redfield relaxation theory (Eq. 3.19). The Hamiltonians of the perturbing interactions, that is, of the quadrupolar coupling and ZFS are given by Eq. 4.10 and 4.32, respectively. Although the matrix representation of the quadrupole Hamiltonian in the Zeemann basis

$\{|m_{\mathrm{I}}\rangle\}$ can be extracted from Eq. 4.12, we write it down explicitly for convenience:

$$[H_{\mathrm{Q}}(t)] = a_{\mathrm{Q}} \begin{bmatrix} \frac{1}{4}V_0^{2(\mathrm{L})}(t) & -\frac{\sqrt{3}}{4}V_1^{2(\mathrm{L})}(t) & \frac{1}{2}\sqrt{\frac{3}{2}}V_2^{2(\mathrm{L})}(t) \\ \frac{\sqrt{3}}{4}V_{-1}^{2(\mathrm{L})}(t) & -\frac{1}{2}V_0^{2(\mathrm{L})}(t) & -\frac{\sqrt{3}}{4}V_1^{2(\mathrm{L})}(t) \\ \frac{1}{2}\sqrt{\frac{3}{2}}V_{-2}^{2(\mathrm{L})}(t) & \frac{\sqrt{3}}{4}V_{-1}^{2(\mathrm{L})}(t) & \frac{1}{4}V_0^{2(\mathrm{L})}(t) \end{bmatrix} \tag{4.38}$$

where the time dependence of the quadrupolar interaction has explicitly been noted (the quadrupolar Hamiltonian in Eq. 4.12 is time independent as it has been assumed that the system is rigid). The quadrupolar relaxation rates are given by the expression:

$$R_{1212}^{\mathrm{Q}} = R_{2323}^{\mathrm{Q}} = T_{2,\mathrm{Q}}^{-1} = \frac{1}{2}a_{\mathrm{Q}}^2\left(3+\eta^2\right)\left[3J(0)+5J(\omega_{\mathrm{I}})+2J(2\omega_{\mathrm{I}})\right] \tag{4.39}$$

For electron spin $S = 1$ one obtains from Eq. 4.34a,b that $D \rightarrow \frac{3}{4}a_{\mathrm{Q}}$ and $E \rightarrow \frac{1}{4}\eta a_{\mathrm{Q}}$; in consequence the corresponding expression for the electron spin relaxation rates takes the form:

$$R_{1212}^{\mathrm{ZFS}} = R_{2323}^{\mathrm{ZFS}} = T_{2,\mathrm{ZFS}}^{-1} = \left(\frac{1}{3}D^2 + E^2\right)\left[3J(0)+5J(\omega_{\mathrm{S}})+2J(2\omega_{\mathrm{S}})\right] \tag{4.40}$$

The off-diagonal relaxation rates (Eq. 4.36) for quadrupolar and ZFS interactions are given as:

$$R_{1223}^{\mathrm{Q}} = \frac{3}{8}a_{\mathrm{Q}}^2\left(3+\frac{\eta^2}{4}\right)J(\omega_{\mathrm{I}}) \tag{4.41}$$

$$R_{1223}^{\mathrm{ZFS}} = \frac{3}{4}\left(\frac{1}{3}D^2+E^2\right)J(\omega_{\mathrm{S}}) \tag{4.42}$$

Let us remind here that for isotropic molecular tumbling the spectral densities $J(\omega)$ are expressed by Eq. 2.36.

Figure 4.3 shows examples of ^{2}H NMR spectra ($I = 1$) calculated for the two limiting cases: Pake doublet (rigid limit) and Lorentzian line (fast motion limit). The shape of the rigid spectrum is similar to the experimental spectra shown in Fig. 4.1.

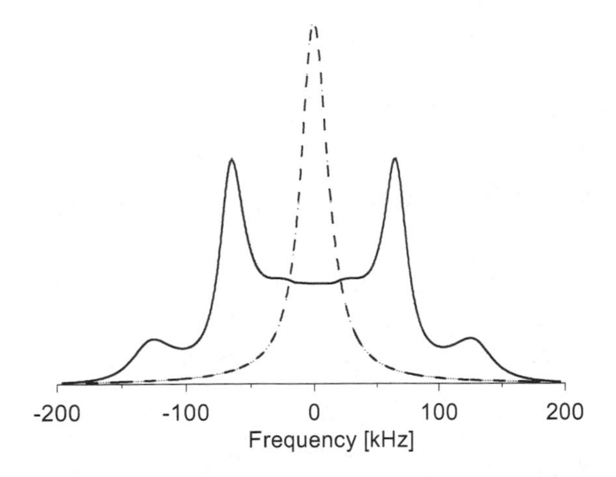

Figure 4.3 ^{2}H NMR spectra at 1T, $a_{\mathrm{Q}} = 220$ kHz; rigid limit $\omega_{\mathrm{Q}}\tau_{\mathrm{rot}} = 15$ (solid line), fast motion limit $\omega_{\mathrm{Q}}\tau_{\mathrm{rot}} = 0.5$(dashed line); ω_{Q} denotes the amplitude of the quadrupole interaction in angular frequency units, $\omega_{\mathrm{Q}} = 1.38 \times 10^{4}$ rad/s.

Now let us consider two-spin spectra, an electron spin $S = 1/2$ coupled by a dipole–dipole interaction to a nuclear spin $I = 1/2$. For this system one can discuss an ESR spectrum (looking from the perspective of the electron spin) and a NMR spectrum (from the perspective of the nuclear spin). The total Hamiltonian contains the Zeeman couplings for the S and I spins, $H_{\mathrm{Z}}(S)$ and $H_{\mathrm{Z}}(I)$, respectively, and the mutual I–S dipole–dipole coupling, $H_{\mathrm{DD}}(I, S)$:

$$H(I, S) = H_{\mathrm{Z}}(I) + H_{\mathrm{Z}}(S) + H_{\mathrm{DD}}(I, S) \tag{4.43}$$

The system is characterized by four states $|m_{\mathrm{I}}, m_{\mathrm{S}}\rangle$, which we shall number, in analogy to the case of two equivalent spins, $I_1 - I_2$, as follows: $|1\rangle = |1/2, 1/2\rangle$, $|2\rangle = |1/2, -1/2\rangle$, $|3\rangle = |-1/2, 1/2\rangle$, and $|4\rangle = |-1/2, -1/2\rangle$.

In this case the Liouville space is of dimension 16×16. As it has already been explained, the crucial factor for the lineshape theory is the time scale of the fluctuations of the dipole–dipole coupling; here we shall consider the case when $\omega_{\mathrm{DD}}\tau_{\mathrm{rot}} << 1$, that is, the I–S dipolar interaction acts as a relaxation mechanisms. To calculate the lineshape function one has to consider the operators S_{+} (for the electron spin) and I_+ (for the nuclear spin):

$$S_+ \propto \left\{ \left| \frac{1}{2}, -\frac{1}{2} \right\rangle \left\langle \frac{1}{2}, \frac{1}{2} \right| + \left| -\frac{1}{2}, -\frac{1}{2} \right\rangle \left\langle -\frac{1}{2}, \frac{1}{2} \right| \right\} = \{|2\rangle\langle 1| + |4\rangle\langle 3|\} \tag{4.44}$$

$$I_+ \propto \left\{ \left| -\frac{1}{2}, \frac{1}{2} \right\rangle \left\langle \frac{1}{2}, \frac{1}{2} \right| + \left| -\frac{1}{2}, -\frac{1}{2} \right\rangle \left\langle \frac{1}{2}, -\frac{1}{2} \right| \right\} = \{|3\rangle\langle 1| + |4\rangle\langle 2|\} \tag{4.45}$$

Equations 4.42 and 4.43 are analogous to Eqs. 4.24 and 4.35. This implies that the electron (ESR) and nuclear (NMR) lineshape functions, $L_S(\omega)$ and $L_I(\omega)$, respectively, are given by the expressions:

$$L_S(\omega) \propto \text{Re} \left\{ [1, 1] \cdot \begin{bmatrix} i(\omega - \omega_S) + R_{1212} & 0 \\ 0 & i(\omega - \omega_S) + R_{3434} \end{bmatrix}^{-1} \cdot \begin{bmatrix} 1 \\ 1 \end{bmatrix} \right\} \tag{4.46}$$

$$L_I(\omega) \propto \text{Re} \left\{ [1, 1] \cdot \begin{bmatrix} i(\omega - \omega_I) + R_{1313} & 0 \\ 0 & i(\omega - \omega_I) + R_{2424} \end{bmatrix}^{-1} \cdot \begin{bmatrix} 1 \\ 1 \end{bmatrix} \right\} \tag{4.47}$$

It is worth to note that in this case the relaxation matrix $[M]$ is diagonal as the energy levels of the I–S spin system are non-degenerated. Using again the matrix representation of the dipole-dipole interaction (Eq. 3.30) and the expression for relaxation rates (Eq. 3.19) one obtains:

$$R_{1212} = R_{3434} = R_{1313} = R_{2424} = T_2^{-1} = \frac{1}{4}\left(\frac{\mu_0}{4\pi}\frac{\gamma_I \gamma_S \hbar}{r^3}\right)^2 \times [4J(0) + 3J(\omega_I) + 3J(\omega_S) + 6J(\omega_I + \omega_S) + J(\omega_S - \omega_I)] \tag{4.48}$$

The only difference between the expressions for the lineshape functions $L_I(\omega)$ and $L_S(\omega)$, is that they are taken at different frequencies $|\omega_I - \omega|$ and $|\omega_S - \omega|$, respectively:

$$L_{I(S)}(\omega) \propto \frac{T_2}{1 + \left(\omega - \omega_{I(S)}\right)^2 T_2^2} \tag{4.49}$$

Nevertheless, it should be noted that the spin–spin relaxation time, T_2, depends on both, the nuclear as well as the electron spin transition frequencies (Eq. 4.47).

4.4 Rigid Spectrum and the Lineshape Theory

One should not think that the principle of the lineshape theory depends on whether the system is rigid or not. Equations 4.20 and 4.23 are special cases of a more general approach, which we shall present in forthcoming chapters. At this stage, it is important to understand how Eq. 4.20 can be derived from Eq. 4.23.

When the dynamics is very slow the energy level structure of the system depends on the molecular orientation which enters the dipole–dipole contribution to the spin Hamiltonian:

$$H_0(I_1, I_2)(\Omega) = H_Z(I_1) + H_Z(I_2) + H_{DD}^{(L)}(I_1, I_2)(\Omega) \qquad (4.50)$$

where the laboratory representation of the dipole–dipole Hamiltonian, $H_{DD}^{(L)}(I_1, I_2)(\Omega)$, is given by Eq. 2.13 and 2.14:

$$H_{DD}^{(L)}(I_1, I_2)(\Omega) = a_{DD} \sum_{m=-2}^{2} (-1)^m D_{0,-m}^2(\Omega)\, T_m^2(I_1, I_2) \qquad (4.51)$$

with $\Omega = (0, \beta_{DDL}, \gamma_{DDL}) = (0, \theta, \varphi)$ describing the molecular orientation. In the Zeeman basis $\{|n\rangle = |m_1, m_2\rangle\}$: $|1\rangle = |1/2, 1/2\rangle\, |2\rangle = |1/2, -1/2\rangle$, $|3\rangle = |-1/2, 1/2\rangle$, and $|4\rangle = |-1/2, -1/2\rangle$ (Section 4.2), the entire Hamiltonian $H_0(I_1, I_2)(\Omega)$ takes the form:

$$H_0(I_1, I_2)(\Omega) = \begin{bmatrix} \frac{1}{2\sqrt{6}} a_{DD} D_{0,0}^2(\Omega) + \omega_I & -\frac{1}{4} a_{DD} D_{0,-1}^2(\Omega) & -\frac{1}{4} a_{DD} D_{0,-1}^2(\Omega) & -\frac{1}{2} a_{DD} D_{0,-2}^2(\Omega) \\ \frac{1}{4} a_{DD} D_{0,1}^2(\Omega) & \frac{1}{4} a_{DD} D_{0,1}^2(\Omega) & -\frac{1}{2\sqrt{6}} a_{DD} D_{0,0}^2(\Omega) & \frac{1}{4} a_{DD} D_{0,-1}^2(\Omega) \\ \frac{1}{4} a_{DD} D_{0,1}^2(\Omega) & -\frac{1}{2\sqrt{6}} a_{DD} D_{0,0}^2(\Omega) & -\frac{1}{2\sqrt{6}} a_{DD} D_{0,0}^2(\Omega) & \frac{1}{4} a_{DD} D_{0,-1}^2(\Omega) \\ \frac{1}{2} a_{DD} D_{0,2}^2(\Omega) & \frac{1}{4} a_{DD} D_{0,1}^2(\Omega) & \frac{1}{4} a_{DD} D_{0,1}^2(\Omega) & \frac{1}{2\sqrt{6}} a_{DD} D_{0,0}^2(\Omega) - \omega_I \end{bmatrix} \qquad (4.52)$$

Some of the elements were calculated in Chapter 2 (Eqs. 2.24 and 2.25). Diagonalization of the matrix of Eq. 4.52 leads to orientation-dependent energy levels (eigenvalues of the matrix), $E_\alpha(\Omega)$, associated with the corresponding eigenfunctions $|\psi_\alpha\rangle$ given as linear combinations of the Zeeman states $|n\rangle$, $|\psi_\alpha\rangle = \sum_{n=1}^{4} C_{\alpha n}(\Omega)\,|n\rangle$, with orientation dependent coefficients $C_{\alpha n}(\Omega)$. As the coefficient matrix $[c_{\alpha n}]$ jest hermitian, one can easily obtain the inverse relationship, $|n\rangle = \sum_{\alpha=1}^{4} c_{n\alpha}(\Omega)|\psi_\alpha\rangle$. Using this relation one can express the operator I_+ (Eq. 4.24) as:

$$
\begin{aligned}
|I_+\rangle &\propto \left|\frac{1}{2},-\frac{1}{2}\right\rangle\left\langle-\frac{1}{2},-\frac{1}{2}\right| + \left|\frac{1}{2},\frac{1}{2}\right\rangle\left\langle-\frac{1}{2},\frac{1}{2}\right| \\
&= \sum_{\alpha,\beta=1}^{4}\left(c_{2\alpha}c_{4\beta}+c_{3\alpha}c_{1\beta}\right)|\psi_\alpha\rangle\left\langle\psi_\beta\right|
\end{aligned}
\tag{4.53}
$$

Equation 4.53 determines the form of the matrix $[I_+]$ as it provides the coefficients associated with the elements $|\psi_\alpha\rangle\left\langle\psi_\beta\right|$; the coefficients are orientation dependent. Thus, the lineshape function $L(\omega)$ can be written as:

$$
L(\omega) \propto \int \mathrm{Re}\left\{[I_+]^+ [M - i\omega 1]^{-1}[I_+]\right\}(\theta,\phi)\sin^2\theta d\theta d\phi \tag{4.54}
$$

Equation 4.54 is a generalized version of Eq. 4.23 in this sense that it includes integration over molecular orientations. As already said, the matrix $[I_+]$ is orientation dependent via the coefficients $c_{n\alpha}(\Omega)$. The matrix $[M - i\omega 1]$ also depends on the molecular orientation via the transition frequencies $\omega_{\alpha\beta}(\Omega) = E_\alpha(\Omega) - E(\Omega)$. The relaxation rates entering the matrix $[M]$ (Eqs. 4.26 and 4.27) should be, in principle, zero as rigid systems do not relax. Nevertheless, to avoid mathematical singularities, one should set them to small, but non-zero values.

In Fig. 4.4 examples of ^{2}H NMR spectra which can be described as a superposition of a rigid spectrum (Pake doublet), discussed above, and a Lorentzian line characterizing fast dynamics, discussed in Section 4.3 are shown. The spectra have been obtained for mixture of partially deuterated lysine (Fig. 4.1) and glycerol (Fig. 4.2).

One can simplify the description of the spectral shape by neglecting the off-diagonal elements in the Hamiltonian matrix of Eq. 4.51. This procedure is often explained by saying that the diagonal elements are much larger than the off-diagonal ones as they include the Zeeman interaction. This is, however, not true. Firstly, for a system of two identical spins only two of the diagonal elements contain the transition frequency ω_I. Secondly, the off-diagonal elements do not need to be comparable with the diagonal

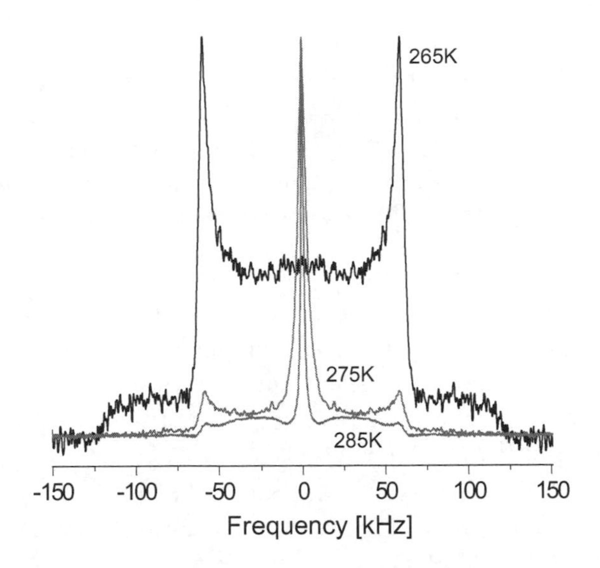

Figure 4.4 ^{2}H NMR spectra for mixture of partially deuterated lysine and glycerol.

values to lead to a considerable mixing of the Zeeman energy levels. In analogy to Eq. 4.19, in the present case the single-quantum transition frequency yields:

$$\omega(\theta) = E_{12}(\theta) = -\omega_I + \left(\frac{\mu_0}{4\pi}\frac{\gamma_I^2\hbar}{r^3}\right)\left(\frac{3\cos^2\theta - 1}{2}\right) \tag{4.55}$$

and, as $[I_+] = [1, 1]$ (the Hamiltonian matrix is diagonal in the Zeeman basis), Eq. 4.54 reduces to Eq. 4.20. This brings us to the conclusion that the rigid spectrum for two identical spins 1/2 also takes the form of Pake doublet, like in the case of $I = 1$ and quadrupole interactions. At the same time, we see how to generalize Eq. 4.23 including off-diagonal elements of the quadrupolar interaction.

4.5 Spin Resonance Spectra and Correlation Functions

Magnetic resonance spectrum (NMR and ESR) represents a "generalized" spectral density which is, by definition, Fourier transform of a "generalized" correlation function. By the "generalization" we understand a correlation function including not only dynamical (motional) degrees of freedom (Eq. 2.29), but also spin variables. As we are interested in a single-quantum spectrum, the correlation function of interest is $\langle I_\pm(0)\, I_\pm(t)\rangle$. The lineshape function, $L(\omega)$, is given by Fourier transform of this quantity taken at the frequency ω:

$$L(\omega) \propto \mathrm{Re} \int_0^\infty \langle I_\pm(0)\, I_\pm(t)\rangle \exp(i\omega t)\, dt \tag{4.56}$$

To calculate the correlation function $\langle I_\pm(0)\, I_\pm(t)\rangle$ one has to consider all spin interactions and motional processes (generally, all degrees of freedom) relevant for the system. They should be encoded into the Liouville operator, L (see Chapter 3). This allows for expressing the $I_\pm(t)$ operator as:

$$I_\pm(t) = [\exp(-i Lt)]\, I_\pm(0) \tag{4.57}$$

One can see here analogy to the "classical" correlation function $\langle A(t)\, A(0)\rangle$ (Eq. 2.30). The quantity $A(x(t))$ is given as:

$$A(x(t)) = \int P(x, x_0, t) A(x_0)\, P_{\mathrm{eq}}(x_0)\, dx_0 \tag{4.58}$$

The operator $I_\pm(0)$ in Eq. 4.56 corresponds to $A(x_0)$ of Eq. 4.58, while the operator $[\exp(-i Lt)]$ plays the role of the conditional probability $P(x, x_0, t)$.

As already stated, the spectrum can be evaluated using the operator I_+ as well as I_-; let us decide for I_+. Then, the spectral density of Eq. 4.56 can be written as:

$$L(\omega) = \mathrm{Re} \int_0^\infty \langle I_+ \exp[-i(L + \omega 1)\, t]\, I_+\rangle\, dt \tag{4.59}$$

where we set for simplicity $I_\pm(0) \equiv I_+$. For the purpose of calculations it is convenient to express Eq. 4.59 in a matrix form. Setting $M = i(L + \omega 1)$ one can write:

$$L(\omega) \infty \mathrm{Re} \int_0^\infty Tr\{[I_+] \exp(-[Mt])\, [I_+]\}\, dt \propto [I_+]^+ \cdot [M]^{-1} \cdot [I_+] \tag{4.60}$$

The averaging, $\langle\rangle$, in Eq. 4.59 is equivalent to taking a trace in the corresponding matrix formulation (Eq. 4.60). The last equality stems from the simple relation $\int_0^\infty \exp(-\alpha t)\, dt = \alpha^{-1}$, where now the matrix $[M]$ plays the role of α.

References

1. Abragam, A. (1961). *The Principles of Nuclear Magnetism* (New York: Oxford University Press).
2. Kimmich, R. (1997). *NMR—Tomography, Diffusometry, Relaxometry* (Berlin: Springer-Verlag).
3. Kowalewski, J. and Mäler, L. (2006). *Nuclear Spin Relaxation in Liquids: Theory, Experiments and Applications* (New York: Taylor & Francis).
4. Kruk, D. (2007). *Theory of Evolution and Relaxation of Multi-Spin Systems* (Bury St Edmunds: Arima).
5. Man, P. P. (1996). *Encyclopedia of Nuclear Magnetic Resonance, Quadrupolar Interactions* (Chichester: Wiley), pp. 3838–3847.
6. Slichter, C. P. (1990). *Principles of Magnetic Resonance* (Berlin: Springer-Verlag).
7. Goldman, M. (2001). Formal theory of spin–lattice relaxation. *J. Magn. Reson.*, **149**, pp. 160–187.
8. Griffith, J. S. (1961). *The Theory of Transition Metal Ions* (Cambridge: Cambridge University Press).
9. Varshalovich, D. A., Moskalev, A. N. and Khersonkii, V. K. (1988). *Quantum Theory of Angular Momentum* (Singapore: Word Scientific Publishing).
10. Pake, G. E. (1948). Fine structure of the proton line. *J. Chem. Phys.*, **16**, pp. 327.
11. Goldman, M. (2001). Formal theory of spin lattice relaxation. *J. Magn. Reson.*, **149**, pp. 160–187.
12. Redfield, A. G. (1996). *Encyclopedia of Nuclear Magnetic Resonance, Relaxation Theory: Density Matrix Formulation* (Chichester: Wiley), pp. 4085–4092.
13. Kruk, D., Kowalewski, J., Tipikin, S., Freed, J. H., Mościcki, M., Mielczarek, A., and Port, M. (2011). Joint analysis of ESR lineshapes and 1H NMRD profiles of DOTA-Gd derivatives by means of the slow motion theory. *J. Chem. Phys.*, **134**, pp. 024508–024523.
14. Kruk, D., Earle, K. A., Mielczarek, A., Kubica, A., Milewska, A. and Moscicki, J. (2011). Nuclear quadrupole resonance lineshape analysis for different motional models: Stochastic Liouville approach. *J. Chem. Phys.*, **135**, pp. 224511–224520.
15. Kruk, D., Mielczarek, A., Korpała, A., Kozłowski, A., Earle, K. A. and Mościcki, J. (2012). Sensitivity of ^{2}H NMR Spectroscopy to motional

models: Proteins and highly viscous liquids as examples. *J. Chem. Phys.* 136, pp. 44509–44517.

16. Abragam, A. and Bleaney, B. (1970). *Electron Paramagnetic Resonance of Transition Ions* (Oxford: Clarendon Press).
17. Slichter, C. P. (1990). *Principles of Magnetic Resonance* (Berlin: Springer-Verlag).
18. Bertini, I., Luchinat, C. and Parigi, G. (2001). *Solution NMR of Paramagnetic Molecules* (Amsterdam: Elsevier).
19. Griffith, J. S. (1961). *The Theory of Transition Metal Ions* (Cambridge: Cambridge University Press).
20. Langhoff, S. R., and Kern, C. W. (1977). *Applications of Electronic Structure Theory* (New York: Plenum Press), pp. 381–437.

Chapter 5

Spin Relaxation: A More General Approach

This chapter explains how to describe spin relaxation phenomena in more complex systems accounting for different kinds of spin interactions. The origin of interference of spin interactions and cross-correlation effects is described and their influence on the spin relaxation discussed in examples.

5.1 Generalized Spectral Densities

In Chapters 2 and 3 we have discussed the relaxation phenomena in a system containing two non-equivalent (possessing different resonance frequencies) spins, $I = 1/2$ and $S = 1/2$. It has been shown that the spin–lattice relaxation process (represented by the population block of the relaxation matrix) is, in general, bi-exponential. When the S spin is in equilibrium, $\langle S_Z \rangle = \langle S_Z \rangle_{\text{eq}}$ (Eq. 2.65), the I spin relaxation is single exponential, with the relaxation rate $R_{1I}(\omega_I) = R_{II}(\omega_I)$ given by Eq. 2.62, which can be rewritten in

Understanding Spin Dynamics
Danuta Kruk

ISBN 978-981-4463-49-2 (Hardcover), 978-981-4463-50-8 (eBook)
www.panstanford.com

the form:

$$R_{1I}(\omega_I) = \frac{2}{3}\left(\frac{\mu_0}{4\pi}\frac{\gamma_I\gamma_S\hbar}{r^3}\right)^2 S(S+1)\left[J(\omega_I-\omega_S) + 3J(\omega_I) + 6J(\omega_I+\omega_S)\right] \quad (5.1)$$

One can write an analogous formula for the S spin relaxation (interchanging the indices I and S) under the assumption that $\langle I_z\rangle = \langle I_z\rangle_{eq}$.

Equation 5.1 is a special case of a more general expression [7, 9–11]:

$$R_{1I}(\omega_I) = \frac{2}{15}\left(\frac{\mu_0}{4\pi}\frac{\gamma_I\gamma_S\hbar}{r^3}\right)^2 S(S+1)\left[S_{1,1}(\omega_I) + 3S_{0,0}(\omega_I) + 6S_{-1,-1}(\omega_I)\right] \quad (5.2)$$

where the quantities $S_{m,m}(\omega_I)$ are referred to as generalized spectral densities. Although they are written as functions of the I spin resonance frequency, ω_I (in analogy to $R_{1I}(\omega_I)$), the spectral densities implicitly depend also on ω_S, as we shall see soon. They can be expressed as a matrix product [7, 9–12]:

$$S_{m,m}(\omega_I) = \frac{1}{2}\mathrm{Re}\left\{\left[S^1_m\right]^+ \left[M - i\omega_I 1\right]^{-1}\left[S^1_m\right]\right\} \quad (5.3)$$

This reminds us the matrix representation of the lineshape function given, for instance by Eq. 4.23. Thus, we should also remember that the I–S spin system is characterized by four Zeeman functions $|n\rangle = |m_I, m_S\rangle$: $|1\rangle = |1/2, 1/2\rangle$, $|2\rangle = |1/2, -1/2\rangle$, $|3\rangle = |-1/2, 1/2\rangle$, and $|4\rangle = |-1/2, -1/2\rangle$, which form the Liouville basis $\{|n\rangle\langle n'|\}$ for which we use the notation $\{|\alpha\rangle\langle\beta|\}$ (see Chapter 3). The matrices $\lfloor S^1_m\rfloor$ represent the S spin operators, $S^1_0 = S_z$ and $S^1_{\pm1} = S_\pm/\sqrt{2}$ in the $\{|\alpha\rangle\langle\beta|\}$ basis. The representation of the S_+ operator is given by Eq. 4.42, in analogy for $S^1_{\pm1}$ and S^1_0 one gets:

$$S_1 = \left\{\left|\frac{1}{2},\frac{1}{2}\right\rangle\left\langle\frac{1}{2},\frac{1}{2}\right| + \left|-\frac{1}{2},-\frac{1}{2}\right\rangle\left\langle-\frac{1}{2},\frac{1}{2}\right|\right\} = \{|2\rangle\langle1| + |4\rangle\langle3|\} \quad (5.4a)$$

$$S_{-1} = \left\{\left|\frac{1}{2},\frac{1}{2}\right\rangle\left\langle\frac{1}{2},-\frac{1}{2}\right| + \left|-\frac{1}{2},\frac{1}{2}\right\rangle\left\langle-\frac{1}{2},-\frac{1}{2}\right|\right\} = \{|1\rangle\langle2| + |3\rangle\langle4|\} \quad (5.4b)$$

$$S_0 = \left\{\left|\frac{1}{2},\frac{1}{2}\right\rangle\left\langle-\frac{1}{2},\frac{1}{2}\right| + \left|\frac{1}{2},-\frac{1}{2}\right\rangle\left\langle-\frac{1}{2},-\frac{1}{2}\right|\right\} = \{|1\rangle\langle3| + |2\rangle\langle4|\} \quad (5.4c)$$

The operator M of Eq. 5.3 encloses, in this case, the Zeeman interaction of the S spin, and the rotational dynamics (assuming that the I–S dipole–dipole coupling fluctuates in time due to molecular rotation), characterized by a correlation time τ_{rot}. The matrix $[M]$ is diagonal with the elements [9, 10]:

$$[M]_{\alpha\beta,\alpha\beta} = i\omega_{\alpha\beta} - i\omega_{\text{I}} + \tau_{\text{rot}}^{-1} \tag{5.5}$$

The frequency $\omega_{\alpha\beta}$ is a transition frequency of the spin S between the levels described by the functions $|\alpha\rangle$ and $|\beta\rangle$, that is, $\omega_{21} = \omega_{43} = -\omega_{12} = -\omega_{34} = \omega_{\text{S}}$, where ω_{S} is the S spin resonance frequency, and $\omega_{13} = \omega_{24} = 0$. Thus, one obtains for the $S_{m,m}(\omega_{\text{I}})$ spectral densities:

$$\begin{aligned} S_{\pm1,\pm1}(\omega_{\text{I}}) &= \frac{1}{2}\text{Re}\left\{[1,1]\begin{bmatrix} i(\pm\omega_{\text{S}} - \omega_{\text{I}}) + \tau_{\text{rot}}^{-1} & 0 \\ 0 & i(\pm\omega_{\text{S}} - \omega_{\text{I}}) + \tau_{\text{rot}}^{-1} \end{bmatrix}^{-1}\begin{bmatrix} 1 \\ 1 \end{bmatrix}\right\} \\ &= \frac{\tau_{\text{rot}}}{1 + (\omega_{\text{S}} \mp \omega_{\text{I}})^2\tau_{\text{rot}}^2} \end{aligned} \tag{5.6}$$

while

$$\begin{aligned} S_{0,0}(\omega_{\text{I}}) &= \frac{1}{2}\text{Re}\left\{[1,1]\begin{bmatrix} -i\omega_{\text{I}} + \tau_{\text{rot}}^{-1} & 0 \\ 0 & i\omega_{\text{I}} + \tau_{\text{rot}}^{-1} \end{bmatrix}^{-1}\begin{bmatrix} 1 \\ 1 \end{bmatrix}\right\} \\ &= \frac{\tau_{\text{rot}}}{1 + \omega_{\text{I}}^2\tau_{\text{rot}}^2} \end{aligned} \tag{5.7}$$

Substituting Eqs. 5.7 and 5.6 into Eq. 5.2 one gets Eq. 5.1.

At this stage one might not fully see the reason for introducing the generalization of Eq. 5.2, but the examples given below should clarify this concept.

5.2 Residual Dipolar Interactions

So far we have discussed dipolar interactions which entirely act as a relaxation mechanism. For this purpose the motion which causes stochastic fluctuations of the dipolar interaction must lead to its zero average, $\langle H_{\text{DD}}(t)\rangle = 0$. This condition is in most cases fulfilled for liquids which can be treated as isotropic systems, but can be (and in most cases is) violated in solids due to anisotropy of the molecular environment and geometrical constrains. In such a case the dipole–dipole coupling can be split into two parts:

$$H_{\text{DD}}(t) = \langle H_{\text{DD}}(t)\rangle + \lfloor H_{\text{DD}}(t) - \langle H_{\text{DD}}(t)\rangle \rfloor = H_{\text{DD}}^{\text{res}} + H'_{\text{DD}}(t) \quad (5.8)$$

The first term, $H_{\text{DD}}^{\text{res}} = \langle H_{\text{DD}}(t)\rangle \neq 0$, is often referred to as a "residual" dipolar coupling and it contributes to the energy level structure of the interacting spins, as described in Chapter 4, while the fluctuations around the residual (averaged) coupling represented by $H'_{\text{DD}}(t) = [H_{\text{DD}}(t) - \langle H_{\text{DD}}(t)\rangle]$ lead to the spin relaxation.

To understand the influence of the residual dipolar coupling on the relaxation process let us consider the total Hamiltonian determining the energy levels of the I–S spin system:

$$H_0(I, S) = \omega_{\text{I}} I_z + \omega_{\text{S}} S_z + H_{\text{DD}}^{\text{res(L)}}(I, S) \quad (5.9)$$

In Eq. 5.9 it has explicitly been stressed that the residual dipolar coupling has to be expressed in the laboratory frame. The laboratory form of the dipole–dipole Hamiltonian has been given in Chapters 2 and 4; it can straightforwardly be adapted to the residual interaction, $H_{\text{DD}}^{\text{res(L)}}(I, S)$. To make the calculations simpler let us limit ourselves at this stage to the diagonal elements of the matrix representation of the dipolar coupling given by Eq. 4.51 (Chapter 4). This leads to the following energy values corresponding to the listed above Zeeman states $|m_{\text{I}}, m_{\text{S}}\rangle$:

$$E_1 = \frac{1}{2}\omega_{\text{I}} + \frac{1}{2}\omega_{\text{S}} + \left(\frac{3\cos^2\beta - 1}{4}\right)\left(\frac{\mu_0}{4\pi}\frac{\gamma_{\text{I}}\gamma_{\text{S}}\hbar}{r_{\text{res}}^3}\right) \quad (5.10\text{a})$$

$$E_2 = \frac{1}{2}\omega_{\text{I}} - \frac{1}{2}\omega_{\text{S}} - \left(\frac{3\cos^2\beta - 1}{4}\right)\left(\frac{\mu_0}{4\pi}\frac{\gamma_{\text{I}}\gamma_{\text{S}}\hbar}{r_{\text{res}}^3}\right) \quad (5.10\text{b})$$

$$E_3 = -\frac{1}{2}\omega_{\text{I}} + \frac{1}{2}\omega_{\text{S}} - \left(\frac{3\cos^2\beta - 1}{4}\right)\left(\frac{\mu_0}{4\pi}\frac{\gamma_{\text{I}}\gamma_{\text{S}}\hbar}{r_{\text{res}}^3}\right) \quad (5.10\text{c})$$

$$E_4 = -\frac{1}{2}\omega_{\text{I}} - \frac{1}{2}\omega_{\text{S}} + \left(\frac{3\cos^2\beta - 1}{4}\right)\left(\frac{\mu_0}{4\pi}\frac{\gamma_{\text{I}}\gamma_{\text{S}}\hbar}{r_{\text{res}}^3}\right) \quad (5.10\text{d})$$

where β describes the orientation of the dipole–dipole axis. In Eq. 5.10 an inter-spin distance r_{res} has been introduced. One should be aware that the amplitude of the residual dipole–dipole coupling is smaller than the amplitude of the total dipolar interaction; this can be reflected by an effective distance between the interacting spins. When the effects of the residual dipolar coupling are restricted only to the "correction" of the energy levels, one can easily modify the expression for the generalized spectral densities:

$$S_{1,1}(\omega_I) = \frac{1}{2}\mathrm{Re}\left\{[1,1]\begin{bmatrix} i(\omega_S+\omega_{DD}-\omega_I)+\tau_{rot}^{-1} & 0 \\ 0 & i(\omega_S-\omega_{DD}-\omega_I)+\tau_{rot}^{-1}\end{bmatrix}^{-1}\begin{bmatrix}1\\1\end{bmatrix}\right\}$$
$$= \frac{1}{2}\frac{\tau_{rot}}{1+(\omega_S+\omega_{DD}-\omega_I)^2\tau_{rot}^2} + \frac{1}{2}\frac{\tau_{rot}}{1+(\omega_S-\omega_{DD}+\omega_I)^2\tau_{rot}^2} \quad (5.11)$$

$$S_{-1,-1}(\omega_I)$$
$$= \frac{1}{2}\mathrm{Re}\left\{[1,1]\begin{bmatrix} i(-\omega_S-\omega_{DD}-\omega_I)+\tau_{rot}^{-1} & 0 \\ 0 & i(-\omega_S+\omega_{DD}-\omega_I)+\tau_{rot}^{-1}\end{bmatrix}^{-1}\begin{bmatrix}1\\1\end{bmatrix}\right\}$$
$$= \frac{1}{2}\frac{\tau_{rot}}{1+(\omega_S+\omega_{DD}+\omega_I)^2\tau_{rot}^2} + \frac{1}{2}\frac{\tau_{rot}}{1+(\omega_S-\omega_{DD}+\omega_I)^2\tau_{rot}^2} \quad (5.12)$$

and

$$S_{0,0}(\omega_I) = \frac{1}{2}\mathrm{Re}\left\{[1,1]\begin{bmatrix} i(\omega_{DD}-\omega_I)+\tau_{rot}^{-1} & 0 \\ 0 & i(\omega_{DD}+\omega_I)+\tau_{rot}^{-1}\end{bmatrix}^{-1}\begin{bmatrix}1\\1\end{bmatrix}\right\}$$
$$= \frac{1}{2}\frac{\tau_{rot}}{1+(\omega_I+\omega_{DD})^2\tau_{rot}^2} + \frac{1}{2}\frac{\tau_{rot}}{1+(\omega_I-\omega_{DD})^2\tau_{rot}^2} \quad (5.13)$$

where we have introduced the dipolar frequency $\omega_{DD} = \left(\frac{3\cos^2\beta-1}{4}\right)\left(\frac{\mu_0}{4\pi}\frac{\gamma_I\gamma_S\hbar}{r_{res}^3}\right)$. Substituting Eqs. 5.11–5.13 into Eq. 5.2 one obtains for the relaxation rate $R_{1I}(\omega_I)$ (for $S = 1/2$):

$$R_{1I}(\omega_I) = \frac{1}{20}\xi_{DD}^2$$
$$\times\left[\frac{2\tau_{rot}}{1+(\omega_S+\omega_I-\omega_{DD})^2\tau_{rot}^2} + \frac{\tau_{rot}}{1+(\omega_S-\omega_I+\omega_{DD})^2\tau_{rot}^2} + \frac{\tau_{rot}}{1+(\omega_S+\omega_I+\omega_{DD})^2\tau_{rot}^2} + \frac{\tau_{rot}}{1+(\omega_I+\omega_{DD})^2\tau_{rot}^2} + \frac{\tau_{rot}}{1+(\omega_I-\omega_{DD})^2\tau_{rot}^2}\right] \quad (5.14)$$

One should note in Eq. 5.14 the dipolar coupling constant, ξ_{DD}, defined as $\xi_{DD} = \left(\frac{\mu_0}{4\pi}\frac{\gamma_I\gamma_S\hbar}{r_{fluct}^3}\right)$, where r_{fluct} is again an effective inter-spin distance reflecting, in this case, the amplitude of the fluctuating part of the dipolar interaction. The residual dipolar coupling is relevant when it is comparable with the resonance frequencies $\omega_{I(S)}$. For solids, the residual coupling is typically of the order of 10th kilo Hertz; thus, it becomes of importance for relaxation experiments performed at low frequencies.

The above description of the influence of the residual dipolar coupling on the spin relaxation is important for didactic purposes; however, it is not complete. If we accepted the fact that ω_{DD} is of the order of $\omega_{I(S)}$ there is no reason to neglect the off-diagonal elements in the matrix representation of the residual dipolar Hamiltonian. However, when the off-diagonal elements are present the Zeeman basis is not anymore the eigenbasis of this spin system, as explained in Chapter 4. In fact, one should take into account all terms of the residual dipolar Hamiltonian, but then the calculations of the relaxation rate $R_{1I}(\omega_I)$ have to be performed in a numerical way. Therefore, we limit the discussion including only the $R_0^2(I, S)$ term. Then, the Hamiltonian yields:

$$H_0(I, S) = \omega_I I_z + \omega_S S_z + 2\omega_{DD}\left[I_z S_z + \frac{1}{2}(I_+ S_- + I_- S_+)\right] \quad (5.15)$$

Its matrix representation in the basis $\{|\alpha\rangle\}$ takes the form:

$$[H_0(I, S)] = \begin{bmatrix} \omega_I + \omega_S + \frac{\omega_{DD}}{2} & 0 & 0 & 0 \\ 0 & \omega_I - \omega_S - \frac{\omega_{DD}}{2} & \omega_{DD} & 0 \\ 0 & \omega_{DD} & -\omega_I + \omega_S - \frac{\omega_{DD}}{2} & 0 \\ 0 & 0 & 0 & -\omega_I - \omega_S + \frac{\omega_{DD}}{2} \end{bmatrix} \quad (5.16)$$

The matrix can be easily diagonalized leading to the following set of eigenvalues E_α, and eigensfunctions, $|\psi_\alpha\rangle$, which are linear combinations of the Zeeman functions, $|\alpha\rangle$:

$$|\psi_1\rangle = |1\rangle = \left|\frac{1}{2}, \frac{1}{2}\right\rangle, \; E_1 = \omega_I + \omega_s + \frac{\omega_{DD}}{2} \quad (5.17a)$$

$$|\psi_2\rangle = a\,|2\rangle + b\,|3\rangle = a\left|\frac{1}{2}, \frac{1}{2}\right\rangle + b\left|-\frac{1}{2}, \frac{1}{2}\right\rangle,$$

$$E_2 = \frac{\sqrt{(\omega_I - \omega_s)^2 + \omega_{DD}^2}}{2} - \frac{\omega_{DD}}{2} \quad (5.17b)$$

$$|\psi_3\rangle = -b\,|2\rangle + a\,|3\rangle = -b\left|\frac{1}{2}, -\frac{1}{2}\right\rangle + a\left|-\frac{1}{2}, \frac{1}{2}\right\rangle,$$

$$E_3 = \frac{\sqrt{(\omega_I - \omega_s)^2 + \omega_{DD}^2}}{2} - \frac{\omega_{DD}}{2} \tag{5.17c}$$

$$|\psi_4\rangle = |4\rangle = \left|-\frac{1}{2}, \frac{1}{2}\right\rangle,\ E_4 = -\omega_I - \omega_s + \frac{\omega_{DD}}{2} \tag{5.17d}$$

where the coefficients a, b are given as:

$$a = \left\{\frac{1}{2}\left[1 + \frac{\omega_I - \omega_s}{\sqrt{(\omega_I + \omega_s)^2 + \omega_{DD}^2}}\right]\right\}^{1/2}$$

$$b = \left\{\frac{1}{2}\left[1 - \frac{\omega_I - \omega_s}{\sqrt{(\omega_I + \omega_s)^2 + \omega_{DD}^2}}\right]\right\}^{1/2} \tag{5.18}$$

Using the inverse relationship, $|\psi_1\rangle = |1\rangle, |\psi_2\rangle = a\,|2\rangle + b\,|3\rangle, |\psi_3\rangle = -b\,|2\rangle + a\,|3\rangle$, and $|\psi_4\rangle = |4\rangle$, one can now set up the representation of the operators S_m^1 in the Liouville space $|\psi_\alpha\rangle\langle\psi_\beta|$ constructed from the eigenfunctions $|\psi_\alpha\rangle$.

$$S_1 = a\left(|\psi_2\rangle\langle\psi_1| + |\psi_4\rangle\langle\psi_3|\right) + b\left(|\psi_3\rangle\langle\psi_1| - |\psi_4\rangle\langle\psi_2|\right) \tag{5.19a}$$

$$S_{-1} = a\left(|\psi_1\rangle\langle\psi_2| + |\psi_3\rangle\langle\psi_4|\right) + b\left(|\psi_1\rangle\langle\psi_3| - |\psi_2\rangle\langle\psi_4|\right) \tag{5.19b}$$

$$S_0 = a\left(|\psi_2\rangle\langle\psi_4| + |\psi_1\rangle\langle\psi_3|\right) + b\left(|\psi_3\rangle\langle\psi_4| - |\psi_1\rangle\langle\psi_2|\right) \tag{5.19c}$$

This implies that the spectral densities $S_{m,m}(\omega_I)$ are now given by somewhat more complex expressions:

$$\begin{aligned} S_{\pm1,\pm1}(\omega_I) &= \mathrm{Re}\left\{[a, a, b, -b]\begin{bmatrix}\Lambda_{21} & 0 & 0 & 0\\ 0 & \Lambda_{43} & 0 & 0\\ 0 & 0 & \Lambda_{31} & 0\\ 0 & 0 & 0 & \Lambda_{42}\end{bmatrix}^{-1}\begin{bmatrix}a\\ a\\ b\\ -b\end{bmatrix}\right\} \\ &= a^2\left[\frac{\tau_{rot}}{1 + (\pm\omega_{21} - \omega_I)^2\,\tau_{rot}^2} + \frac{\tau_{rot}}{1 + (\pm\omega_{43} - \omega_I)^2\,\tau_{rot}^2}\right] \\ &\quad + b^2\left[\frac{\tau_{rot}}{1 + (\pm\omega_{31} - \omega_I)^2\,\tau_{rot}^2} + \frac{\tau_{rot}}{1 + (\pm\omega_{42} - \omega_I)^2\,\tau_{rot}^2}\right] \end{aligned} \tag{5.20}$$

where $\Lambda_{\alpha\beta} = i\left(\pm\omega_{\alpha\beta} - \omega_{\mathrm{I}}\right) + \tau_{\mathrm{rot}}^{-1}$, $\omega_{\alpha\beta} = E_{\alpha} - E_{\beta}$ and

$$S_{0,0}\left(\omega_{\mathrm{I}}\right) = \mathrm{Re}\left\{[a, a, b, -b]\begin{bmatrix}\Lambda_{24} & 0 & 0 & 0\\ 0 & \Lambda_{13} & 0 & 0\\ 0 & 0 & \Lambda_{34} & 0\\ 0 & 0 & 0 & \Lambda_{12}\end{bmatrix}^{-1}\begin{bmatrix}a\\ a\\ b\\ -b\end{bmatrix}\right\}$$
$$= a^{2}\left[\frac{\tau_{\mathrm{rot}}}{1+\left(\omega_{24}-\omega_{\mathrm{I}}\right)^{2}\tau_{\mathrm{rot}}^{2}} + \frac{\tau_{\mathrm{rot}}}{1+\left(\omega_{13}-\omega_{\mathrm{I}}\right)^{2}\tau_{\mathrm{rot}}^{2}}\right]$$
$$+ b^{2}\left[\frac{\tau_{\mathrm{rot}}}{1+\left(\omega_{34}-\omega_{\mathrm{I}}\right)^{2}\tau_{\mathrm{rot}}^{2}} + \frac{\tau_{\mathrm{rot}}}{1+\left(\omega_{12}-\omega_{\mathrm{I}}\right)^{2}\tau_{\mathrm{rot}}^{2}}\right] \tag{5.21}$$

Then, substituting Eqs. 5.20 and 5.21 into Eq. 5.2 one obtains an expression for the relaxation rate $R_{1\mathrm{I}}\left(\omega_{\mathrm{I}}\right)$ which indeed reflects the fact that the residual dipolar coupling leads to a non-Zeeman set of eigenfunctions. The essential difference between Eqs. 5.11–5.13 and Eqs. 5.20 and 5.21 is that in the last case the residual dipolar coupling affects not only the transition frequencies, but also the 'weigh-factors' a^2 and b^2 of the contributions to $R_{1\mathrm{I}}\left(\omega_{\mathrm{I}}\right)$ associated with the individual transition frequencies $\omega_{\alpha\beta}$. This description of the influence of the residual dipolar coupling on $R_{1\mathrm{I}}\left(\omega_{\mathrm{I}}\right)$ is, however, also simplified. A full treatment requires taking into account all terms of the dipolar Hamiltonian, according to Eq. 4.51.

5.3 Interference Effects

In Chapters 2 and 3, it has been discussed that the spin relaxation processes in a system of mutually coupled (by a dipole–dipole interaction) non-equivalent spins, I–S, is bi-exponential and one should distinguish between auto-relaxation and cross-relaxation rates. The auto-relaxation rate, R_{II} links the changes in the I spin magnetization, $\frac{dM_z(I)(t)}{dt}$, with its current value, $M_z(I)(t)$; analogously R_{SS} links $\frac{dM_z(S)(t)}{dt}$ with $M_z(S)(t)$. The cross-relaxation rate R_{IS} links $\frac{dM_z(I)(t)}{dt}$ with $M_z(S)(t)$ (and analogously for R_{SI}). The expressions for the relaxation rates are given by Eqs. 2.62–2.64.

In this section we remain with a two spins I–S interacting by a dipole–dipole coupling, but we shall take into account that one of the spins (let us say S) is subject to chemical shift anisotropy (CSA) interaction (often referred to as chemical shielding). The term "chemical shielding" means that the local magnetic field $\vec{B}_{\mathrm{loc}}$ differs from the external magnetic field $\vec{B}_0$ due to an induced magnetic field $\vec{B}_{\mathrm{ind}}$ [8, 13–15]:

$$\vec{B}_{\mathrm{loc}} = \vec{B}_0 + \vec{B}_{\mathrm{ind}} = \vec{B}_0 \left(1 - \sigma\right) \tag{5.22}$$

where σ denotes the chemical shielding tensor. The tensor can be decomposed into symmetric and asymmetric parts. The asymmetric part is modulated by molecular rotation (like the I–S dipole–dipole coupling) and, in consequence, acts as a relaxation mechanism for the S spin. As the S and I spins are coupled, it is of interest to inquire into whether the CSA interaction affects also the I spin dynamics. For this purpose we have to consider the Hamiltonian form of the CSA interaction distinguishing between its principal axis frame (P) (this is a frame in which the symmetric part of the chemical shielding tensor is diagonal) and the laboratory (L) representations. The form of the CSA Hamiltonian in the (P) frame, $H_{\mathrm{CSA}}^{(\mathrm{P})}(S)$, yields [1, 2, 8, 13–15]:

$$H_{\mathrm{CSA}}^{(\mathrm{P})}(S) = \gamma_S \sum_{m=-2}^{2} (-1)^m \, \Phi_{-m}^{2(\mathrm{P})} T_m^2 \left(S, B_0\right) \tag{5.23}$$

where the functions $\Phi_{-m}^{2(\mathrm{P})}$ are defined as:

$$\Phi_{-m}^{2(\mathrm{P})} = \sqrt{\frac{2}{3}} \left[\sigma_{zz} - \frac{1}{2}\left(\sigma_{xx} + \sigma_{yy}\right)\right], \; \Phi_{\pm 1}^{2(\mathrm{P})} = 0, \; \Phi_{\pm 2}^{2(\mathrm{P})} = \frac{1}{2}\left(\sigma_{xx} + \sigma_{yy}\right) \tag{5.24}$$

while the tensor components $T_m^2(S, B_0)$ are:

$$T_0^2(S, B_0) = \sqrt{\frac{2}{3}} S_z B_0, \; T_{\pm 1}^2(S, B_0) = \mp \frac{1}{2} S_{\pm} B_0, \; T_{\pm 2}^2(S, B_0) = 0 \tag{5.25}$$

Transforming the Hamiltonian of Eq. 5.23 into the laboratory frame one obtains:

$$\Phi_{-m}^{2(L)}(t) = \sum_{k=-2}^{2} D_{k,-m}^2 \left(\Omega(t)\right) \Phi_{-m}^{2(\mathrm{P})} \tag{5.26}$$

Where $\Omega(t) \equiv \Omega_{\mathrm{PL}}(t)$ describes the orientation of the molecule (the (P) frame is fixed in the molecule) relative to the laboratory frame. The transformation to the laboratory frame leads to:

$$B_0\Phi^2_{-m}(t) = \omega_S\left\{\sqrt{\frac{2}{3}}\left[\sigma_{zz} - \frac{1}{2}\left(\sigma_{xx}+\sigma_{yy}\right)\right]D^2_{0,-m}(\Omega) + \frac{1}{2}\left(\sigma_{xx}+\sigma_{yy}\right)\left[D^2_{2,-m}(\Omega) + D^2_{-2,-m}(\Omega)\right]\right\} \quad (5.27)$$

In Eq. 5.27 the relationship $\omega_S = \gamma_S B_0$ has been applied. In the already introduced basis $\{|n\rangle = |m_I, m_S\rangle\}$: $(|1\rangle = |1/2, 1/2\rangle$, $|2\rangle = |1/2, -1/2\rangle$, $|3\rangle = |-1/2, 1/2\rangle$, and $|4\rangle = |-1/2, -1/2\rangle)$, the Hamiltonian $H'(I, S)(t)$ which is now responsible for the relaxation process:

$$H'(I, S)(t) = H^{(L)}_{DD}(I, S)(t) + H^{(L)}_{CSA}(S)(t) \quad (5.28)$$

has the following form (for the representation of the dipole-dipole part, $H^{(L)}_{DD}(I, S)(t)$, Eq. 4.52 has been used):

$$\left[H'(I,S)\right] = a_{DD}\begin{bmatrix} \frac{1}{2\sqrt{6}}D^2_{0,0} & \frac{1}{4}D^2_{0,-1} & \frac{1}{4}D^2_{0,-1} & \frac{1}{2}D^2_{0,-2} \\ -\frac{1}{4}D^2_{0,-1} & -\frac{1}{2\sqrt{6}}D^2_{0,0} & -\frac{1}{2\sqrt{6}}D^2_{0,0} & -\frac{1}{4}D^2_{0,-1} \\ -\frac{1}{4}D^2_{0,1} & \frac{1}{2\sqrt{6}}D^2_{0,0} & \frac{1}{2\sqrt{6}}D^2_{0,0} & -\frac{1}{4}D^2_{0,-1} \\ -\frac{1}{2}D^2_{0,2} & \frac{1}{4}D^2_{0,1} & \frac{1}{4}D^2_{0,1} & \frac{1}{2\sqrt{6}}D^2_{0,0} \end{bmatrix} + \begin{bmatrix} \frac{1}{\sqrt{6}}\Phi^2_0 & \frac{1}{2}\Phi^2_{-1} & 0 & 0 \\ -\frac{1}{2}\Phi^2_1 & -\frac{1}{\sqrt{6}}\Phi^2_0 & 0 & 0 \\ 0 & 0 & \frac{1}{\sqrt{6}}\Phi^2_0 & -\frac{1}{2}\Phi^2_{-1} \\ 0 & 0 & \frac{1}{2}\Phi^2_1 & -\frac{1}{\sqrt{6}}\Phi^2_0 \end{bmatrix} \quad (5.29)$$

In Eq. 5.29 the index (L) has been omitted for simplicity. The extended form of the relaxation Hamiltonian leads to the following set of equations for the populations, $\rho_{\alpha\alpha}(t)$:

$$\frac{d}{dt}\begin{bmatrix} \rho_{11} \\ \rho_{22} \\ \rho_{33} \\ \rho_{44} \end{bmatrix}(t) = \left\{\left[R^{DD}\right] + \left[R^{CSA}\right] + \left[R^{DD\text{-}CSA}\right]\right\} \times \begin{bmatrix} \rho_{11} \\ \rho_{22} \\ \rho_{33} \\ \rho_{44} \end{bmatrix}(t) \quad (5.30)$$

Equation 5.30 contains three relaxation matrices $\lfloor R^{DD} \rfloor$, $\lfloor R^{CSA} \rfloor$, and $\lfloor R^{DD\text{-}CSA} \rfloor$, referred to as dipolar, CSA and interference relaxation matrices, respectively. The dipolar relaxation matrix has already been given in Chapter 3 (Eq. 3.48). Analogously, using the Redfield relaxation formula (Eq. 3.19), one can calculate the $\lfloor R^{CSA} \rfloor$ matrix:

$$\left[R^{CSA}\right] = \begin{bmatrix} R^{CSA}_{1111} & \frac{1}{2}J^{CSA}(\omega_S) & 0 & 0 \\ \frac{1}{2}J^{CSA}(\omega_S) & R^{CSA}_{2222} & 0 & 0 \\ 0 & 0 & R^{CSA}_{3333} & \frac{1}{2}J^{CSA}(\omega_S) \\ 0 & 0 & \frac{1}{2}J^{CSA}(\omega_S) & R^{CSA}_{4444} \end{bmatrix} \tag{5.31}$$

where $R^{CSA}_{\alpha\alpha\alpha\alpha} = -\frac{1}{2}J^{CSA}(\omega_S)$ [2, 8]. The spectral density $J^{CSA}(\omega_S)$ is defined as (assuming isotropic molecular tumbling with a correlation time, τ_{rot}):

$$J^{CSA}(\omega_S) = \omega_S^2 \left\{ \frac{2}{3}\left[\sigma_{zz} - \frac{1}{2}\left(\sigma_{xx} + \sigma_{yy}\right)\right]^2 + \frac{1}{2}\left(\sigma_{xx} - \sigma_{yy}\right)^2 \right\} \frac{1}{5}\frac{\tau_{rot}}{1 + \omega_S^2\tau_{rot}^2} \tag{5.32}$$

One should note that the amplitude of CSA interaction increases linearly with ω_S. Thus, one can expect that the CSA relaxation mechanism becomes more efficient at higher frequencies (magnetic fields). Nevertheless with increasing frequency, the amplitude of CSA competes with the decreasing spectral density (a Lorentzian function in the simplest case). When $\omega_S\tau_{rot} \gg 1$ one obtains:

$$J^{CSA}(\omega_S) \to \frac{1}{5\tau_{rot}} \left\{ \frac{2}{3}\left[\sigma_{zz} - \frac{1}{2}\left(\sigma_{xx} + \sigma_{yy}\right)\right]^2 + \frac{1}{2}\left(\sigma_{xx} - \sigma_{yy}\right)^2 \right\} \tag{5.33}$$

that is, the spectral density reaches a constant (frequency independent) value which is proportional to τ_{rot}^{-1}. In Fig. 5.1, the frequency dependence of the $J^{CSA}(\omega_S)$ spectral density, $J^{CSA}(\omega_S) \propto \frac{1}{\tau_{rot}} \frac{(\omega_S\tau_{rot})^2}{1+(\omega_S\tau_{rot})^2}$ is compared with the spectral shape of a Lorentzian function $J(\omega_S) \propto \frac{\tau_{rot}}{1+(\omega_S\tau_{rot})^2}$.

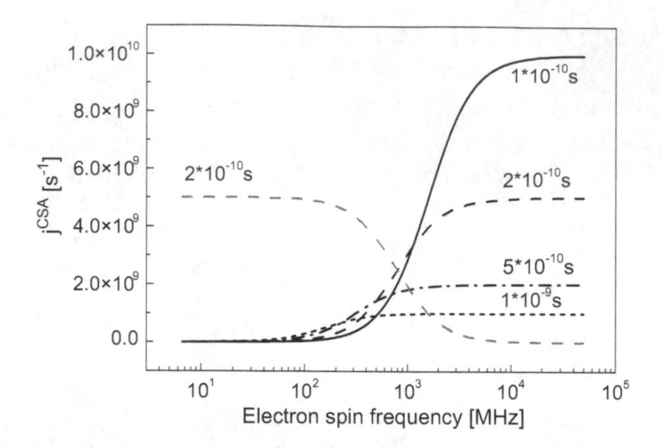

Figure 5.1 Frequency dependences of the correlation functions $J^{\mathrm{CSA}}(\omega_S) = \tau_{\mathrm{rot}}^{-1}(\omega_S\tau_{\mathrm{rot}})^2/[1+(\omega_S\tau_{\mathrm{rot}})^2]$ for different correlation times τ_{rot} compared with $\tau_{\mathrm{rot}}/[1+(\omega_S\tau_{\mathrm{rot}})^2]$ scaled to reach the value of τ_{rot}^{-1} at low frequencies.

The third contribution to the relaxation stems from interference between the dipole–dipole and CSA interactions as both of them are modulated by the same motional process (rotational dynamics in this case) [8, 18, 19]. To understand the concept of interference effects let us consider a correlation function $C(\tau) = \langle H'(\tau)\, H'(0)\rangle$ for a Hamiltonian $H'(\tau)$ being a sum of two terms, $H'(t) = H_1(t) + H_2(t)$:

$$\begin{aligned} C(\tau) &= \langle [H_1(\tau) + H_2(\tau)]\,[H_1(\tau) + H_2(\tau)]\rangle \\ &= \langle H_1(\tau)\, H_1(0)\rangle + \langle H_2(\tau)\, H_2(0)\rangle \\ &\quad + (\langle H_1(\tau) + H_2(0)\rangle + \langle H_2(\tau)\, H_1(0)\rangle) \end{aligned} \tag{5.34}$$

The last two terms

$$\tilde{C}(\tau) = \langle H_1(\tau)\, H_2(0)\rangle + \langle H_2(\tau)\, H_1(0)\rangle = 2\,\langle H_1(\tau)\, H_2(0)\rangle \tag{5.35}$$

give a correlation function between two different spin interactions. When one considers the general form of a Hamiltonian, $H(t) \propto \sum_{m=-1}^{l} (-1)^m F_{-m}^1(t)\, T_{m'}^1$, the function $\tilde{C}(\tau)$ represents, in fact, the correlation between the time dependent components $\left\langle F_{-m(1)}^{1*}(\tau)\, F_{m(2)}^1(0)\right\rangle$ of the Hamiltonians (the indices (1), (2) refer to $H_1(t)$ and $H_2(t)$, respectively). These functions for the dipole–dipole and CSA interactions both depend on time via the same angle $\Omega(\tau)$ describing the molecular orientation so obviously they are fully correlated. From the Hamiltonian matrix of Eq. 5.29 and the relaxation formula (Eq. 3.19), one can evaluate the $\lfloor R^{\text{DD - CSA}} \rfloor$ relaxation matrix:

$$\left[R^{\text{DD - CSA}}\right] = \begin{bmatrix} R_{1111}^{\text{DD - CSA}} & \frac{1}{2} J^{\text{DD - CSA}}(\omega_S) & 0 & 0 \\ \frac{1}{2} J^{\text{DD - CSA}}(\omega_S) & R_{2222}^{\text{DD - CSA}} & 0 & 0 \\ 0 & 0 & R_{3333}^{\text{DD - CSA}} & \frac{1}{2} J^{\text{DD - CSA}}(\omega_S) \\ 0 & 0 & \frac{1}{2} J^{\text{DD - CSA}}(\omega_S) & R_{4444}^{\text{DD - CSA}} \end{bmatrix} \tag{5.36}$$

where $\left[R_{\alpha\alpha\alpha\alpha}^{\text{DD - CSA}}\right] = -\frac{1}{2} J^{\text{DD - CSA}}(\omega_S)$ [8, 18, 19]. The spectral density $J^{\text{DD - CSA}}(\omega_S)$ is defined as:

$$J^{\text{DD - CSA}}(\omega_S) = \sqrt{\frac{2}{3}} \left[\sigma_{zz} - \frac{1}{2}\left(\sigma_{xx} + \sigma_{yy}\right)\right] a_{\text{DD}} \omega_S \frac{1}{5} \frac{\tau_{\text{rot}}}{1 + \omega_S^2 \tau_{\text{rot}}^2} \tag{5.37}$$

Using the relaxation matrices of Eq. 5.31 and Eq. 5.37 one can write a set of equations analogous to Eqs. 3.49a,b, but also including the CSA and DD-CSA contributions (this is reflected by using the symbol $\tilde{R}$ instead of R):

$$\frac{d}{dt}\left(\rho_{11} + \rho_{22} - \rho_{33} - \rho_{44}\right) = -\tilde{R}_{\text{II}}\left(\rho_{11} + \rho_{22} - \rho_{33} - \rho_{44}\right) - \tilde{R}_{\text{IS}}\left(\rho_{11} + \rho_{33} - \rho_{22} - \rho_{44}\right) \tag{5.38a}$$

$$\frac{d}{dt}\left(\rho_{11} + \rho_{33} - \rho_{22} - \rho_{44}\right) = -\tilde{R}_{\text{SS}}\left(\rho_{11} + \rho_{33} - \rho_{22} - \rho_{44}\right) - \overset{\leftrightarrow}{R}_{\text{SI}}\left(\rho_{11} + \rho_{22} - \rho_{33} - \rho_{44}\right) \tag{5.38b}$$

It is worth to remain that $\langle I_Z \rangle = \rho_{11} + \rho_{22} - \rho_{33} - \rho_{44}$, while $\langle S_Z \rangle = \rho_{11} + \rho_{33} - \rho_{22} - \rho_{44}$. After simple calculations, one can conclude that the CSA and DD-CSA terms do not affect the relaxation rates R_{II} and R_{IS}, that is, $\tilde{R}_{\text{II}} = R_{\text{II}}$ and $\tilde{R}_{\text{IS}} = R_{\text{IS}}$ (the relaxation rates R_{II} and R_{IS} are given by Eq. 2.62 and Eq. 2.64, respectively), while the relaxation rate $\tilde{R}_{\text{SS}}$ yields:

$$\tilde{R}_{\text{SS}} = R_{\text{SS}} + J^{\text{CSA}}(\omega_S) + J^{\text{DD - CSA}}(\omega_S) \tag{5.39}$$

where R_{SS} is given by Eq. 2.63. As the $\tilde{R}_{\text{SS}}$ relaxation rate enters the expressions for $R_1^{\pm}$ (Eq. 2.61) it affects the magnetization evolution of both spins, I and S.

5.4 Cross-Correlation Effects

In the previous section we have discussed the subject of interference between different spin interactions which are modulated by the same motional process. Here, we shall discuss correlation between spin interactions of the same type, but involving different nuclei [5, 8, 18].

To explain the origin of such correlation a system including more than two spins is needed. Let us thus consider a systems consisting of three spins 1/2 denoted as S, I_1, and I_2, as shown in Fig. 5.2.

The dipole–dipole Hamiltonian consists now of three terms:

$$H_{\mathrm{DD}}(S, I_1, I_2) = H_{\mathrm{DD}}(S, I_1) + H_{\mathrm{DD}}(S, I_2) + H_{\mathrm{DD}}(I_1, I_2) \quad (5.40)$$

However, when one assumes that S represents an electron spin, while I_1 and I_2 correspond to nuclei, one can safely neglect the last term, $H_{\mathrm{DD}}(I_1, I_2)$ in Eq. 5.40. The basis $\{|\alpha\rangle\}$ is now constructed from eight functions $\left\{\left|m_S, m_{I_1}, m_{I_2}\right\rangle = |m_S, m_1, m_2\rangle\right\}$, labeled as $|1\rangle = |1/2, 1/2, 1/2\rangle$, $|2\rangle = |1/2, 1/2, -1/2\rangle$, $|3\rangle =$

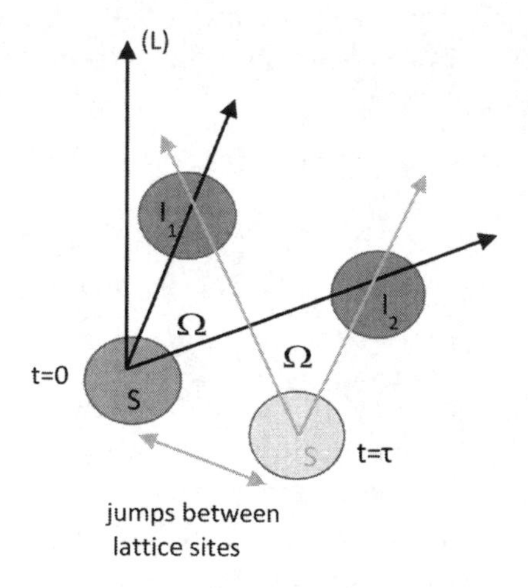

Figure 5.2 A system of three spins, I_1–S–I_2, connected by dipole–dipole interactions. Assuming that S is an electron spin, while I_1 and I_2 represent nuclear spins, the I_1–I_2 dipolar coupling can be neglected. When the nuclear spins are immobilized (i.e., the dipolar interactions fluctuate in time entirely due to the motion of the S spin), the relative orientation of the S–I_1 and S–I_2 dipole–dipole axes remain unchanged in time.

$|1/2, -1/2, 1/2\rangle$, $|4\rangle = |1/2, -1/2, -1/2\rangle$, $|5\rangle = |-1/2, 1/2, 1/2\rangle$, $|6\rangle = |-1/2, 1/2, -1/2\rangle$, $|7\rangle = |1/2, -1/2, 1/2\rangle$, and $|8\rangle = |-1/2, -1/2, -1/2\rangle$. The representation of the Hamiltonian $H_{\mathrm{DD}}(S, I_1, I_2)$ in this basis takes the form:

$$[H_{\mathrm{DD}}(S, I_1, I_2)] = [H_{\mathrm{DD}}(S, I_1, I_2)]_0 + \left\{[H_{\mathrm{DD}}(S, I_1, I_2)]_1 + [H_{\mathrm{DD}}(S, I_1, I_2)]_2\right\}$$

$$= \begin{bmatrix} a_0+b_0 & 0 & 0 & 0 & 0 & 0 & 0 & 0 \\ 0 & a_0-b_0 & 0 & 0 & -b_0 & 0 & 0 & 0 \\ 0 & 0 & -a_0+b_0 & 0 & -a_0 & 0 & 0 & 0 \\ 0 & 0 & 0 & -a_0-b_0 & 0 & -a_0 & -b_0 & 0 \\ 0 & -b_0 & -a_0 & 0 & -a_0-b_0 & 0 & 0 & 0 \\ 0 & 0 & 0 & -a_0 & 0 & -a_0+b_0 & 0 & 0 \\ 0 & 0 & 0 & -b_0 & 0 & 0 & a_0-b_0 & 0 \\ 0 & 0 & 0 & 0 & 0 & 0 & 0 & a_0+b_0 \end{bmatrix}$$

$$+ \begin{bmatrix} 0 & d_{-1} & c_{-1} & 0 & c_{-1}+d_{-1} & f_{-2} & e_{-2} & 0 \\ -d_1 & 0 & 0 & c_{-1} & 0 & c_{-1} & 0 & e_{-2} \\ -c_{-1} & 0 & 0 & d_{-1} & 0 & 0 & -c_{-1} & f_{-2} \\ 0 & -c_{-1} & -d_1 & 0 & 0 & 0 & 0 & -c_{-1}-d_{-1} \\ -c_1-d_1 & 0 & 0 & 0 & 0 & -d_1 & -c_{-1} & -f_{-2} \\ f_2 & -c_1 & 0 & 0 & -d_1 & 0 & 0 & -c_{-1} \\ e_2 & 0 & c_1 & 0 & c_{-1} & 0 & 0 & -d_{-1} \\ 0 & e_2 & f_2 & c_1+d_1 & f_2 & c_1 & d_1 & 0 \end{bmatrix} \tag{5.41}$$

where the zero-quantum part, $[H_{\mathrm{DD}}(S, I_1, I_2)]_0$ is represented by the first matrix, while the sum of the single- and double-quantum parts $\left\{[H_{\mathrm{DD}}(S, I_1, I_2)]_1 + [H_{\mathrm{DD}}(S, I_1, I_2)]_2\right\}$ is given by the second matrix. The symbols used have the following meaning

$$a_0 = \frac{1}{2\sqrt{6}} a_{\mathrm{DD}}^{S-I_1} D^2_{0,0}(\Omega_1),\ b_0 = \frac{1}{2\sqrt{6}} a_{\mathrm{DD}}^{S-I_2} D^2_{0,0}(\Omega_2),$$

$$c_{-1} = -\frac{1}{4} a_{\mathrm{DD}}^{S-I_1} D^2_{0,-1}(\Omega_1),\ C_1 = \frac{1}{4} a_{\mathrm{DD}}^{S-I_1} D^2_{0,1}(\Omega_1),$$

$$d_{-1} = -\frac{1}{4} a_{\mathrm{DD}}^{S-I_2} D^2_{0,-1}(\Omega_2),\ d_1 = \frac{1}{4} a_{\mathrm{DD}}^{S-I_2} D^2_{0,1}(\Omega_2),$$

$$e_{-2} = -\frac{1}{2} a_{\mathrm{DD}}^{S-I_1} D^2_{0,-2}(\Omega_1),\ e_2 = -\frac{1}{2} a_{\mathrm{DD}}^{S-I_1} D^2_{0,2}(\Omega_1),$$

$$f_{-2} = -\frac{1}{2} a_{\mathrm{DD}}^{S-I_2} D^2_{0,-2}(\Omega_2),\ \text{and}\ f_2 = -\frac{1}{2} a_{\mathrm{DD}}^{S-I_2} D^2_{0,2}(\Omega_2),$$

where $a_{\mathrm{DD}}^{S-I_1}$ and $a_{\mathrm{DD}}^{S-I_2}$ are dipole–dipole coupling constants for the $S - I_1$ and S–I_2 pairs of spins, respectively, and Ω_1 and Ω_2 describe the orientations of the S–I_1 and S–I_2, respectively. When the I_1 and I_2 spins are fixed at their positions (the nuclei are immobilized), the fluctuations of the S–I_1 and S–I_2 dipole–dipole axes are solely caused by the motion of the S spin (it can be, for instance, a diffusing paramagnetic molecule added to a crystal). This implies that the orientations of these axes are completely correlated, that is, the orientation of the S–I_1 axis determines the orientation of the S–I_2 axis. Keeping this in mind and inspecting the matrix of Eq. 5.41, one can conclude that some of its elements contain terms originating from both S–I_1 and S–I_2 dipole–dipole couplings. Thus, according to Eq. 5.35, the corresponding relaxation matrix elements R_{1155} and R_{4488} contains terms originating from the correlation between the orientations of the S–I_1 and S–I_2 dipole–dipole axes:

$$R_{1155} = R_{4488} = -\frac{1}{8}J^{(1)}(2\omega_{\mathrm{I}}) - \frac{1}{8}J^{(2)}(2\omega_{\mathrm{I}}) - \frac{1}{4}J^{(1-2)}(2\omega_{\mathrm{I}}) \quad (5.42)$$

with $J^{(\mathrm{i})}(\omega) = \left(a_{\mathrm{DD}}^{\mathrm{S-I_i}}\right)^2 J(\omega)$, and $J^{(1-2)}(\omega) = a_{\mathrm{DD}}^{\mathrm{S-I_1}} a_{\mathrm{DD}}^{\mathrm{S-I_2}} J(\omega)$, where the form of the spectral density $J(\omega)$ depends on the assumed model of motion; the frequency $2\omega_{\mathrm{I}}$ corresponds to the transition between the states $|1\rangle - |5\rangle$ and $|4\rangle - |8\rangle$.

Finishing this example some points should be stressed. The first one is that one should not mistake the cross-relaxation effects represented by the relaxation rates R_{IS} and R_{SI} of Eq. 2.59 with the cross-correlation. The second one is that in the present case the spin operator $\langle S_{\mathrm{Z}}\rangle$ and $\langle I_{\mathrm{Z}}\rangle$ are represented by the combinations:

$$\langle S_{\mathrm{Z}}\rangle \propto \rho_{11} + \rho_{22} + \rho_{33} + \rho_{44} - \rho_{55} - \rho_{66} - \rho_{77} - \rho_{88} \quad (5.43)$$

$$\langle I_{\mathrm{Z}}\rangle \propto \rho_{11} + \rho_{55} - \rho_{44} - \rho_{44} - \rho_{88} \quad (5.44)$$

Eventually, when the I_1 and I_2 undergo a motion the correlation is lost (there are no cross-correlation terms), but we face the problem of bi-exponential relaxation in a system containing more than two spins.

Consider an ensemble of N_{I} identical spins $I = 1/2$ and N_{S} identical spins $S = 1/2$; both group of species undergo dynamics characterized by the correlation times $\tau_{\mathrm{c,I}}$ and $\tau_{\mathrm{c,S}}$, respectively.

The relaxation process is bi-exponential as described by the set of equations, Eq. 2.59, but the relaxation rate R_{II} contains contributions originating from dipole–dipole interactions within the I spin sub-ensemble and between the I and S spins. Analogously, the R_{SS} relaxation rate consists of S–S and S–I parts. The corresponding expressions yield:

$$R_{\mathrm{II}}Z = K_{\mathrm{IS}}\left\lfloor J^{\mathrm{IS}}(\omega_{\mathrm{I}}-\omega_{\mathrm{S}}) + 3J^{\mathrm{IS}}(\omega_{\mathrm{I}}) + 6J^{\mathrm{IS}}(\omega_{\mathrm{I}}+\omega_{\mathrm{S}})\right\rfloor + K_{\mathrm{II}}\left\lfloor J^{\mathrm{II}}(\omega_{\mathrm{I}}) + 4J^{\mathrm{II}}(2\omega_{\mathrm{I}})\right\rfloor \quad (5.45a)$$

$$R_{\mathrm{SS}} = K_{\mathrm{SI}}\left\lfloor J^{\mathrm{IS}}(\omega_{\mathrm{I}}-\omega_{\mathrm{S}}) + 3J^{\mathrm{IS}}(\omega_{\mathrm{I}}) + 6J^{\mathrm{IS}}(\omega_{\mathrm{I}}+\omega_{\mathrm{S}})\right\rfloor + K_{\mathrm{SS}}\left\lfloor J^{\mathrm{SS}}(\omega_{\mathrm{S}}) + 4J^{\mathrm{SS}}(2\omega_{\mathrm{S}})\right\rfloor \quad (5.45b)$$

where K_{IS}, K_{SI}, K_{II}, and K_{SS} are effective dipolar coupling constants. The spectral densities $J^{\mathrm{II(SS)}}(\omega)$ and $J^{\mathrm{IS}}(\omega)$ depend on the applied motional model, but they also depend upon effective correlation times defined as $\tau_{\mathrm{eff,I}} = \tau_{\mathrm{c,I}}/2$ $(\tau_{\mathrm{eff,S}} = \tau_{\mathrm{c,S}}/2)$ and $\tau_{\mathrm{eff,IS}}^{-1} = \tau_{\mathrm{c,I}}^{-1} + \tau_{\mathrm{c,S}}^{-1}$. The cross-relaxation rates R_{IS} and R_{SI} are now given as:

$$R_{\mathrm{IS}} = K_{\mathrm{IS}}\left[6J^{\mathrm{IS}}(\omega_{\mathrm{I}}+\omega_{\mathrm{S}}) - J^{\mathrm{IS}}(\omega_{\mathrm{S}}-\omega_{\mathrm{I}})\right]\frac{N_{\mathrm{S}}}{N_{\mathrm{I}}} \quad (5.46a)$$

$$R_{\mathrm{SI}} = K_{\mathrm{SI}}\left[6J^{\mathrm{IS}}(\omega_{\mathrm{I}}+\omega_{\mathrm{S}}) - J^{\mathrm{IS}}(\omega_{\mathrm{S}}-\omega_{\mathrm{I}})\right]\frac{N_{\mathrm{I}}}{N_{\mathrm{S}}} \quad (5.46b)$$

For a more general case of the spin quantum number, $S \geq 1$, the reader is referred to [9].

5.5 Hierarchy of Spin Relaxation Processes

In Section 2.4 a set of equations describing bi-exponential relaxation in a system of two non-equivalent spins has been derived. The description has been extended in Section 5.4 to a system composed of N_{I} and N_{S} fractions of spins I and S, respectively. When one of the spins (let us say the S spin) relaxes very fast, that is, $R_{\mathrm{SS}} \gg R_{\mathrm{II}}$ (Eq. 2.59), the I spin relaxation becomes single exponential as then the condition $\left(\langle S_{\mathrm{Z}}\rangle - \langle S_{\mathrm{Z}}\rangle_{\mathrm{eq}}\right) = 0$ is reached very fast [4, 6]. Then, Eq. 2.59a reduces to the simple form:

$$\frac{d\left(\langle I_{\mathrm{Z}}\rangle - \langle I_{\mathrm{Z}}\rangle_{\mathrm{eq}}\right)}{dt} = -R_{\mathrm{II}}\left(\langle I_{\mathrm{Z}}\rangle - \langle I_{\mathrm{Z}}\rangle_{\mathrm{eq}}\right) \quad (5.47)$$

with a single exponential solution with the exponential factor R_{II}. Following Eq. 2.62 and assuming rotational dynamics described by Lorentzian spectral densities (Eq. 2.36), one could then write:

$$R_{II} = R_{1I} = \frac{1}{10}\left(\frac{\mu_0}{4\pi}\frac{\gamma_I\gamma_S\hbar}{r^3}\right)^2\left[\frac{\tau_{rot}}{1+(\omega_I-\omega_S)^2\tau_{rot}^2} + \frac{3\tau_{rot}}{1+\omega_I^2\tau_{rot}^2} + \frac{6\tau_{rot}}{1+(\omega_I+\omega_S)^2\tau_{rot}^2}\right] \quad (5.48)$$

Equation 5.48 is, however, not complete as the S spin relaxation acts as an additional source of fluctuations of the I–S dipole–dipole coupling, contributing to the spectral densities of Eq. 5.48. When the S spin relaxes fast due to spin interactions which do not involve the spin I, one talks about hierarchy of events. The S spin relaxation is independent of the presence of the spin I, while the I spin relaxation is influenced by the S spin relaxation. Such situations are typical for electron (S)–nuclear (I) spin systems. There are several sources of fast electronic relaxation (depending on the spin quantum number), for instance g-tensor anisotropy ($S = 1/2$) or zero-field splitting interactions (for $S \geq 1$); at this stage we shall not inquire into their detailed mechanisms.

Nevertheless, in the simplest case when the electron spin relaxation can be just described by two relaxation rates, R_{1S} and R_{2S}, for the spin–lattice and spin–spin relaxation, respectively, Eq. 5.48 can be straightforwardly modified accounting for the influence of the electronic relaxation on the modulations of the I–S dipolar coupling [2, 3, 7–9, 16, 17]:

$$R_{1I} = \frac{2}{15}\left(\frac{\mu_0}{4\pi}\frac{\gamma_I\gamma_S\hbar}{r^3}\right)^2 S(S+1)\left[\frac{\tau_{c2}}{1+(\omega_I-\omega_S)^2\tau_{c2}^2} + \frac{3\tau_{c1}}{1+\omega_I^2\tau_{c1}^2} + \frac{6\tau_{c2}}{1+(\omega_I+\omega_S)^2\tau_{c2}^2}\right] \quad (5.49)$$

where $\tau_{c,i}^{-1} = \tau_{rot}^{-1} + R_{i,S}$ ($i = 1, 2$).

Equation 5.49 has also been generalized to an arbitrary spin quantum number, S. The expression of Eq. 5.49 is referred to as Solomon–Bloembergen–Morgan (SBM) formula (originally derived for $S = 1/2$) [2, 3, 16, 17]. The underlying concept of this expression is illustrated in Fig. 5.3. In Eq. 5.49, the spectral density which does not include the electron spin transition frequency, ω_S, is associated

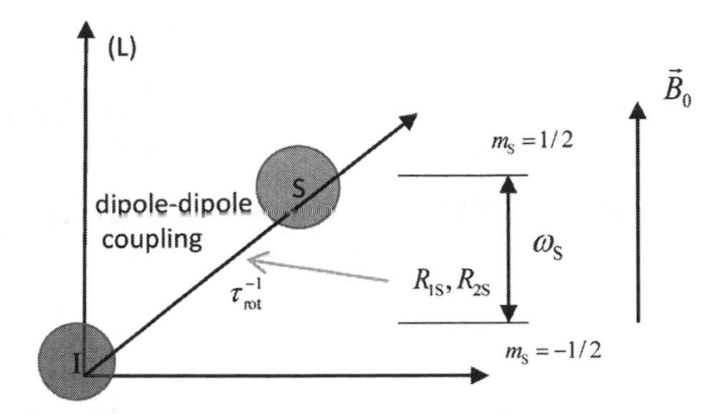

Figure 5.3 Relaxation processes in a system in which the *S* spin relaxes due to an interaction which is independent of the presence of the *I* spin. The *S* spin relaxation contributes to the modulations of the *I*–*S* dipole–dipole coupling.

with the spin–lattice electronic relaxation rate, R_{1S}, as it corresponds to the population part of the relaxation matrix ($\Delta m_S = 0$; Eq. 3.25). The spectral densities including the electronic frequency ω_S are associated with the spin–spin relaxation rate, R_{2S}, as in this case $|\Delta m_S| = 1$.

References

1. Anet, F.A.L and O'Leary, D.J. (1992). The shielding tensor. Part I: Understanding its symmetry properties. *Concepts Magn. Reson.*, **3**, pp. 193–214.
2. Anet, F.A.L and O'Leary, D.J. (1992). The shielding tensor. Part II: Understanding its strange effects on relaxation. *Concepts Magn. Reson.*, **4**, pp. 35–52.
3. Bloembergen, N. (1957). Proton relaxation times in paramagnetic solutions. *J. Chem. Phys.*, **27**, pp. 572–573.
4. Bloembergen, N. and Morgan, L.O. (1961). Proton relaxation times in paramagnetic solutions: Effects of electron spin relaxation. *J. Chem. Phys.*, **34**, pp. 842–850.
5. Hubbard, P.S. (1958). Nuclear magnetic relaxation of three or four molecules in liquid. *Phys. Rev.*, **109**, pp. 1153–1158.

6. Goldman, M. (1984). Interference effects in the relaxation of a pair of unlike spin-1/2 nuclei. *J. Magn. Reson.*, **60**, pp. 437–452.
7. Kowalewski, J., Kruk, D. and Parigi, G. (2005). NMR relaxation in solution of paramagnetic complexes: Recent theoretical progress for $S \geq 1$. *Adv. Inorg. Chem.*, **57**, pp. 41–104.
8. Kowalewski, J. and Mäler, L. (2006). *Nuclear Spin Relaxation in Liquids: Theory, Experiments and Applications* (New York: Taylor & Francis).
9. Kruk, D. (2007). *Theory of Evolution and Relaxation of Multi-Spin Systems* (Bury St Edmunds: Arima).
10. Kruk, D., Kubica, A., Masierak, W., Privalov, A.F., M. Wojciechowski, M. and Medycki, W. (2011). Quadrupole relaxation enhancement: Application to molecular crystals. *Solid StateNucl. Magn. Reson.*, **40**, pp. 114–125.
11. Kumar, A., Grace, R.C.R. and Madhu, P.K. (2000). Cross-correlations in NMR. *Prog. Nucl. Magn. Reson. Spectr.*, **37**, pp. 191–319.
12. Nilsson, T. and Kowalewski, J. (2000). Slow-motion theory of nuclear spin relaxation in paramagnetic low-symmetry complexes: A generalization to high electron spin. *J. Magn. Reson.*, **146**, pp. 345–358.
13. Smith, S.A., Palke, W.E. and Gerig, J.T. (1992). The Hamiltonians of NMR. Part I. *Concepts Magn. Reson.*, **4**, pp. 107–144.
14. Smith, S.A., Palke, W.E. and Gerig, J.T. (1992). The Hamiltonians of NMR. Part II. *Concepts Magn. Reson.*, **4**, pp. 181–204.
15. Smith, S.A., Palke, W.E. and Gerig, J.T. (1993). The Hamiltonians of NMR. Part III. *Concepts Magn. Reson.*, **4**, pp. 151–177.
16. Solomon, I. (1955). Relaxation processes in a system of two spins. *Phys. Rev.*, **99**, pp. 559–565.
17. Solomon, I. and Bloembergen, N. (1956). Nuclear magnetic interactions in the HF molecule. *J. Chem. Phys.*, **25**, pp. 261–266.
18. Werbelow, L.G. (1996). Encyclopedia of nuclear magnetic resonance, relaxation processes: Cross correlation and interference terms (Chichester: Wiley), pp. 4072–4078.
19. Werbelow, L.G. (1978). The use of interference terms to separate dipolar and chemical shift anisotropy contributions to nuclear spin relaxation. *J. Magn. Reson.*, **71**, pp. 151–153.

Chapter 6

Electron Spin Resonance of Spins 1/2

In Chapter 4, we have explained the concept of spin resonance lineshape and presented the formal theory of lineshape analysis. We have put special attention to the issue of how the spectral shape is related to dynamical properties of molecules currying the spins of interest. In this chapter we shall apply this theory to electron spin resonance (ESR) spectra of nitroxide radicals.

6.1 ESR Spectra and Scalar Interactions for ^{15}N Systems

Interesting examples of lineshape analysis, very important for biological applications, are provided by ESR spectroscopy of nitroxide radicals. Figure 6.1 shows 4-oxo-TEMPO-h_{16}-$^{14(15)}$N molecule which is one of nitroxide radicals typically applied as spin labels [1, 2].

In nitroxide radicals, the electron spin $S = 1/2$ is coupled to the nuclear spin $P = 1/2$ (^{15}N) or $P = 1$(^{14}N) by a hyperfine scalar coupling. The hyperfine coupling results from dipole–dipole interactions between the electron- and nitrogen spins. This coupling can be split into a time-independent, scalar part (referred to as isotropic hyperfine coupling) and anisotropic, time-dependent part.

Understanding Spin Dynamics
Danuta Kruk

ISBN 978-981-4463-49-2 (Hardcover), 978-981-4463-50-8 (eBook)
www.panstanford.com

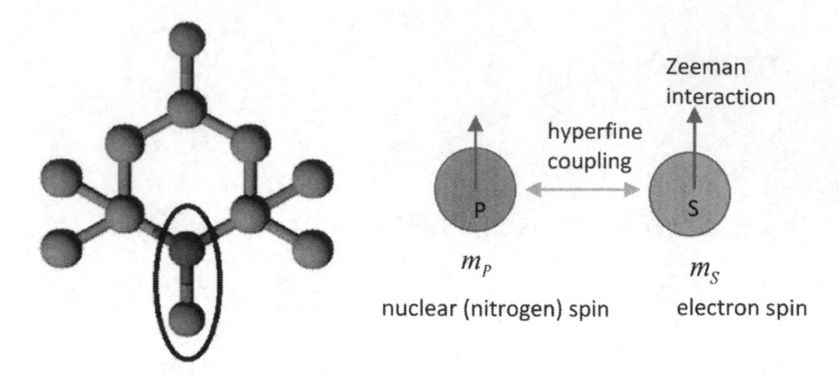

Figure 6.1 Structure of 4-oxo-TEMPO-h_{16}-$^{14(15)}$N molecule (4-oxo-2,2,6,6-tetramethyl-1-piperidinyloxy–h_{16}-$^{14(15)}$N), ^{1}H not shown, and its schematic view from the perspective of a spin system.

The Hamiltonian of the scalar (isotropic) hyperfine interaction, $H_{\mathrm{sc}}(P, S)$, has the form [3–5]:

$$H_{\mathrm{A,sc}}(P, S) = AP \cdot S = A\left[P_z S_z + \frac{1}{2}(P_+S_- + P_-S_+)\right] \quad (6.1)$$

where A is the hyperfine coupling constant. This means that the static interaction determining the energy level structure of the I–S spin system is now given as a sum of the Zeeman and scalar couplings:

$$\begin{aligned} H_0(P, S) &= H_z(S) + H_z(P) + H_{\mathrm{A,sc}}(P, S) \\ &= \omega_S S_z + \omega_P P_z + A\left[P_z S_z + \frac{1}{2}(P_+S_- + P_-S_+)\right] \end{aligned} \quad (6.2)$$

The Zeeman interaction of the nucleus is much weaker than that of the electron and therefore, it can be neglected.

The lineshape analysis is considerable easier for ^{15}N radicals due to the nitrogen spin quantum number $P = 1/2$. Using for the $P - S$ system the same numbering of the Zeeman states, $|m_S, m_P\rangle$, as in the previous chapters ($|1\rangle = |1/2, 1/2\rangle$, $|2\rangle = |1/2, -1/2\rangle$, $|3\rangle = |-1/2, 1/2\rangle$, and $|4\rangle = |-1/2, -1/2\rangle$), the matrix representation of the main Hamiltonian, $H_0(P, S)$, is as follows (neglecting the

Zeeman interaction of ^{15}N):

$$H_0(P,S) = \begin{bmatrix} \frac{\omega_S}{2} + \frac{A}{4} & 0 & 0 & 0 \\ 0 & \frac{\omega_S}{2} - \frac{A}{4} & \frac{A}{2} & 0 \\ 0 & \frac{A}{2} & -\frac{\omega_S}{2} - \frac{A}{4} & 0 \\ 0 & 0 & 0 & -\frac{\omega_S}{2} + \frac{A}{4} \end{bmatrix} \quad (6.3)$$

Because of the simple form of the Hamiltonian matrix one can easily find analytical expressions for the eigenvalues, E_α, and eigenstates, $|\psi_\alpha\rangle$ [3, 4, 6]:

$$|\psi_1\rangle = |1\rangle = \left|\frac{1}{2}, \frac{1}{2}\right\rangle, \ E_1 = \frac{\omega_S}{2} + \frac{A}{4} \quad (6.4a)$$

$$|\psi_2\rangle = a\,|2\rangle + b\,|3\rangle = a\left|\frac{1}{2}, \frac{1}{2}\right\rangle + b\left|-\frac{1}{2}, \frac{1}{2}\right\rangle,$$

$$E_2 = \frac{\sqrt{\omega_S^2 + A^2}}{2} - \frac{A}{4} \quad (6.4b)$$

$$|\psi_3\rangle = -b\,|2\rangle + a\,|3\rangle = -b\left|\frac{1}{2}, -\frac{1}{2}\right\rangle + a\left|-\frac{1}{2}, \frac{1}{2}\right\rangle,$$

$$E_3 = \frac{\sqrt{\omega_S^2 + A^2}}{2} - \frac{A}{4} \quad (6.4c)$$

$$|\psi_4\rangle = |4\rangle = \left|-\frac{1}{2}, -\frac{1}{2}\right\rangle, \ E_4 = -\frac{\omega_S}{2} + \frac{A}{4} \quad (6.4d)$$

with the coefficients:

$$a = \left\{\frac{1}{2}\left[1 + \frac{\omega_S}{\sqrt{\omega_S^2 + A^2}}\right]\right\}^{1/2}, \ b = \left\{\frac{1}{2}\left[1 - \frac{\omega_S}{\sqrt{\omega_S^2 + A^2}}\right]\right\}^{1/2} \quad (6.5)$$

The energy level structure of the P–S system is shown in Fig. 6.2. Equations 6.6a,b give the inverse relationship:

$$|2\rangle = a\,|\psi_2\rangle - b\,|\psi_3\rangle \quad (6.6a)$$

$$|3\rangle = b\,|\psi_2\rangle + a\,|\psi_3\rangle \quad (6.6b)$$

Then, the operator $S_+ \propto (|1\rangle\langle 2| + |3\rangle\langle 4|)$ is represented in the Liouville space $\{|\psi_\alpha\rangle\langle\psi_\beta|\}$ as:

$$\langle S_+\rangle \propto a\,(|\psi_1\rangle\langle\psi_3| + |\psi_2\rangle\langle\psi_4|) + b\,(|\psi_1\rangle\langle\psi_2| - |\psi_3\rangle\langle\psi_4|) \quad (6.7)$$

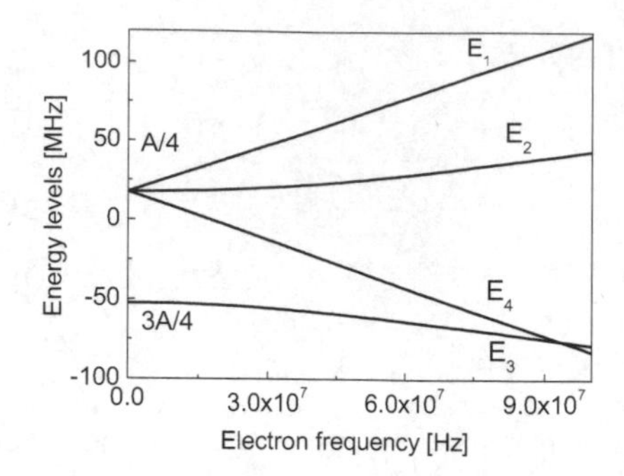

Figure 6.2 Energy level structure of electron spin–nuclear ^{15}N spin system versus the electron spin frequency, the hyperfine coupling constant $A = 70$ MHz. At zero magnetic field the energy levels yield $E_1 = E_2 = E_4 = A/4$, $E_3 = -3A/4$.

This implies that the lineshape function, $L_S(\omega)$, can be calculated from the matrix expression:

$$L_S(\omega) \propto \mathrm{Re}\left\{[a, a, b, -b] \cdot \begin{bmatrix} \Lambda_{1313} & 0 & 0 & 0 \\ 0 & \Lambda_{2424} & 0 & 0 \\ 0 & 0 & \Lambda_{1212} & 0 \\ 0 & 0 & 0 & \Lambda_{3434} \end{bmatrix}^{-1} \cdot \begin{bmatrix} a \\ a \\ b \\ -b \end{bmatrix}\right\} \tag{6.8}$$

where $\Lambda_{\alpha\beta\alpha\beta} = i\Delta\omega_{\alpha\beta} + R_{\alpha\beta\alpha\beta}$, $\Delta\omega_{\alpha\beta} = \omega + \omega_{\alpha\beta}$. The transition frequencies yield [3, 4, 6]:

$$\omega_{12} = -\frac{\omega_s + A}{2} + \frac{\sqrt{\omega_S^2 + A^2}}{2} \tag{6.9a}$$

$$\omega_{34} = -\frac{\omega_s - A}{2} + \frac{\sqrt{\omega_S^2 + A^2}}{2} \tag{6.9b}$$

$$\omega_{13} = -\frac{\omega_s + A}{2} - \frac{\sqrt{\omega_S^2 + A^2}}{2} \tag{6.9c}$$

$$\omega_{24} = -\frac{\omega_s - A}{2} - \frac{\sqrt{\omega_S^2 + A^2}}{2} \tag{6.9d}$$

The spectrum consists of two sets of lines. The first set contains two lines corresponding to the frequencies ω_{12} and ω_{34} which are separated by A, while the second pair of lines (ω_{13} and ω_{24}) is shifted with respect to the first one by $\sqrt{\omega_S^2 + A^2}$. As the first pair is weighted by the factor b which tends to zero for $\omega_S \gg A$, it vanishes anyway (ESR experiments are performed at frequencies for which $\omega_S \gg A$). Thus, the ESR lineshape, $L_S(\omega)$, is given as a sum of two Lorentzian functions:

$$L_S(\omega) \propto \frac{T_{13}}{1 + (\omega - \omega_{13})^2 \tau_{13}^2} + \frac{T_{24}}{1 + (\omega - \omega_{24})^2 \tau_{24}^2} \tag{6.10}$$

where $T_{13} = R_{1313}^{-1}$ and $T_{24} = R_{2424}^{-1}$

To evaluate the relaxation rates one has to identify the electron spin relaxation mechanism. There are several contributions to the electronic relaxation, but the dominating one is due to the anisotropic part of the hyperfine interaction, $H_{\mathrm{A,aniso}}$. The Hamiltonian has, in its principal axis system (P) (this should not be mistaken with the nitrogen spin quantum number, P), the form:

$$H_{\mathrm{A,aniso}}^{(P)}(P, S) = \sum_{m=-2}^{2} (-1)^m G_{-m}^{2(P)} T_m^2 (P, S) \tag{6.11}$$

with

$$G_0^{2(P)} = \sqrt{\frac{2}{3}} \left[A_{zz} - \frac{1}{2} (A_{xx} + A_{yy}) \right],$$
$$G_{\pm 1}^{2(P)} = 0, \; G_{\pm 2}^{2(P)} = \frac{1}{2} (A_{xx} - A_{yy}) \tag{6.12}$$

where A_{xx}, A_{yy}, A_{zz} are the Cartesian components of the hyperfine tensor; $A = \frac{1}{3}(A_{xx} + A_{yy} + A_{zz})$. The tensor operators, $T_m^2(P, S)$, are given by Eqs. 2.11, 2.18, and 2.19.

The anisotropic part of the hyperfine coupling fluctuates in time due to rotational dynamics of the paramagnetic molecule. The transformation of Eq. 2.13 now yields:

$$G_{-m}^{2(\mathrm{L})} = \sum_{k=-2}^{2} G_k^{2(\mathrm{P})} D_{k,-m}^2 (\Omega_{\mathrm{PL}}) \tag{6.13}$$

Applying it one obtains the laboratory-frame form of the Hamiltonian of the isotropic hyperfine coupling:

$$H_{\mathrm{A,aniso}}^{(\mathrm{L})}(P, S) = \sum_{m=-2}^{2} (-1)^m G_{-m}^{2(\mathrm{L})} T_m^2 (P, S) \tag{6.14}$$

As usually, the angle Ω_{PL} describes the orientation of the principal axis system (P) of the isotropic part of the hyperfine coupling with respect to the laboratory frame (L). Then, the $G_m^{2(\mathrm{L})}$ quantities are given as:

$$G_{-m}^{2(\mathrm{L})} = D_{0,-m}^{2}\left(\Omega_{\mathrm{PL}}\right)\sqrt{\frac{2}{3}}\left[A_{zz} - \frac{1}{2}\left(A_{xx} + A_{yy}\right)\right] + \left(D_{-2,-m}^{2}\left(\Omega_{\mathrm{PL}}\right) + D_{2,-m}^{2}\left(\Omega_{\mathrm{PL}}\right)\right)\frac{1}{2}\left(A_{xx} - A_{yy}\right) \quad (6.15)$$

The Hamiltonian $H_{\mathrm{A,aniso}}^{(\mathrm{L})}(P,S)$ represented in the Zeeman basis $\{|m_{\mathrm{P}}, m_{\mathrm{S}}\rangle\}$ takes the form:

$$\left[H_{\mathrm{A,aniso}}^{(\mathrm{L})}(P,S)\right](t) = \begin{bmatrix} \frac{1}{2\sqrt{6}}G_0^{2(\mathrm{L})} & \frac{1}{4}G_{-1}^{2(\mathrm{L})} & \frac{1}{4}G_{-1}^{2(\mathrm{L})} & \frac{1}{2}G_{-2}^{2(\mathrm{L})} \\ -\frac{1}{4}G_1^{2(\mathrm{L})} & -\frac{1}{2\sqrt{6}}G_0^{2(\mathrm{L})} & -\frac{1}{2\sqrt{6}}G_0^{2(\mathrm{L})} & -\frac{1}{4}G_{-1}^{2(\mathrm{L})} \\ -\frac{1}{4}G_1^{2(\mathrm{L})} & -\frac{1}{2\sqrt{6}}G_0^{2(\mathrm{L})} & -\frac{1}{2\sqrt{6}}G_0^{2(\mathrm{L})} & -\frac{1}{4}G_{-1}^{2(\mathrm{L})} \\ \frac{1}{2}G_2^{2(\mathrm{L})} & \frac{1}{4}G_1^{2(\mathrm{L})} & \frac{1}{4}G_1^{2(\mathrm{L})} & \frac{1}{2\sqrt{6}}G_0^{2(\mathrm{L})} \end{bmatrix} \quad (6.16)$$

However, the Hamiltonian has to be expressed in the basis $\{|\psi_\alpha\rangle\}$ which is the eigenbasis of the system (the Zeeman basis is not). Using the relationship $|\psi_1\rangle = |1\rangle, |\psi_2\rangle = a|2\rangle + b|3\rangle, |\psi_3\rangle = -b|2\rangle + a|3\rangle$, and $|\psi_4\rangle = |4\rangle$, one obtains:

$$\left[H_{\mathrm{A,aniso}}^{(\mathrm{L})}(P,S)\right](t) = \begin{bmatrix} \frac{1}{2\sqrt{6}}G_0^{2(\mathrm{L})} & \frac{1}{4}(a+b)G_{-1}^{2(\mathrm{L})} & \frac{1}{4}(a-b)G_{-1}^{2(\mathrm{L})} & \frac{1}{2}G_{-2}^{2(\mathrm{L})} \\ -\frac{1}{4}(a+b)G_1^{2(\mathrm{L})} & -(a+b)^2\frac{1}{2\sqrt{6}}G_0^{2(\mathrm{L})} & -\left(a^2-b^2\right)\frac{1}{2\sqrt{6}}G_0^{2(\mathrm{L})} & -\frac{1}{4}(a+b)G_{-1}^{2(\mathrm{L})} \\ -\frac{1}{4}(a-b)G_1^{2(\mathrm{L})} & -\left(a^2-b^2\right)\frac{1}{2\sqrt{6}}G_0^{2(\mathrm{L})} & -(a-b)^2\frac{1}{2\sqrt{6}}G_0^{2(\mathrm{L})} & -\frac{1}{4}(a-b)G_{-1}^{2(\mathrm{L})} \\ \frac{1}{2}G_2^{2(\mathrm{L})} & \frac{1}{4}(a+b)G_1^{2(\mathrm{L})} & \frac{1}{4}(a-b)G_1^{2(\mathrm{L})} & \frac{1}{2\sqrt{6}}G_0^{2(\mathrm{L})} \end{bmatrix} \quad (6.17)$$

Knowing the representation of the Hamiltonian which is responsible for the relaxation process, one can apply Eq. 3.19, to calculate the electron spin relaxation rates R_{1313} and R_{2424} which are relevant for the ESR lineshape [7]:

$$R_{1313} = \frac{1}{24}\left[(a-b)^2+1\right]^2 J^A(0) + \frac{1}{16}(a+b)^2 J^A(\omega_{12}) + \frac{1}{8}(a-b)^2 J^A(\omega_{13}) + \frac{1}{4} J^A(\omega_{14}) + \frac{1}{24}\left(a^2-b^2\right)^2 J^A(\omega_{23}) + \frac{1}{16}(a-b)^2 J^A(\omega_{34}) \quad (6.18a)$$

$$R_{2424} = \frac{1}{24}\left[(a+b)^2+1\right]^2 J^A(0) + \frac{1}{16}(a+b)^2 J^A(\omega_{12}) + \frac{1}{8}(a+b)^2 J^A(\omega_{24}) + \frac{1}{4} J^A(\omega_{14}) + \frac{1}{24}\left(a^2-b^2\right)^2 J^A(\omega_{23}) + \frac{1}{16}(a-b)^2 J^A(\omega_{34}) \quad (6.18b)$$

For completeness one can also calculate the other relaxation rates [7]:

$$R_{1212} = \frac{1}{24}\left[(a+b)^2+1\right]^2 J^A(0) + \frac{1}{8}(a+b)^2 J^A(\omega_{12}) + \frac{1}{16}(a-b)^2 J^A(\omega_{13}) + \frac{1}{4} J^A(\omega_{14}) + \frac{1}{24}\left(a^2-b^2\right)^2 J^A(\omega_{23}) + \frac{1}{16}(a+b)^2 J^A(\omega_{24}) \quad (6.19a)$$

$$R_{3434} = \frac{1}{24}\left[(a-b)^2+1\right]^2 J^A(0) + \frac{1}{8}(a-b)^2 J^A(\omega_{34}) + \frac{1}{16}(a-b)^2 J^A(\omega_{13}) + \frac{1}{4} J^A(\omega_{14}) + \frac{1}{24}\left(a^2-b^2\right)^2 J^A(\omega_{23}) + \frac{1}{16}(a+b)^2 J^A(\omega_{24}) \quad (6.19b)$$

At high magnetic fields, when $a \to 1$ and $b \to 0$, Eq. 6.18 and Eq. 6.19 converge to

$$R_{1313} = \frac{1}{6} J^A(0) + \frac{1}{8} J^A \frac{A}{2} + \frac{1}{8} J^A\left(\omega_S + \frac{A}{2}\right) + \frac{7}{24} J^A(\omega_S) \quad (6.20a)$$

$$R_{2424} = \frac{1}{6} J^A(0) + \frac{1}{8} J^A \frac{A}{2} + \frac{1}{8} J^A\left(\omega_S - \frac{A}{2}\right) + \frac{7}{24} J^A(\omega_S) \quad (6.20b)$$

while

$$R_{1212} = R_{3434} = \frac{1}{6} J^A(0) + \frac{1}{8} J^A \frac{A}{2} + \frac{1}{16} J^A\left(\omega_S + \frac{A}{2}\right) + \frac{1}{16} J^A\left(\omega_S - \frac{A}{2}\right) + \frac{7}{24} J^A(\omega_S) \quad (6.21)$$

As $J^A(\omega_S \pm \frac{A}{2}) \cong J^A(\omega_S)$, eventually all relaxation rates converge to:

$$R_{1212} = R_{1313} = R_{2424} = R_{3434} \cong \frac{1}{6}J^A(0) + \frac{1}{8}J^A\left(\frac{A}{2}\right) + \frac{5}{12}J^A(\omega_S) \tag{6.22}$$

Taking into account, the orthogonality properties of Wigner rotation matrices one gets from Eq. 6.15, the following expression for the spectral density:

$$J^A(\omega) = \left\{\frac{2}{3}\left[A_{zz} - \frac{1}{2}(A_{xx} + A_{yy})\right]^2 + \frac{1}{2}(A_{xx} - A_{yy})^2\right\}\frac{1}{5}\frac{\tau_{rot}}{1 + \omega^2\tau_{rot}^2} \tag{6.23}$$

In Eq. 6.23 isotropic molecular tumbling (i.e., a Lorentzian spectral density) has been assumed.

In Fig. 6.3, illustrative ESR spectra for 4-oxo-TEMPO-d_{16}-^{15}N dissolved in glycerol are shown. The spectrum at 294 K has been fitted (calculated) by means of the described theory. Figure 6.4 presents examples of calculated ESR spectra for ^{15}N radicals. The

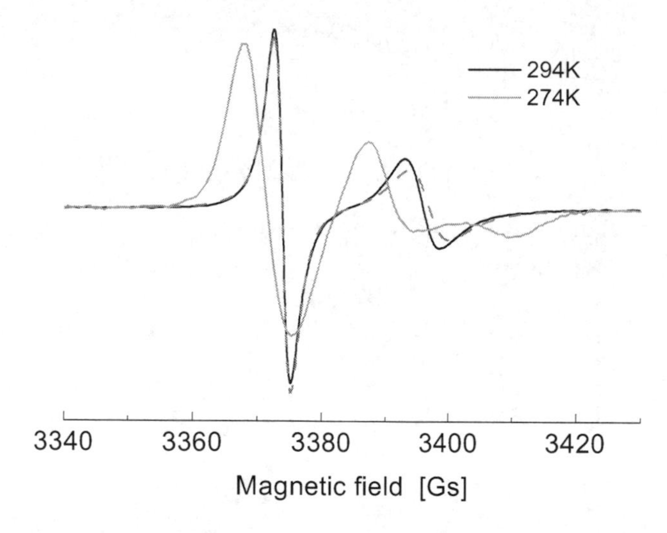

Figure 6.3 ESR spectra for 4-oxo-TEMPO-d_{16}-^{15}N in glycerol at 9.53 GHz (0.34T); a theoretical spectrum calculated in terms of the described theory (dashed line).

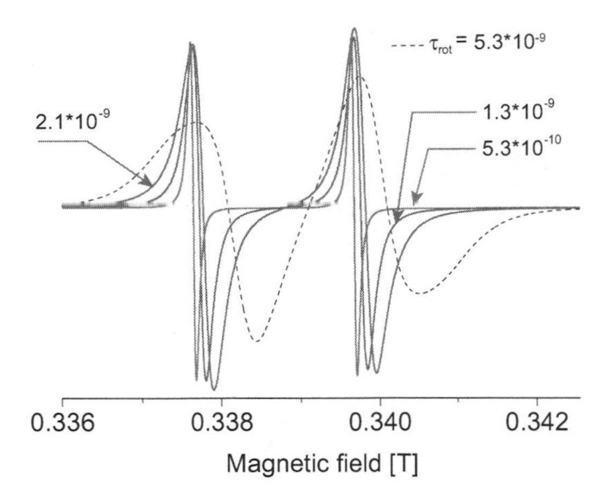

Figure 6.4 Simulated ESR spectra for a ^{15}N radical at 0.34T for different rotational correlation times, τ_{rot} (in sec).

dynamics is again modeled as isotropic rotation characterized by Lorentzian spectral densities.

As it has already been pointed out, spin relaxation processes can be described as linear combinations of spectral densities only when the perturbation theory applies [8, 9]; in this case this means that the relationship $\omega_{\mathrm{aniso}}\tau_{\mathrm{rot}} < 1$, where ω_{aniso} is the amplitude of the anisotropic part of the hyperfine coupling (in angular frequency units) holds. This implies (taking into account typical values of the components of the hyperfine coupling tensor, which for ^{15}N radicals are about $A_{xx} \cong A_{yy} \cong$ (15–25) MHz, $A_{xx} \cong$ (120–140) MHz [10, 11]), that the rotational correlation time should not exceed few nanoseconds. Therefore, the spectrum for $\tau_{\mathrm{rot}} = 5.3$ ns has been calculated using a more general approach (discussed in the forthcoming chapters). This also explains the moderate agreement between the experimental and calculated spectra shown in Fig. 6.3; the validity limit of the presented theory has been approached in this case.

Before finishing this section it is worth to inspect more closely the energy level structure shown in Fig. 6.5 plotted as transition frequencies $\omega_{\alpha\beta} = E_{\alpha\beta} = E_{\beta} - E_{\alpha}$ (Fig. 6.5a). The transition frequencies associated with different spin coherences are not equal, except of some specific magnetic fields (electron frequencies; see the crossings of the transition frequencies in Fig. 6.5a).

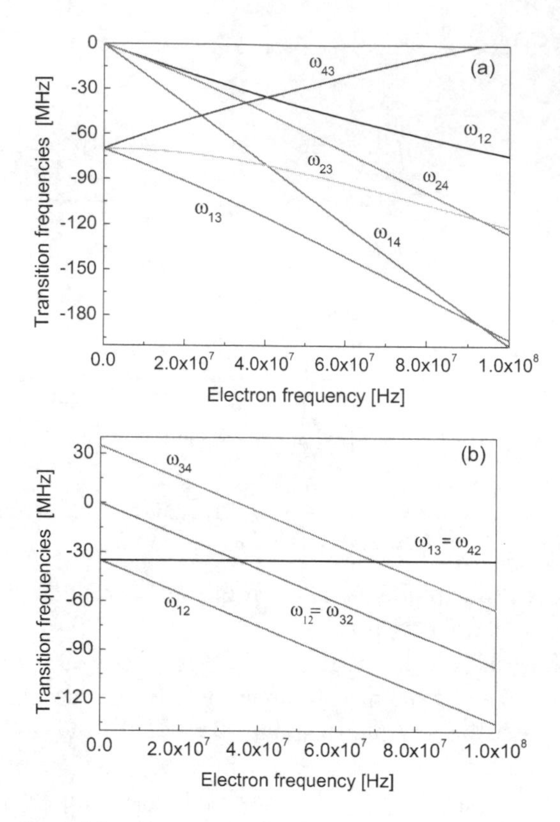

Figure 6.5 Transition frequencies for an electron spin–nitrogen (^{15}N) spin system versus the electron spin frequency, ω_S. (a) Exact energy level structure has been taken into account and (b) high field approximation of the energy levels.

One should note that the highest frequency considered in Fig. 6.5, $\omega_S = 100$ MHz, is by two orders of magnitude smaller than the frequency (referred to as X-band).

Comparing Fig. 6.5a with 6.5b in which the transition frequencies in high field approximation (i.e., neglecting the off-diagonal elements of the Hamiltonian matrix) are presented, one clearly sees that the approximation may not be applied at lower magnetic fields as the obtained transition frequencies are significantly different. At high frequencies Fig. 6.5b becomes correct, but then one has to face the problem of coupled spin coherences as some of the transition frequencies are equal.

Nevertheless, neglecting for a while, the coupling between different coherences one gets at high fields a simple expression for the ESR lineshape function:

$$L_S(\omega) = \mathrm{Re}\left\{[1,\,1]\cdot\begin{bmatrix} i\Delta\omega_{13}+R_{1313} & 0 \\ 0 & i\Delta\omega_{24}+R_{2424}\end{bmatrix}^{-1}\cdot\begin{bmatrix}1\\1\end{bmatrix}\right\} \tag{6.24}$$

which gives:

$$L_S(\omega) \propto \frac{T_2}{1+(\omega-\omega_{13})^2 T_2^2} + \frac{T_2}{1+(\omega-\omega_{24})^2 T_2^2} \tag{6.25}$$

where $T_2 = R_{1313}^{-1} = R_{2424}^{-1}$ (the relaxation rates are given by Eq. 6.20). Now, taking into account that $\omega_{13} = \omega_{42}$ (and then, $\omega_{31} = \omega_{24}$), one should consider the coupled evolution of the coherences ρ_{13} and ρ_{42}(ρ_{13} and ρ_{42}). However, due to orthogonality properties of Wigner rotation matrices encoded in the quantities G_m^2, one gets $R_{1342} = R_{3124} = 0$.

6.2 ESR Spectra and Scalar Interactions for ^{14}N Systems

It is of interest to compare the ESR lineshape for ^{15}N radicals with the case of ^{14}N as in the last case the nitrogen spin quantum number is $P = 1$. Then, the system is characterized by six spin functions $|m_S, m_P\rangle$, that is, $|1\rangle = |1/2, 1\rangle$, $|2\rangle = |1/2, 0\rangle$, $|3\rangle = |1/2, -1\rangle$, $|4\rangle = |-1/2, 1\rangle$, $|5\rangle = |-1/2, 0\rangle$, and $|6\rangle = |-1/2, -1\rangle$. The matrix representation of the Hamiltonian $H_0(P, S)$ takes in this basis the form:

$$[H_0(P,S)] = \begin{bmatrix} \frac{\omega_S}{2}+\frac{A}{2} & 0 & 0 & 0 & 0 & 0 \\ 0 & \frac{\omega_S}{2} & \frac{A}{\sqrt{2}} & 0 & 0 & 0 \\ 0 & \frac{A}{\sqrt{2}} & \frac{\omega_S}{2}-\frac{A}{2} & 0 & 0 & 0 \\ 0 & 0 & 0 & -\frac{\omega_S}{2}-\frac{A}{2} & \frac{A}{\sqrt{2}} & 0 \\ 0 & 0 & 0 & \frac{A}{\sqrt{2}} & -\frac{\omega_S}{2} & 0 \\ 0 & 0 & 0 & 0 & 0 & -\frac{\omega_S}{2}+\frac{A}{2} \end{bmatrix} \tag{6.26}$$

Now, it is somewhat more difficult to calculate the eigenvalues (energy levels), E_α, and eigenvectors, $|\psi_\alpha\rangle$, but after some manipulations one gets [6]:

$$|\psi_1\rangle = \left|\frac{1}{2}, 1\right\rangle,\ E_1 = \frac{\omega_S}{2} + \frac{A}{2} \tag{6.27a}$$

$$|\psi_2\rangle = a\left|\frac{1}{2}, 0\right\rangle + \mathrm{b}\left|-\frac{1}{2}, 1\right\rangle,\ E_2 = \frac{1}{2}\sqrt{\frac{9A^2}{2} + \omega_S^2 + \omega_S A} - \frac{A}{4} \tag{6.27b}$$

$$|\psi_3\rangle = c\left|\frac{1}{2}, 0\right\rangle + d\left|-\frac{1}{2}, 1\right\rangle,\ E_3 = -\frac{1}{2}\sqrt{\frac{9A^2}{4} + \omega_S^2 + \omega_S A} - \frac{A}{4} \tag{6.27c}$$

$$|\psi_4\rangle = e\left|\frac{1}{2}, -1\right\rangle + f\left|-\frac{1}{2}, 0\right\rangle,\ E_4 = \frac{1}{2}\sqrt{\frac{9A^2}{4} + \omega_S^2 - \omega_S A} - \frac{A}{4} \tag{6.27d}$$

$$|\psi_5\rangle = g\left|\frac{1}{2}, -1\right\rangle + h\left|-\frac{1}{2}, 0\right\rangle,\ E_5 = -\frac{1}{2}\sqrt{\frac{9A^2}{4} + \omega_S^2 - \omega_S A} - \frac{A}{4} \tag{6.27e}$$

$$|\psi_6\rangle = \left|-\frac{1}{2}, -1\right\rangle,\ E_6 = -\frac{\omega_S}{2} + \frac{A}{2} \tag{6.27f}$$

with:

$$\begin{aligned} a &= \left(1+\alpha^2\right)^{-1/2},\ b = \alpha a,\ c = \left(1+\beta^2\right)^{-1/2},\ d = \beta c, \\ e &= \left(1+\gamma^2\right)^{-1/2},\ f = \gamma e,\ g = \left(1+\delta^2\right)^{-1/2},\ h = \delta g \end{aligned} \tag{6.28}$$

where

$$\begin{aligned} \alpha &= -\frac{A + 2\omega_S - \sqrt{9A^2 + 4A\omega_S + 4\omega_S^2}}{2\sqrt{2}A}, \\ \beta &= -\frac{A + 2\omega_S + \sqrt{9A^2 + 4A\omega_S + 4\omega_S^2}}{2\sqrt{2}A} \\ \gamma &= -\frac{A - 2\omega_S + \sqrt{9A^2 - 4A\omega_S + 4\omega_S^2}}{2\sqrt{2}A}, \\ \delta &= -\frac{A - 2\omega_S - \sqrt{9A^2 - 4A\omega_S + 4\omega_S^2}}{2\sqrt{2}A} \end{aligned} \tag{6.29}$$

In Fig. 6.6, the energy level structure of the system is shown.

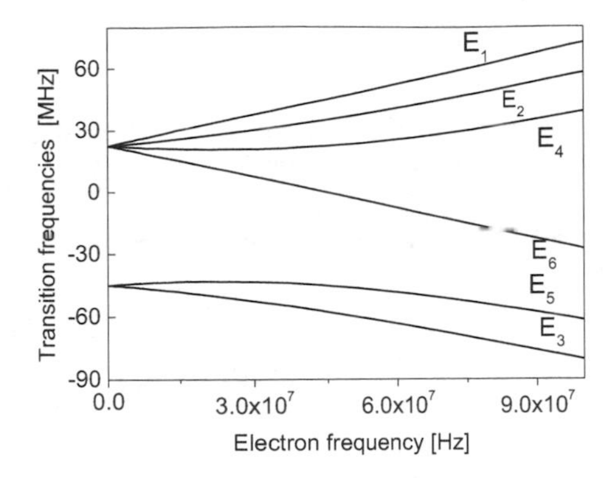

Figure 6.6 Energy level structure of electron spin–nuclear ^{14}N spin system versus the electron spin frequency, $A = 45$ MHz. At zero magnetic field $E_1 = E_2 = E_4 = E_6 = A/2$ and $E_3 = E_5 = -A$.

The scalar coupling is proportional to the gyromagnetic factor of the nucleus; the ratio between the values for ^{15}N and ^{14}N corresponds to the ratio between the ^{15}N and ^{14}N gyromagnetic factors.

The operator S_+, which in the Zeeman basis can be expressed as $S_+ \propto |1\rangle\langle 4| + |2\rangle\langle 5| + |3\rangle\langle 6|$, in the Liouville space $\{|\psi_\alpha\rangle\langle\psi_\beta|\}$, constructed from the eigenfunction of the system yields:

$$\langle S_+\rangle \propto b\,|\psi_1\rangle\langle\psi_2| + d\,|\psi_1\rangle\langle\psi_3| + e\,|\psi_4\rangle\langle\psi_6| + g\,|\psi_5\rangle\langle\psi_6|$$
$$+ af\,|\psi_2\rangle\langle\psi_4| + ah\,|\psi_2\rangle\langle\psi_5| + cf\,|\psi_3\rangle\langle\psi_4| + ch\,|\psi_3\rangle\langle\psi_5| \quad (6.30)$$

where the inverse relationship $|1\rangle = |\psi_1\rangle$, $|2\rangle = a\,|\psi_2\rangle + c\,|\psi_3\rangle$, $|3\rangle = e\,|\psi_4\rangle + g\,|\psi_5\rangle$, $|4\rangle = b\,|\psi_2\rangle + d\,|\psi_3\rangle$, $|5\rangle = f\,|\psi_4\rangle + h\,|\psi_5\rangle$, and $|6\rangle = |\psi_6\rangle$ has been applied. As the expressions are quite cumbersome, it is useful to discuss at the first stage a simplified description of the problem.

Neglecting the off-diagonal terms in the matrix representation of the static Hamiltonian of Eq. 6.26 (this is justified only at high magnetic fields), one obtains for the energy levels $\tilde{E}_1 = \frac{\omega_s}{2} + \frac{A}{2}$, $\tilde{E}_2 = \frac{\omega_s}{2}$, $\tilde{E}_3 = \frac{\omega_s}{2} - \frac{A}{2}$, $\tilde{E}_4 = \frac{\omega_s}{2} - \frac{A}{2}$, $\tilde{E}_5 = -\frac{\omega_s}{2}$, and $\tilde{E}_6 = -\frac{\omega_s}{2} - \frac{A}{2}$, while $a \to d \to e \to h \to 1$ and $b \to c \to f \to g \to 0$. One should note that when considering the high field (frequency) limit, $A \ll \omega_S$, the energy levels E_3 and E_4 are interchanged with respect to the general case of Eq. 6.27; therefore, the symbols, $\tilde{E}$, have been introduced. Neglecting the off-diagonal elements of the main Hamiltonian, $H_0(P, S)$, implies that the Zeeman basis $\{|m_S, m_P\rangle\}$ becomes the eigenbasis of the S–P spin system.

The representation of the anisotropic part of the hyperfine coupling, $H^{(L)}_{A,aniso}(P, S)(t)$ in the $\{|m_S, m_P\rangle\}$ basis is as follows (it might be of help to remind that $S_\pm |S, m_s\rangle = \sqrt{S(S+1) - m_s(m_s \pm 1)}\,|S, m_S \pm 1\rangle$, $S_z |S, m_S\rangle = m_S |S, m_S\rangle$, $T^2_0(P, S) = \frac{1}{\sqrt{6}}\left[2P_+S_Z - \frac{1}{2}(P_+S_- + P_-S_+)\right]$, $T^2_{\pm 1}(P, S) = \mp\frac{1}{2}[P_ZS_\pm + P_\pm S_Z]$, and $T^2_{\pm 2}(P, S) = \frac{1}{2}P_\pm S_\pm$):

$$\left[H^{(L)}_{A,aniso}(P, S)\right](t) = \begin{bmatrix} \frac{1}{\sqrt{6}}G^{2(L)}_0 & \frac{1}{2\sqrt{2}}G^{2(L)}_{-1} & 0 & \frac{1}{2}G^{2(L)}_{-1} & \frac{1}{\sqrt{2}}G^{2(L)}_{-1} & 0 \\ -\frac{1}{2\sqrt{2}}G^{2(L)}_1 & 0 & \frac{1}{2\sqrt{2}}G^{2(L)}_{-1} & -\frac{1}{2\sqrt{3}}G^{2(L)}_0 & 0 & \frac{1}{\sqrt{2}}G^{2(L)}_{-2} \\ 0 & -\frac{1}{2\sqrt{2}}G^{2(L)}_1 & -\frac{1}{\sqrt{6}}G^{2(L)}_0 & 0 & -\frac{1}{2\sqrt{3}}G^{2(L)}_0 & -\frac{1}{2}G^{2(L)}_{-1} \\ -\frac{1}{2}G^{2(L)}_1 & -\frac{1}{2\sqrt{3}}G^{2(L)}_0 & 0 & -\frac{1}{\sqrt{6}}G^{2(L)}_0 & -\frac{1}{2\sqrt{2}}G^{2(L)}_{-1} & 0 \\ \frac{1}{\sqrt{2}}G^{2(L)}_2 & 0 & -\frac{1}{2\sqrt{3}}G^{2(L)}_0 & \frac{1}{2\sqrt{2}}G^{2(L)}_1 & 0 & -\frac{1}{2\sqrt{2}}G^{2(L)}_{-1} \\ 0 & \frac{1}{\sqrt{2}}G^{2(L)}_2 & \frac{1}{2}G^{2(L)}_1 & 0 & \frac{1}{2\sqrt{2}}G^{2(L)}_1 & \frac{1}{\sqrt{6}}G^{2(L)}_0 \end{bmatrix}(t) \tag{6.31}$$

where the functions $G^{2(L)}_m(t)$ are defined in Eq. 6.15. The lineshape function, $L_S(\omega)$, for $\omega_S \gg A$ is described by the matrix expression:

$$L_S(\omega) \propto Re \left\{ \begin{bmatrix} 1 & 1 & 1 \end{bmatrix} \cdot \begin{bmatrix} i(\omega + \omega_{14}) + R_{1414} & 0 & 0 \\ 0 & i(\omega + \omega_{25}) + R_{2525} & 0 \\ 0 & 0 & i(\omega + \omega_{36}) + R_{3636} \end{bmatrix}^{-1} \cdot \begin{bmatrix} 1 \\ 1 \\ 1 \end{bmatrix} \right\} \tag{6.32}$$

where $\omega_{14} = -\omega_S + A$, $\omega_{36} = -\omega_S - A$ and $\omega_{25} = -\omega_S$. These transition frequencies are indicated in Fig. 6.7. It is of importance to compare the energy level structure obtained in the "high field

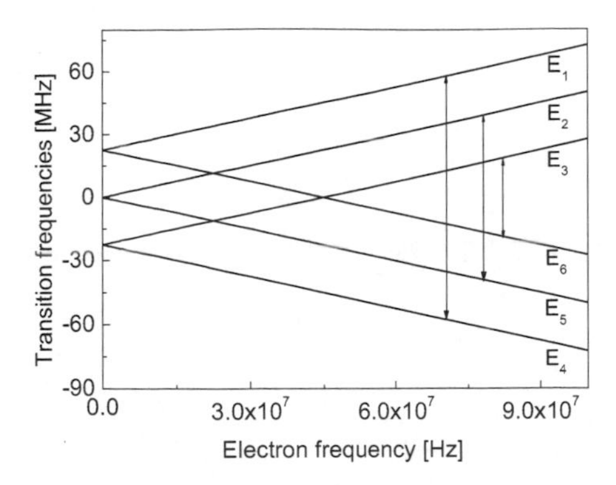

Figure 6.7 Energy level structure of an electron spin–nitrogen (^{14}N) spin system in the "high field approximation".

approximation", that is, when the off-diagonal elements of the Hamiltonian matrix are neglected (Fig. 6.7) with the energy levels shown in Fig. 6.6, which account for the influence of the isotropic hyperfine coupling, they are significantly different.

Equation 6.32 implies that the spectrum consists of three lines separated by A:

$$L_S(\omega) \propto \frac{T_{14}}{1+(\omega-\omega_S+A)^2 T_{14}^2} + \frac{T_{25}}{1+(\omega-\omega_S)^2 T_{25}^2} + \frac{T_{36}}{1+(\omega-\omega_S+A)^2 T_{36}^2} \tag{6.33}$$

with $T_{\alpha\beta} = (R_{\alpha\beta\alpha\beta})^{-1}$. The relaxation rates associated with the individual lines are given as the following combinations of spectral densities:

$$R_{1414} = \frac{2}{3}J^A(0) + \frac{1}{4}J^A\left(\frac{A}{2}\right) + \frac{7}{12}J^A\left(\omega_S + \frac{A}{2}\right) + \frac{1}{2}J^A(\omega_S + A) \tag{6.34}$$

$$R_{2525} = \frac{7}{12}J^A\left(\omega_S + \frac{A}{2}\right) + \frac{7}{12}J^A\left(\omega_S - \frac{A}{2}\right) + \frac{3}{8}J^A\left(\frac{A}{2}\right) \tag{6.35}$$

$$R_{3636} = \frac{2}{3}J^A(0) + \frac{1}{4}J^A\left(\frac{A}{2}\right) + \frac{7}{12}J^A\left(\omega_S - \frac{A}{2}\right) + \frac{1}{2}J^A(\omega_S + A) \tag{6.36}$$

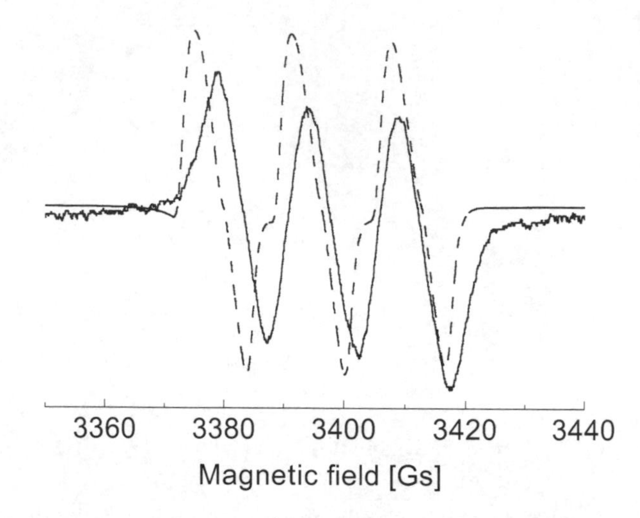

Figure 6.8 ESR spectra of 4-oxo-TEMPO-d_{16}-^{14}N in decalin (solid line) and propylene glycol (dashed line) at room temperature, 9.53 GHz (0.34T).

Illustrative ESR spectra for 4-oxo-TEMPO-d_{16}-^{14}N are shown in Fig. 6.8. One can see how dynamical properties of the solution change the ESR lineshape. Viscosity of the solvent determines rotational correlation times of the paramagnetic molecule, which then determine the electron spin relaxation times. In Fig. 6.9 two further examples of ESR spectra are shown. The first one is for 4-oxo-TEMPO-d_{16}-^{14}N dissolved in toluene. Due to relatively high viscosity of toluene (long rotational correlation time), the electron spin relaxation becomes so fast that the lines for $m_\mathrm{p} = -1,0,1$ are not resolved. The second radical, di-tert-butyl nitroxide (this is a pure compound, not a solution, which is liquid at room temperature) contains two paramagnetic centers (unpaired electron spins). Strong interactions between them lead to electron spin relaxation and, in consequence, overlapping of the lines for $m_\mathrm{p} = -1,0,1$.

In analogy to the case of ^{15}N, for ^{14}N systems one should also analyze the energy level structure in the high field range (Fig. 6.7), examining the relevant transition frequencies. One sees from Fig. 6.10 that some of them are equal which implies that a coupled evolution of the corresponding spin coherences should be carefully considered.

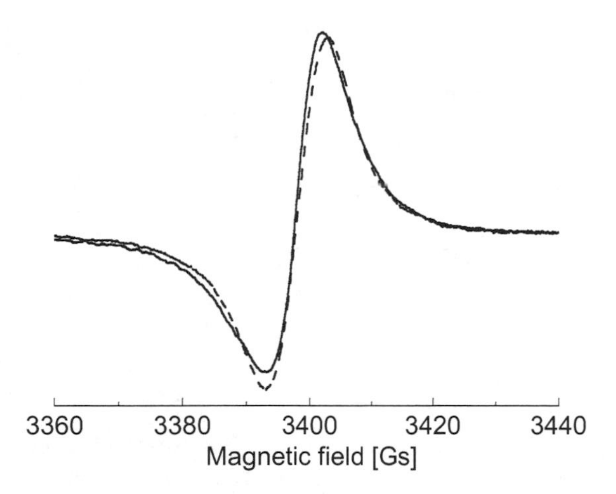

Figure 6.9 ESR spectrum of 4-oxo-TEMPO-d_{16}-^{14}N in toluene (solid line) and of di-tert-butyl nitroxide (dashed line) at room temperature, 9.53 GHz (0.34T).

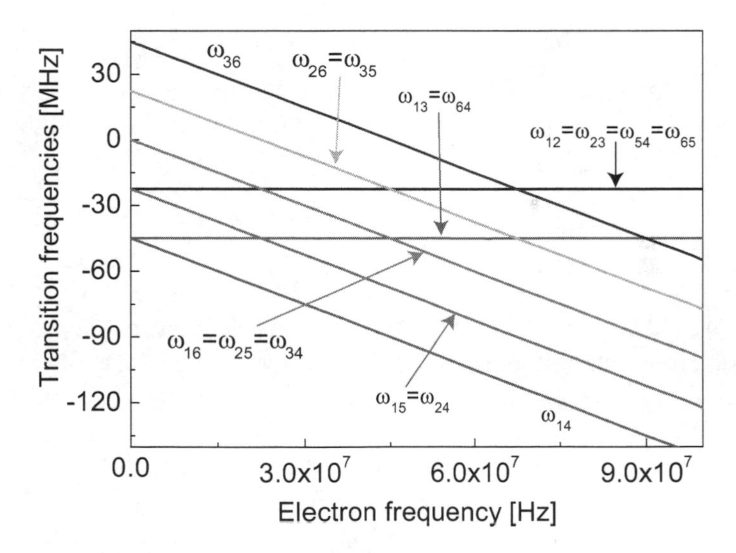

Figure 6.10 Transition frequencies for a coupled electron spin–^{14}N spin system in the high frequency range.

We are interested in the spin coherences ρ_{14}, ρ_{36}, and ρ_{25}; while the first two coherences are isolated, the third one, ρ_{25}, is coupled to ρ_{16} and ρ_{34}. In consequence, the lineshape function of Eq. 6.32 should be, in principle, extended to the form:

$$L_S(\omega) \propto \mathrm{Re}\left\{[1, 0, 1, 0, 1] \cdot \begin{bmatrix} \Lambda_{1414} & 0 & 0 & 0 & 0 \\ 0 & \Lambda_{1616} & R_{1625} & R_{1634} & 0 \\ 0 & R_{1625} & \Lambda_{2525} & R_{2534} & 0 \\ 0 & R_{1634} & R_{2534} & \Lambda_{3434} & 0 \\ 0 & 0 & 0 & 0 & \Lambda_{3636} \end{bmatrix}^{-1} \cdot \begin{bmatrix} 1 \\ 0 \\ 1 \\ 0 \\ 1 \end{bmatrix}\right\} \tag{6.37}$$

where $\Lambda_{\alpha\beta} = i\left(\omega + \omega_{\alpha\beta}\right) + R_{\alpha\beta\alpha\beta}$. Fortunately, the additional relaxation rates, R_{1634}, R_{2534}, and R_{1625} are again zero due to orthogonality of the Wigner rotation matrices.

In the general case, the off-diagonal elements of the Hamiltonian, $H_0(P, S)$(Eq. 6.26), lead to a much more complex expression for the lineshape function, $L_S(\omega)$:

$$L_S(\omega) \propto \mathrm{Re}\left\{[S_+]^+ \cdot [i\Delta\omega + M]^{-1} \cdot [S_+]\right\} \tag{6.38}$$

where the vector $[S_+]^+$ (referred in the literature as the "projection vector" [8, 12–14]) contains the representation of the S_+ operator in the basis $\left\{|\psi_\alpha\rangle\left\langle\psi_\beta\right|\right\}$; $[S_+]^+$ denotes matrix transposition:

$$[S_+]^+ = [b, d, e, g, af, ah, cf, ch] \tag{6.39}$$

The matrix $[i\Delta\omega + M]$ is diagonal, with the elements $\left\{i\left(\omega + \omega_{\alpha\beta}\right) + R_{\alpha\beta\alpha\beta}\right\}$, where $(\alpha, \beta) = (1,2), (1,3); (4,6), (5,6); (2,4), (2,5);$ and $(3,4), (3,5)$ (they are written in the order corresponding to the elements of the projection vector). The pairs (α, β) correspond to the elements (functions) $|\psi_\alpha\rangle\left\langle\psi_\beta\right|$ of Eq. 6.30 (i.e., the representation of the S_+ operator in the eigenbasis $\left\{|\psi_\alpha\rangle\left\langle\psi_\beta\right|\right\}$). In the high field range this representation converges to the already discussed form $\langle S_+\rangle \propto |\psi_1\rangle\langle\psi_3| + |\psi_4\rangle\langle\psi_6| + |\psi_2\rangle\langle\psi_5| = |1\rangle\langle 4| + |3\rangle\langle 6| + |2\rangle\langle 5|$.

To evaluate the relevant relaxation rates one needs to, in analogy to the case of ^{15}N, transform the Zeeman representation of the Hamiltonian $H^{(L)}_{\mathrm{aniso}}(P, S)$ to the basis $\left\{|\psi_\alpha\rangle\left\langle\psi_\beta\right|\right\}$; one obtains:

$$H^{(\mathrm{L})}_{\mathrm{A,aniso}}(P,S)(t) =$$

$$\begin{bmatrix}
\frac{1}{\sqrt{6}}G^2_0 & \frac{a+b\sqrt{2}}{2\sqrt{2}}G^2_{-1} & \frac{c+d\sqrt{2}}{2\sqrt{2}}G^2_{-1} & \frac{f}{\sqrt{2}}G^2_{-2} & \frac{h}{\sqrt{2}}G^2_{-2} & 0 \\
-\frac{a+b\sqrt{2}}{2\sqrt{2}}G^2_1 & -\frac{b}{\sqrt{6}}\left(b+a\sqrt{2}\right)G^2_0 & -\frac{ad+cb+bd\sqrt{2}}{2\sqrt{3}}G^2_0 & \frac{ae+bf}{2\sqrt{2}}G^1_{-1} & \frac{ag-bh}{2\sqrt{2}}G^1_{-1} & \frac{a}{2\sqrt{2}}G^2_{-2} \\
-\frac{c+d\sqrt{2}}{2\sqrt{2}}G^2_1 & -\frac{ad+cb+bd\sqrt{2}}{2\sqrt{3}}G^2_0 & -\frac{d}{\sqrt{6}}\left(d+c\sqrt{2}\right)G^2_0 & \frac{ce-df}{2\sqrt{2}}G^1_{-1} & \frac{cg-dh}{2\sqrt{2}}G^1_{-1} & \frac{c}{\sqrt{2}}G^2_{-2} \\
\frac{f}{\sqrt{2}}G^2_2 & -\frac{ae-bf}{2\sqrt{2}}G^1_1 & -\frac{ce-df}{2\sqrt{2}}G^1_1 & -\frac{e}{\sqrt{6}}\left(e+f\sqrt{2}\right)G^2_0 & -\frac{eh+fg+eg\sqrt{2}}{2\sqrt{3}}G^2_0 & -\frac{f+e\sqrt{2}}{2\sqrt{2}}G^2_{-1} \\
\frac{h}{\sqrt{2}}G^2_2 & -\frac{ag-bh}{2\sqrt{2}}G^1_1 & -\frac{cg-dh}{2\sqrt{2}}G^1_1 & -\frac{eh+fg+eg\sqrt{2}}{2\sqrt{3}}G^2_0 & -\frac{g}{\sqrt{6}}\left(g+\sqrt{2}h\right)G^2_0 & -\frac{h+g\sqrt{2}}{2\sqrt{2}}G^2_{-1} \\
0 & \frac{a}{\sqrt{2}}G^2_{-2} & \frac{c}{\sqrt{2}}G^2_{-2} & \frac{f+e\sqrt{2}}{2\sqrt{2}}G^2_1 & \frac{h+g\sqrt{2}}{2\sqrt{2}}G^2_1 & \frac{1}{\sqrt{6}}G^2_0
\end{bmatrix} \tag{6.40}$$

Although this representation is quite complicated, Eq. 3.19 allows for explicit calculations of the relaxation matrix elements. Some of them are given below as an example [7]:

$$\begin{aligned} R_{1212} = & \frac{1}{6}\left[1+b\left(b+a\sqrt{2}\right)\right]^2 J^A(0) + \frac{1}{4}\left(a+b\sqrt{2}\right)^2 J^A(\omega_{12}) \\ & + \frac{1}{8}\left(c+d\sqrt{2}\right)^2 J^A(\omega_{13}) \frac{1}{2} f^2 J^A(\omega_{14}) + \frac{1}{2} h^2 J^A(\omega_{15}) \\ & + \frac{1}{12}\left(ad+bc+bd\sqrt{2}\right)^2 J^A(\omega_{23}) + \frac{1}{8}(bf-ae)^2 J^A(\omega_{24}) \\ & + \frac{1}{8}(bh-ag)^2 J^A(\omega_{25}) + \frac{1}{2} a^2 J^A(\omega_{26}) \end{aligned} \tag{6.41a}$$

$$\begin{aligned} R_{1313} = & \frac{1}{6}\left[1+d\left(d+c\sqrt{2}\right)\right]^2 J^A(0) + \frac{1}{8}\left(a+b\sqrt{2}\right)^2 J^A(\omega_{12}) \\ & + \frac{1}{4}\left(c+d\sqrt{2}\right)^2 J^A(\omega_{13}) \frac{1}{2} f^2 J^A(\omega_{14}) + \frac{1}{2} h^2 J^A(\omega_{15}) \\ & + \frac{1}{12}\left(ad+bc+bd\sqrt{2}\right)^2 J^A(\omega_{23}) + \frac{1}{8}(df-ce)^2 J^A(\omega_{34}) \\ & + \frac{1}{8}(dh-cg)^2 J^A(\omega_{35}) + \frac{1}{2} c^2 J^A(\omega_{36}) \end{aligned} \tag{6.41b}$$

$$\begin{aligned} R_{2424} = & \frac{1}{6}\left[b\left(b+a\sqrt{2}\right) - e\left(e+f\sqrt{2}\right)\right]^2 J^A(0) \\ & + \frac{1}{8}\left(a+b\sqrt{2}\right)^2 J^A(\omega_{12}) \\ & + \frac{1}{12}\left(ad+bc+bd\sqrt{2}\right)^2 J^A(\omega_{23}) \\ & + \frac{1}{4}(bf-ae) J^A(\omega_{24}) + \frac{1}{8}(bh-ag)^2 J^A(\omega_{25}) \\ & + \frac{1}{2} a^2 J^A(\omega_{26}) + \frac{1}{2} f^2 J^A(\omega_{14}) + \frac{1}{8}(df-ce)^2 J^A(\omega_{34}) \\ & + \frac{1}{12}\left(eh+fg+eg\sqrt{2}\right)^2 J^A(\omega_{45}) \\ & + \frac{1}{8}\left(f+e\sqrt{2}\right)^2 J^A(\omega_{46}) \end{aligned} \tag{6.41c}$$

while the remaining elements needed for calculating the ESR lineshape, the reader can find in [7].

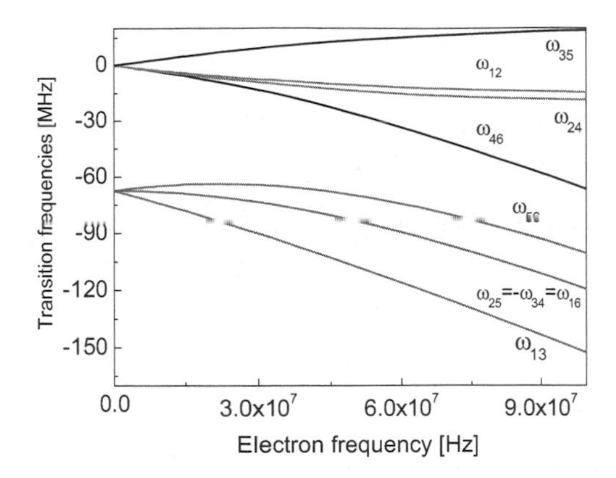

Figure 6.11 Transition frequencies for an electron spin—^{14}N spin system versus the electron spin frequency.

In Fig. 6.11, the relevant transition frequencies for the electron spin–^{14}N nuclear spin are shown. One sees that $\omega_{25} = -\omega_{34} = \omega_{16}$, that is, $\tilde{\omega}_{25} = \tilde{\omega}_{34} = \tilde{\omega}_{16}$ (the energy levels are inverted as indicated before $E_3 = \tilde{E}_4$, $E_4 = \tilde{E}_3$). This degeneracy has already been discussed in the context of the high field (frequency) spectrum. Other transition frequencies do not overlap, but ω_{12} and ω_{24} are very close. This means that the matrix $[i\Delta\omega + M]$ is not really diagonal. The off-diagonal relaxation coefficient R_{1224} should be introduced. This element is equal to:

$$R_{1224} = \frac{1}{8}\left(a + b\sqrt{2}\right)(bf - ae)\left(J^A(\omega_{12}) + J^A(\omega_{24})\right) \qquad (6.42)$$

At high magnetic fields, the contributions associated with the frequencies ω_{12} and ω_{24} vanish.

6.3 ESR Spectra at Low Frequencies

After this quite complex example, let us investigate more closely the opposite limit of magnetic fields, that is, the situation when $\omega_S \ll 0$, or even when $\omega_S = 0$. For the case of ^{15}N ($P = 1/2$) one gets in this limit $a \to b \to 1/\sqrt{2}$, $\omega_{13} \to \omega_{23} \to \omega_{34} \to -A$, $\omega_{12} \to \omega_{14} \to$

$\omega_{24} \to 0$. The set of eigenfunctions (Eq. 6.4) converges to:

$$|\psi_1\rangle = |1\rangle = \left|\frac{1}{2}, \frac{1}{2}\right\rangle, E_1 = \frac{A}{4} \tag{6.43a}$$

$$|\psi_2\rangle = \frac{1}{\sqrt{2}}|2\rangle + \frac{1}{\sqrt{2}}|3\rangle = \frac{1}{\sqrt{2}}\left|\frac{1}{2}, -\frac{1}{2}\right\rangle + \frac{1}{\sqrt{2}}\left|-\frac{1}{2}, \frac{1}{2}\right\rangle, E_2 = \frac{A}{4} \tag{6.43b}$$

$$|\psi_3\rangle = -\frac{1}{\sqrt{2}}|2\rangle + \frac{1}{\sqrt{2}}|3\rangle = -\frac{1}{\sqrt{2}}\left|\frac{1}{2}, -\frac{1}{2}\right\rangle + \frac{1}{\sqrt{2}}\left|-\frac{1}{2}, \frac{1}{2}\right\rangle,$$
$$E_3 = -\frac{3A}{4} \tag{6.43c}$$

$$|\psi_4\rangle = |4\rangle = \left|-\frac{1}{2}, \frac{1}{2}\right\rangle, E_4 = \frac{A}{4} \tag{6.43d}$$

Although Eq. 6.43 looks quite simple (at least compared to the previous calculations), one sees that the energy levels are degenerated, and this requires, as we already know, high caution. The opposite relationship $|1\rangle = |\psi_1\rangle$, $|2\rangle = \frac{1}{2}|\psi_2\rangle - \frac{1}{\sqrt{2}}|\psi_3\rangle$, $|3\rangle = \frac{1}{\sqrt{2}}|\psi_2\rangle + \frac{1}{\sqrt{2}}|\psi_3\rangle$ and $|4\rangle = |\psi_4\rangle$ leads to the following representation of the S_+ operator:

$$\begin{aligned}&\langle S_+\rangle \propto |1\rangle\langle 2| + |3\rangle\langle 4| \propto \\ &|\psi_1\rangle\langle\psi_2| - |\psi_1\rangle\langle\psi_3| + |\psi_2\rangle\langle\psi_4| + |\psi_3\rangle\langle\psi_4|\end{aligned} \tag{6.44}$$

Setting now the matrix expression for the lineshape function, $L_S(\omega)$ one should take into account that there is a coupling between the coherences associated with the same transition frequency:

$$L_S(\omega) \propto \mathrm{Re}\left\{[1, 1, -1, 1] \cdot \begin{bmatrix} \Lambda_{1212} & R_{1224} & 0 & 0 \\ R_{1224} & \Lambda_{2424} & 0 & 0 \\ 0 & 0 & \Lambda_{1313} & R_{1334} \\ 0 & 0 & R_{1334} & \Lambda_{3434} \end{bmatrix}^{-1} \cdot \begin{bmatrix} 1 \\ 1 \\ -1 \\ 1 \end{bmatrix}\right\} \tag{6.45}$$

The relaxation rates R_{1212} and R_{3434} can be obtained from Eqs. 6.18b and 6.19b, respectively; they yield $R_{1212} = \frac{3}{4}J^A(0)$, $R_{3434} = \frac{5}{12}J^A(0)$, while $R_{1224} = \frac{1}{4}J^A(0)$, and $R_{1334} = 0$.

For the case of ^{14}N, the energy level structure in the low field range looks as follows:

$$|\psi_1\rangle = \left|\frac{1}{2}, 1\right\rangle, \quad E_1 = \frac{A}{2} \tag{6.46a}$$

$$|\psi_2\rangle = \sqrt{\frac{2}{3}}\left|\frac{1}{2}, 0\right\rangle - \sqrt{\frac{1}{3}}\left|-\frac{1}{2}, 1\right\rangle, \quad E_2 = \frac{A}{2} \tag{6.46b}$$

$$|\psi_3\rangle = \sqrt{\frac{1}{3}}\left|\frac{1}{2}, 0\right\rangle + \sqrt{\frac{2}{3}}\left|-\frac{1}{2}, 1\right\rangle, \quad E_3 = -A \tag{6.46c}$$

$$|\psi_4\rangle = \sqrt{\frac{1}{3}}\left|\frac{1}{2}, -1\right\rangle + \sqrt{\frac{2}{3}}\left|-\frac{1}{2}, 0\right\rangle, \quad E_4 = \frac{A}{2} \tag{6.46d}$$

$$|\psi_5\rangle = \sqrt{\frac{2}{3}}\left|\frac{1}{2}, -1\right\rangle - \sqrt{\frac{1}{3}}\left|-\frac{1}{2}, 0\right\rangle, \quad E_5 = -A \tag{6.46e}$$

$$|\psi_6\rangle = \left|-\frac{1}{2}, -1\right\rangle, E_6 = \frac{A}{2} \tag{6.46f}$$

One can expect that the degeneracy of the energy levels influences the ESR lineshape. First, one has to, however, set up the low field representation of the S_+ operator, in analogy to Eq. 6.44. As $a = d = f = g = \sqrt{2/3}$ and $c = e = -b = -h = \sqrt{1/3}$, one gets: $[S_+]^+ \propto \left[-\sqrt{\frac{1}{3}}, \sqrt{\frac{1}{3}}, \sqrt{\frac{1}{3}}, \sqrt{\frac{2}{3}}, \frac{2}{3}, -\frac{\sqrt{2}}{3}, \frac{\sqrt{2}}{3}, -\frac{1}{3}\right]$, which corresponds to the coherences $(\alpha,\beta) = (1,2), (1,3); (4,6), (5,6); (2,4), (2,5)$; and $(3,4), (3,5)$. Taking into account that $\omega_{12} = \omega_{24} = \omega_{35} = \omega_{46} = 0$, $\omega_{34} = \omega_{56} = 3A/2$, and $\omega_{25} = \omega_{13} = -3A/2$, the matrix $[i\Delta\omega + M]$ takes the form:

$$[i\Delta\omega + M] = \begin{bmatrix} \Lambda_{12} & 0 & R_{1246} & 0 & R_{1224} & 0 & 0 & R_{1235} \\ 0 & \Lambda_{13} & 0 & 0 & 0 & R_{1325} & 0 & 0 \\ R_{1246} & 0 & \Lambda_{46} & 0 & R_{2446} & 0 & 0 & R_{3546} \\ 0 & 0 & 0 & \Lambda_{56} & 0 & 0 & R_{3456} & 0 \\ R_{1224} & 0 & R_{2446} & 0 & \Lambda_{24} & 0 & 0 & R_{2435} \\ 0 & R_{1325} & 0 & 0 & 0 & \Lambda_{25} & 0 & 0 \\ 0 & 0 & 0 & R_{3456} & 0 & 0 & \Lambda_{34} & 0 \\ R_{1235} & 0 & R_{3546} & 0 & R_{2435} & 0 & 0 & \Lambda_{35} \end{bmatrix} \tag{6.47}$$

The relaxation rates are

$$R_{1245} = \frac{f\,(ag - bh)}{2} J^{A}(0),$$

$$R_{1224} = \frac{(ae - bf)\left(a + b\sqrt{2}\right)}{4} J^{A}(0),$$

$$R_{1235} = \frac{(ag - bh)\left(c + d\sqrt{2}\right)}{4} J^{A}\left(\frac{3A}{2}\right),$$

$$R_{1325} = \frac{(cg - dh)\left(a + b\sqrt{2}\right)}{4} J^{A}(0),$$

$$R_{2446} = \frac{(bf - ae)\left(f + e\sqrt{2}\right)}{4} J^{A}(0),$$

$$R_{3546} = \frac{(ce - df)\left(f + e\sqrt{2}\right)}{4} J^{A}\left(\frac{3A}{2}\right),$$

$$R_{3456} = \frac{(dh - cg)\left(f + e\sqrt{2}\right)}{4} J^{A}(0), \text{ and}$$

$$R_{2435} = \frac{(ad + cb + bd)\left(eh + fg + eg\sqrt{2}\right)}{6} J^{A}\left(\frac{3A}{2}\right) \text{ [7].}$$

6.4 *g*-Tensor Anisotropy

At higher magnetic fields there is another contribution to the electron spin relaxation; it is associated with anisotropy of the g-tensor. The g-tensor anisotropy interaction can be treated as an electronic counterpart of the chemical shift anisotropy for nuclei (Section 5.3). The simple form of the Zeeman interaction corresponds to a symmetrical g-tensor. In analogy to the hyperfine tensor, the g-tensor is characterized by three components g_{xx}, g_{yy}, and g_{zz}. The Zeeman Hamiltonian can be rewritten as [3, 5, 13]:

$$H_{Z}(S) = \omega_{S} S_{Z} = \gamma_{S} B_{0} S_{Z} = \frac{\beta_e}{\hbar}\left(\frac{g_{xx} + g_{yy} + g_{zz}}{3}\right) B_{0} S_{z} = \frac{\beta_e}{\hbar} g_{\mathrm{eff}} B_{0} S \tag{6.48}$$

where β_e is Bohr magneton and g_{eff} is defined as $g_{\mathrm{eff}} = (g_{xx} + g_{yy} + g_{zz})/3$. When the electron is put into a local magnetic surrounding in a molecule or crystal, then Eq. 6.48 should be written in a more general form:

$$H_Z(S) = \frac{\omega_S}{2}\left(\frac{g_{xx}+g_{yy}+g_{zz}}{3}\right) S_Z \tag{6.49}$$

and it is likely that $g_{xx} \neq g_{yy} \neq g_{zz}$. This asymmetry gives rise to additional terms in the Zeeman Hamiltonian, which then yields, $H_Z = H_{g,iso} + H_{g,aniso}$, with the isotropic part, $H_{g,iso}$, given by Eq. 6.49. The anisotropic part, $H_{g,aniso}$, is modulated by molecular tumbling and causes electron spin relaxation. In the laboratory frame (determined by the direction of the external magnetic field, that is, by the isotropic part of the electron Zeeman coupling), the Hamiltonian $H_{g,aniso}$ takes the form (compare Eqs. 5.23 and 5.26) for the chemical shift anisotropy Hamiltonian):

$$H^{(L)}_{g,aniso}(t) = \frac{\gamma_S}{2} g_{eff} \sum_{m-2}^{2} (-1)^m X^{2(L)}_{-m}(t)\, T^2_m(S, B_0) \tag{6.50}$$

The rank-two tensor components, $T^2_m(S, B_0)$ are defined by Eq. 5.25. In the principal axis system (P), the tensor components $X^{2(P)}_{-m}$ are defined as

$$X^{2(P)}_0 = \sqrt{\frac{2}{3}}\left[g_{zz} - \frac{1}{2}\left(g_{xx}+g_{yy}\right)\right],$$

$$X^{2(P)}_{\pm 1} = 0,\; X^{2(P)}_{\pm 1} = \frac{1}{2}\left(g_{xx}-g_{yy}\right) \tag{6.51}$$

in close analogy to Eq. 5.24 for the functions $\Phi^{2(P)}_{-m}$ describing chemical shift anisotropy.

One should again note the rule that spin interactions are time-independent in their principal axis systems. Applying the transformation of Eq. 2.13 to the tensor components $X^{2(P)}_{-m}$ one gets (see Eq. 5.26):

$$X^{2(L)}_{-m}(t) = \sum_{k=-2}^{2} D^2_{k,-m}(\Omega(t))\, X^{2(P)}_{-m} \tag{6.52}$$

which explicitly gives:

$$B_0 X^2_{-m}(t) = \omega_S \frac{g_{eff}}{2}\left\{\sqrt{\frac{2}{3}}\left[g_{zz} - \frac{1}{2}\left(g_{xx}+g_{yy}\right)\right] D^2_{0,-m}(\Omega) + \frac{1}{2}\left(g_{xx}-g_{yy}\right)\left[D^2_{1,-m}(\Omega) + D^2_{-1,-m}(\Omega)\right]\right\} \tag{6.53}$$

where the angle $\Omega = \Omega(t)$ describes molecular orientation. The contribution of the g-tensor anisotropy interaction to the electron spin relaxation is reflected by the form of the matrix $\lfloor H_{\text{A,aniso}} + H_{\text{g,aniso}} \rfloor$ which is an extension of Eq. 6.16:

$$\lfloor H_{\text{A,aniso}}(P,S) + H_{\text{g,aniso}}(S) \rfloor (t) =$$
$$\begin{bmatrix} \frac{1}{2\sqrt{6}} G_0^2 + \frac{1}{\sqrt{6}} X_0^2 & \frac{1}{4} G_{-1}^2 & \frac{1}{4} G_{-1}^2 + \frac{1}{4} X_{-1}^2 & \frac{1}{2} G_{-2}^2 \\ \frac{1}{4} G_1^2 & -\frac{1}{2\sqrt{6}} G_0^2 + \frac{1}{\sqrt{6}} X_0^2 & -\frac{1}{2\sqrt{6}} G_0^2 & -\frac{1}{4} G_{-1}^2 - \frac{1}{4} X_{-1}^2 \\ -\frac{1}{4} G_1^2 - \frac{1}{4} X_1^2 & -\frac{1}{2\sqrt{6}} G_0^2 & -\frac{1}{2\sqrt{6}} G_0^2 - \frac{1}{\sqrt{6}} X_0^2 & -\frac{1}{4} G_{-1}^2 \\ \frac{1}{2} G_2^2 & \frac{1}{4} G_1^2 + \frac{1}{4} X_1^2 & \frac{1}{4} G_1^2 & \frac{1}{2\sqrt{6}} G_0^2 - \frac{1}{\sqrt{6}} X_0^2 \end{bmatrix} (t) \tag{6.54}$$

In Eq. 6.54 we have omitted the index (L) for simplicity. As the g-tensor anisotropy interaction is of importance in the high frequency range, it is enough to consider its contribution to Eq. 6.20 (which gives the single-quantum relaxation rates important for the lineshape at high frequencies). Now, the expression reads:

$$R_{1313} = R_{2424} = \frac{1}{6} J^A(0) + \frac{1}{8} J^A\left(\frac{A}{2}\right) + \frac{5}{12} J^A(\omega_S) + \frac{1}{8} J^g(\omega_S) \tag{6.55}$$

where (see for analogy Eq. 5.32):

$$J^g(\omega) = \omega_S^2 \left(\frac{g_{\text{eff}}}{2}\right)^2 \left\{ \frac{2}{3} \left[g_{zz} - \frac{1}{2}(g_{xx} + g_{yy}) \right]^2 + \frac{1}{2}(g_{xx} - g_{yy})^2 \right\} \frac{1}{5} \frac{\tau_{\text{rot}}}{1 + \omega^2 \tau_{\text{rot}}} \tag{6.56}$$

Although extended, Eq. 6.56 still does not completely describe the electron spin relaxation at higher frequencies as it does not take into account interference effects between the anisotropic hyperfine interaction and the g-tensor anisotropy coupling [15, 16]. Both interactions are modulated by the same motional process, that is, molecular tumbling. This implies that Eq. 6.56 should contain the term:

$$R_{1313}^{A-g} = R_{2424}^{A-g} = \frac{1}{12} J^{A-g}(0) + \frac{1}{8} J^{A-g}(\omega_S) \tag{6.57}$$

where the spectral density $J^{A-g}(\omega)$ takes the form:

$$J^{A-g}(\omega) = \frac{2}{3}\omega_S\left(\frac{g_{\mathrm{eff}}}{2}\right)\left[g_{zz} - \frac{1}{2}(g_{xx} + g_{yy})\right] \cdot \left[A_{zz} - \frac{1}{2}(A_{xx} + A_{yy})\right]\frac{1}{5}\frac{\tau_{\mathrm{rot}}}{1 + \omega^2\tau_{\mathrm{rot}}} \quad (6.58)$$

in which one can see close analogy to Eq. 5.37 for the spectral density $J^{\mathrm{DD-CSA}}(\omega)$.

References

1. Borbat, P. P., Mchaourab, H. S. and Freed, J. H. (2002). Protein structure determination using long-distance constraints from double-quantum coherence ESR: Study of T4 Lysozyme. *J. Am. Chem. Soc.*, **124**, pp. 5304–5314.
2. Freed, J. H. (1976). *Spin Labeling Theory and Applications*, ed. Berliner, L. J. (New York: Academic Press).
3. Barci, L., Bertini, I. and Luchinat, C. (1991). *Nuclear and Electronic Relaxation* (Weinheim: Wiley-VCH).
4. Belorizky, E., Gilies, D. G., Gorecki, W., Lang, K., Noack, F., Roux, C., Struppe, J., Suteliffe, L. H., Travers, J. P. and Wu, X. (1998). Translational diffusion constants and intermolecular relaxation in paramagnetic solutions with hyperfine coupling on the electronic site. *J. Phys. Chem. A*, **102**, pp. 3674–3680.
5. Bertini, I., Luchinat, C. and Parigi, G. (2001). *Solution NMR of Paramagnetic Molecules* (Amsterdam: Elsevier).
6. Kruk, D., Korpala, A., Kowalewski, J., Rossler, E. A. and Moscicki, J. (2012). ^{1}H relaxation dispersion in solutions of nitroxide radicals: Effects of hyperfine interactions with ^{14}N and ^{15}N nuclei. *J. Chem. Phys.*, **137**, pp. 044512–044524.
7. Kruk, D., Korpala, A., Kubica, A., Kowalewski J., Rossler, E. A. and Moscicki, J. (2013). ^{1}H relaxation dispersion in solutions of nitroxide radicals: Influence of electron spin relaxation. *J. Chem. Phys.*, **138**, pp. 124506–124521.
8. Kruk, D. (2007). *Theory of Evolution and Relaxation of Multi-Spin Systems* (Bury St Edmunds: Arima).

9. Redfield, A. G. (1996). *Encyclopedia of Nuclear Magnetic Resonance, Relaxation Theory: Density Matrix Formulation* (Chichester: Wiley), pp. 4085–4092.
10. Sato, H., Bottle, S. E., Blinco, J. P., Micallef, A. S., Eaton, G. R. and Eaton, S. S. (2008). Electron spin-lattice relaxation of nitroxyl radicals in temperature ranges that span glassy solutions to low-viscosity liquids. *J. Magn. Res.*, **191**, pp. 66–77.
11. Sato, H., Kathirvelu, V., Fielding, A., Blinco, J. P., Micallef, A. S., Bottle, S. E., Eaton, S. S. and Eaton, G. R. (2007). Impact of molecular size on electron spin relaxation rates of nitroxyl radicals in glassy solvents between 100 and 300 K. *Mol. Phys.*, **105**, pp. 2137–2151.
12. Kowalewski, J., Kruk, D. and Parigi, G. (2005). NMR relaxation in solution of paramagnetic complexes: Recent theoretical progress for $S \geq 1$. *Adv. Inorg. Chem.*, **57**, pp. 41–104.
13. Kruk, D., Kowalewski, J., Tipikin, S., Freed, J. H., Mocicki, M., Mielczarek, A., and Port, M. (2011). Joint analysis of ESR lineshapes and ^{1}H NMRD profiles of DOTA-Gd derivatives by means of the slow motion theory. *J. Chem. Phys.*, **134**, pp. 024508–024523.
14. Nilsson, T. and Kowalewski, J. (2000). Slow-motion theory of nuclear spin relaxation in paramagnetic low-symmetry complexes: A generalization to high electron spin, *J. Magn. Reson.*, **146**, pp. 345–358.
15. Goldman, M. (2001). Formal theory of spin–lattice relaxation. *J. Magn. Reson.*, **149**, pp. 160–187.
16. Kumar, A., Grace, R. C. R. and Madhu, P. K. (2000). Cross-correlations in NMR. *Prog. Nucl. Magn. Reson. Spectr.*, **37**, pp. 191–319.

Chapter 7

Nuclear Spin Relaxation in Paramagnetic Liquids

In Chapter 6 we have discussed electron spin resonance (ESR) lineshapes for nitroxide radicals containing ^{14}N and ^{15}N isotopes. For this purpose we have considered a two-spin system consisting of an electron spin and a nuclear spin of ^{15}N or ^{14}N. In this chapter we shall go a step further and enquire how the isotropic hyperfine coupling between the electron spin and the nitrogen spin affects proton (^{1}H) spin relaxation when the paramagnetic species are dissolved in proton containing liquids.

7.1 Proton Relaxation and Hyperfine Coupling

To discuss proton relaxation in liquids containing paramagnetic species (nitroxide radicals in this case) a system composed of three spins: proton spin ($I = 1/2$), electron spin ($S = 1/2$), and spin of ^{15}N or ^{14}N ($P = 1/2$ or $P = 1$) should be considered. The spin system and the relevant spin interactions are shown in Fig. 7.1. The proton spin relaxes due to dipole–dipole interactions with the electron spin which is, in turn, coupled to the nitrogen spin by the hyperfine

Understanding Spin Dynamics
Danuta Kruk

ISBN 978-981-4463-49-2 (Hardcover), 978-981-4463-50-8 (eBook)
www.panstanford.com

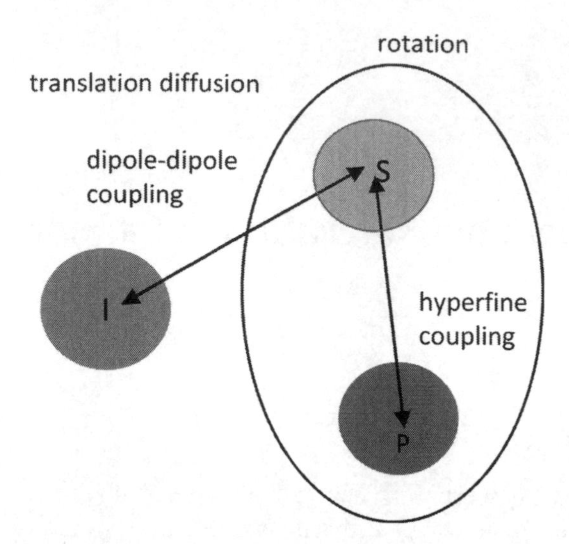

Figure 7.1 Mutually interacting proton ($I = 1/2$), electron ($S = 1/2$), and nitrogen ($P = 1/2$ or $P = 1$) spins in solutions of nitroxide radicals. Rotational dynamics modulates the S–P hyperfine coupling, while the I–S dipole–dipole interaction is modulated by relative translational motion of the solvent and solute molecules.

interaction. The I–S dipole–dipole coupling fluctuates in time due to translational diffusion of the solvent and solute molecules. This is an example of nuclear spin relaxation in systems of mutually interacting nuclei (of different spin quantum numbers) and electrons.

Equation 2.62 in Chapter 2 describes the proton relaxation rate, $R_{1I}(\omega_I)$, induced by dipole–dipole coupling to an electron spin, assuming Zeeman energy level structure for the electron spin. The expression is repeated below for convenience, with explicitly introduced electron spin quantum number S (Eq. 5.57 and Eq. 5.58) [1–7]:

$$R_{1I}(\omega_I) = \frac{2}{3} S(S+1) \left(\frac{\mu_0}{4\pi} \frac{\gamma_I \gamma_S \hbar}{r^3} \right)^2 \times \left[J(\omega_I - \omega_S) + 3J(\omega_I) + 6J(\omega_I + \omega_S) \right] \quad (7.1)$$

This equation assumes that both spins are placed in the same molecule as the inter-spin distance, r, is assumed constant. For translational dynamics (inter-molecular relaxation) this expression has to be modified to the form:

$$R_{1I}^{\text{trans}}(\omega_I) = \frac{2}{3}S(S+1)\left(\frac{\mu_0}{4\pi}\gamma_I\gamma_S\hbar\right)^2 \times \left[J_{\text{trans}}(\omega_S - \omega_I) + 3J_{\text{trans}}(\omega_I) + 6J_{\text{trans}}(\omega_I + \omega_S)\right] \quad (7.2)$$

where the inter-molecular spectral density, $J_{\text{trans}}(\omega)$, has been defined by Eq. 2.45. The electron spin S does not exhibit a Zeeman energy level structure as it is coupled to the nuclear spin P. As explained in Chapter 5, in such a case, the expressions of Eqs. 7.1 and 7.2 have to be extended by introducing generalized spectral densities, $S_{m,m}(\omega)$ (with $m = -1, 0, 1$) [4, 5, 8, 9]; Eq. 7.2 should be rewritten as:

$$R_{1I}^{\text{trans}}(\omega_I) = \frac{2}{3}S(S+1)\left(\frac{\mu_0}{4\pi}\gamma_I\gamma_S\hbar\right)^2 \times \left[S_{1,1}^{\text{trans}}(\omega_I) + 3S_{0,0}^{\text{trans}}(\omega_I) + 6S_{-1,-1}^{\text{trans}}(\omega_I)\right] \quad (7.3)$$

The form of the spectral densities $S_{m,m}^{\text{trans}}(\omega)$ associated with translational diffusion stems from generalization of Eq. 2.45, and it yields [4, 5, 10]:

$$S_{m,m}^{\text{trans}}(\omega_I) = \frac{72}{5}\frac{N_S}{d^3}\text{Re}\int_0^\infty \frac{u^2}{81 + 9u^2 - 2u^4 + u^6} \times \left\{\frac{1}{4}\left[S_m^1\right]^+ [M]^{-1}\left[S_m^1\right]\right\} du \quad (7.4)$$

where N_S denotes (in this case) the number of electron spins per unit volume. Comparing Eq. 7.4 and Eq. 2.45 one sees that the function $\frac{u^2\tau_{\text{trans}}}{u^2+(\omega\tau_{\text{trans}})^2}$ in Eq. 2.45 has now been replaced by the matrix product $\left\{\left[S_m^1\right]^+ [M]^{-1}\left[S_m^1\right]\right\}$; it is worth to remind that the operators S_m^1 are defined as $S_0^1 = S_z$ and $S_{\pm 1}^1 = S_\pm/\sqrt{2}$.

We already know that the matrix $[M]$ is set up in the Liouville space constructed from the eigenvectors of the S spin. However, in this case, the electron spin S is coupled to the nitrogen spin P via the isotropic hyperfine interaction, and the spins cannot be considered separately (the hyperfine coupling is the reason while the energy level structure of the electron spin is not Zeeman, see Chapter 6). As in the first step we neglect the electron spin relaxation, the matrix $[M]$ is diagonal (Eq. 5.5), but now the rotational correlation time, τ_{rot}, is replaced by $u^{-2}\tau_{\text{trans}}$:

$$[M]_{\alpha\beta,\alpha\beta} = i\left(\omega_I + \omega_{\alpha\beta}\right) + u^2\tau_{\text{trans}}^{-1} \tag{7.5}$$

In analogy to ESR lineshape analysis (Chapter 6), we begin the discussion with the case of $P = 1/2$ (^{15}N). In the first step we need to set up the representations of the S_m^1 operators in the basis $\{|\psi_\alpha\rangle\langle\psi_\beta|\}$ defined by Eq. 6.4 [10, 11]. For $S = 1/2$ the general representation of these operators in the Zeeman basis $\{|m_S, m_P\rangle\}$ yields:

$$S_0 = \sum_{m_S>0} \left(|m_S, m_P\rangle\langle m_S, m_P| - |-m_S, m_P\rangle\langle -m_S, m_P|\right) \tag{7.6a}$$

$$S_1 = -\sum_{m_S} \left(|m_S, m_P\rangle\langle m_S - 1, m_P|\right) \tag{7.6b}$$

$$S_{-1} = \sum_{m_S} |m_S - 1, m_P\rangle\langle m_S, m_P| \tag{7.6c}$$

This implies, that taking into account the relationship between the Zeeman states $|n\rangle = |m_S, m_P\rangle$ and the eigenstates $|\psi_\alpha\rangle$ for the case of ^{15}N (Eq. 6.4): $|1\rangle = |1/2, 1/2\rangle = |\psi_1\rangle$, $|2\rangle = |1/2\rangle = a|\psi_2\rangle - b|\psi_3\rangle$, $|3\rangle = |-1/2, 1/2\rangle = b|\psi_2\rangle + a|\psi_3\rangle$, $|4\rangle = |-1/2, -1/2\rangle = |\psi_4\rangle$, Eq. 7.6 can be expressed in the Liouville basis $\{|\psi_\alpha\rangle\langle\psi_\beta|\}$ as:

$$\begin{aligned} S_0 &= |1\rangle\langle 1| + |2\rangle\langle 3| - |3\rangle\langle 3| - |4\rangle\langle 4| \\ &= |\psi_1\rangle\langle\psi_1| + \left(a^2 - b^2\right)|\psi_2\rangle\langle\psi_2| \\ &\quad + \left(b^2 - a^2\right)|\psi_3\rangle\langle\psi_3| + |\psi_4\rangle\langle\psi_4| + 2ab\left(|\psi_2\rangle\langle\psi_3| + |\psi_3\rangle\langle\psi_2|\right) \end{aligned} \tag{7.7a}$$

$$\begin{aligned} S_1 &= \left(|1\rangle\langle 3| + |2\rangle\langle 4|\right) = -a\left(|\psi_1\rangle\langle\psi_3| + |\psi_2\rangle\langle\psi_4|\right) \\ &\quad - b\left(|\psi_1\rangle\langle\psi_2| - |\psi_3\rangle\langle\psi_4|\right) \end{aligned} \tag{7.7b}$$

$$\begin{aligned} S_{-1} &= |3\rangle\langle 1| + |4\rangle\langle 2| = a\left(|\psi_3\rangle\langle\psi_1| + |\psi_4\rangle\langle\psi_2|\right) \\ &\quad + b\left(|\psi_2\rangle\langle\psi_1| - |\psi_4\rangle\langle\psi_3|\right) \end{aligned} \tag{7.7c}$$

Thus, the explicit form of the generalized spectral densities, $S_{m,m}^{\text{trans}}(\omega_I)$ (see Eq. 5.3), is:

$$S_{\pm1,\pm1}^{\text{trans}}(\omega_{\text{I}}) = \frac{1}{4}\text{Re}\left\{[a, a, -b, b] \cdot \begin{bmatrix} \Lambda_{13} & 0 & 0 & 0 \\ 0 & \Lambda_{24} & 0 & 0 \\ 0 & 0 & \Lambda_{12} & 0 \\ 0 & 0 & 0 & \Lambda_{34} \end{bmatrix}^{-1} \begin{bmatrix} a \\ a \\ -b \\ b \end{bmatrix}\right\} \tag{7.8}$$

where $\Lambda_{\alpha\beta} = i\left(\omega_I \pm \omega_{\alpha\beta}\right) + u^2/\tau_{\text{trans}}$. Combining Eq. 7.8 with Eq. 7.4 and the definition of the translational spectral density, $J_{\text{trans}}(\omega)$

(Eq. 2.45), one obtains:

$$\begin{aligned} S^{\text{trans}}_{\pm 1,\pm 1}(\omega_{\text{I}}) &= a^2\left[J_{\text{trans}}(\omega_{\text{I}} \pm \omega_{13}) + J_{\text{trans}}(\omega_{\text{I}} \pm \omega_{24})\right] \\ &\quad + b^2\left[J_{\text{trans}}(\omega_{\text{I}} \pm \omega_{12}) + J_{\text{trans}}(\omega_{\text{I}} \pm \omega_{34})\right] \\ &\cong a^2\left[J_{\text{trans}}(\omega_{13}) + J_{\text{trans}}(\omega_{24})\right] \\ &\quad + b^2\left[J_{\text{trans}}(\omega_{12}) + J_{\text{trans}}(\omega_{34})\right] \end{aligned} \quad (7.9)$$

For the last equality we have used the fact that $\omega \ll \omega_{\alpha\beta}$. Analogously, one gets for $S^{\text{trans}}_{0,0}(\omega_I)$:

$$S^{\text{trans}}_{0,0}(\omega_{\text{I}}) = \left\lfloor 1 + \left(a^2 - b^2\right)\right\rfloor J_{\text{trans}}(\omega_{\text{I}}) + 4a^2b^2 J_{\text{trans}}(\omega_{23}) \quad (7.10)$$

Substituting these spectral densities into Eq. 7.3 one gets the final expression for the proton spin–lattice relaxation rate, $R^{\text{trans}}_{1\text{I}}(\omega_{\text{I}})$, which is valid for an arbitrary magnetic field (resonance frequency) [10–12]:

$$\begin{aligned} R^{\text{trans}}_{1\text{I}}(\omega_{\text{I}}) &= \frac{2}{3}S(S+1)\left(\frac{\mu_0}{4\pi}\gamma_{\text{I}}\gamma_{\text{S}}\hbar\right)^2 \times \\ &\left\{\begin{array}{l} \frac{3}{2}\left[1 + \left(a^2 - b^2\right)\right] J_{\text{trans}}(\omega_I) + 6(ab)^2 J_{\text{trans}}(\omega_{23}) + \\ \frac{7}{2}a^2\left[J_{\text{trans}}(\omega_{13}) + J_{\text{trans}}(\omega_{24})\right] + \frac{7}{2}b^2\left[J_{\text{trans}}(\omega_{12}) + J_{\text{trans}}(\omega_{34})\right] \end{array}\right\} \\ &= \{(\text{I}) + (\text{II}) + (\text{III}) + (\text{IV})\} \end{aligned} \quad (7.11)$$

In Fig. 7.2 examples of the total proton relaxation rate, $R^{\text{trans}}_{1\text{I}}(\omega_{\text{I}})$, as well as the individual contributions are shown ("frequency" denotes in this figure and next figures of this chapter, the proton resonance frequency). The isotropic part of the electron spin–nitrogen spin hyperfine coupling reduces at low frequencies the relaxation rate, $R^{\text{trans}}_{1\text{I}}(\omega_{\text{I}})$ (the relaxation becomes slower) with respect to the case of $A = 0$. For $A = 0$ (i.e., Zeeman energy level structure of the electron spin), the second (II) and forth (IV) terms in Eq. 7.11 vanish ($b = 0$), while the first (I) and third (III) terms converge to $3J_{\text{trans}}(\omega_{\text{I}})$ and $7J_{\text{trans}}(\omega_{\text{S}})$, respectively, that is, Eq. 7.11 converges to Eq. 7.2.

In Fig. 7.3 the relaxation rate, $R^{\text{trans}}_{1\text{I}}(\omega_{\text{I}})$ obtained from Eq. 7.11 is compared with those for $A = 0$ for several correlation times, τ_{trans}; the effect of the hyperfine coupling increases when the dynamics slows down.

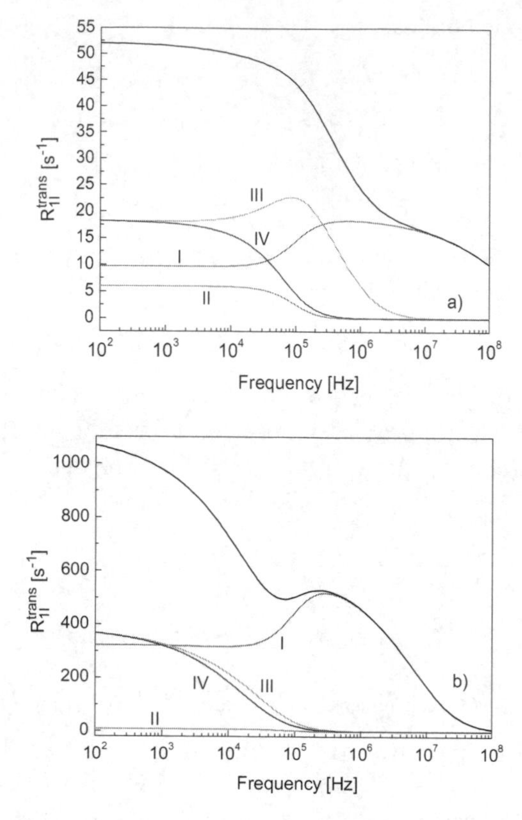

Figure 7.2 Total relaxation rate, $R_{1I}^{\text{trans}}(\omega_I)$ (black line) and individual contributions (I), (II), (III), and (IV) according to Eq. 7.11; $A = 63$ MHz, $d = 4$ Å. (a) $\tau_{\text{trans}} = 1.5 \times 10^{-9}$ s and (b) $\tau_{\text{trans}} = 5 \times 10^{-8}$ s. The concentration of the paramagnetic molecules is assumed to be 10 mMol.

The description of the proton relaxation is more complex when the electron spin relaxation cannot be neglected. As explained in Chapter 5, electronic relaxation plays the role of an additional source of modulations of the I–S dipole–dipole coupling. We already know that the electronic relaxation is primarily caused by the anisotropic part of the S–P hyperfine coupling modulated by the rotational dynamics of the paramagnetic molecule. As mentioned in Chapter 2, the translational correlation time is longer than the rotational one (according to the Stoke's equation, $\tau_{\text{trans}}/\tau_{\text{rot}} = 9$) [1]. This means that although the electron spin relaxation is slower

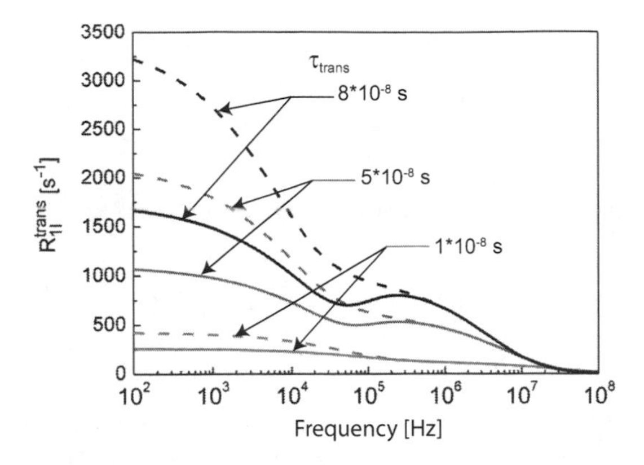

Figure 7.3 $R_{1I}^{\text{trans}}(\omega_I)$ according to Eq. 7.11; $A = 63$ MHz, $d = 4$ Å (solid lines), corresponding $R_{1I}^{\text{trans}}(\omega_I)$ for $A = 0$ (Eq. 7.2, dashed lines). The concentration 10 mMol of the paramagnetic molecules is assumed.

than the rotational dynamics (this is the necessary condition for the presented description to be valid, see Section 3.4 regarding the Redfield condition [1, 13–16]), it might be faster (or comparable) with the translational dynamics (in terms of the correlation time, τ_{trans}). Then, the electron spin relaxation becomes an important factor influencing the proton relaxation.

The matrix representation of the anisotropic hyperfine interaction, $H_{\text{A,aniso}}^{(L)}(P, S)$, in the eigenbasis $\{|\psi_\alpha\rangle\}$ (Eq. 6.4) is given by Eq. 6.17. Using it, one can calculate the relevant electron spin relaxation rates by means of Eq. 3.19. To figure out how to incorporate the electron spin relaxation into Eq. 7.11 one has to closely investigate the origin of the individual terms contributing to this expression. The spectral densities taken at ω_{13}, ω_{24}, ω_{12}, and ω_{34} stem from the $S_{\pm1,\pm1}^{\text{trans}}$ quantities. As all the transition frequencies are different there is no coupling between these coherences. This implies that in order to include the electron spin relaxation it is sufficient to add on the diagonal of the matrix $[M]$ of Eq. 7.8, the corresponding relaxation rates, R_{1313}, R_{2424}, R_{1212}, and R_{3434}. In consequence, the spectral densities $J_{\text{trans}}(\omega_{\alpha\beta})$ (for ω_{13}, ω_{24}, ω_{12}, and ω_{34}) can be rather straightforwardly generalized to the form [5, 10]:

$$J_{\text{trans}}\left(\omega_{\alpha\beta},\tau_{\alpha\beta}\right)=\frac{72}{5}\frac{N_S}{d^3}\int_0^\infty \frac{u^2}{81-9u^2+u^6}\frac{\tau_{\alpha\beta,u}}{1+\left(\omega_{\alpha\beta}\tau_{\alpha\beta,u}\right)^2}du \tag{7.12}$$

where

$$\tau_{\alpha\beta,u}^{-1}=u^2\tau_{\text{trans}}^{-1}+R_{\alpha\beta\alpha\beta} \tag{7.13}$$

The electron spin relaxation rates are given by Eqs. 6.18 and 6.19.

Although the spectral density, $J_{\text{trans}}\left(\omega_{23}\right)$ stems from the $S_{0,0}^{\text{trans}}$ quantity, the coherence ρ_{23} is not coupled to any of the populations $\rho_{\alpha\alpha}$ as $\omega_{23} \neq 0$. Thus, $J_{\text{trans}}\left(\omega_{23}\right)$ can also be generalized to $J_{\text{trans}}\left(\omega_{23},\tau_{23}\right)$ according to Eq. 7.12, with the relaxation rate [17]:

$$\begin{aligned}R_{2323}=&\frac{1}{24}\left[(a+b)^2+(a-b)^2\right]^2 J^A(0)+\frac{1}{16}(a+b)^2 J^A(\omega_{24})\\&+\frac{1}{16}(a+b)^2 J^A(\omega_{12})+\frac{1}{16}(a-b)^2 J^A(\omega_{13})\\&+\frac{1}{12}\left(a^2-b^2\right)^2 J^A(\omega_{23})+\frac{1}{16}(a-b)^2 J^A(\omega_{34})\end{aligned} \tag{7.14}$$

For the last spectral density, $J_{\text{trans}}\left(\omega_{\text{I}}\right)$, the calculations are more cumbersome as it is associated with the population part of the electron spin relaxation matrix involving the coupled coherences $\rho_{\alpha\alpha}$. Thus, the quantity $S_{0,0}^{\text{trans}}\left(\omega_{\text{I}}\right)$ is given as the matrix product [17]:

$$S_{0,0}^{\text{trans}}\left(\omega_I\right)=\frac{1}{2}\text{Re}\left\{[c]^+\begin{bmatrix}\Lambda_{11} & R_{1122} & R_{1133} & R_{1144}\\ R_{1122} & \Lambda_{22} & R_{2233} & R_{2244}\\ R_{1133} & R_{2233} & \Lambda_{33} & R_{3344}\\ R_{1144} & R_{2244} & R_{3344} & \Lambda_{44}\end{bmatrix}^{-1}[c]\right\} \tag{7.15}$$

where, $\Lambda_{\alpha\alpha} = i\omega + u^2/\tau_{\text{trans}} + R_{\alpha\alpha\alpha\alpha}$, while $[c]^+ = \left[1, a^2-b^2, b^2-a^2, 1\right]$. The relaxation rates are equal to $R_{\alpha\alpha\alpha\alpha} = -\sum_{\beta\neq\alpha} R_{\alpha\alpha\beta\beta}$, where $R_{\alpha\alpha\beta\beta}$ are given as $R_{1122} = \frac{1}{8}(a+b)^2 J^A(\omega_{12})$, $R_{1133} = \frac{1}{8}(a-b)^2 J^A(\omega_{13})$, $R_{1144} = \frac{1}{2}J^A(\omega_{13})$, $R_{2233} = \frac{1}{12}\left(a^2-b^2\right)^2 J^A(\omega_{23})$, $R_{2244} = \frac{1}{8}(a+b)^2 J^A(\omega_{24})$, and $R_{3344} = \frac{1}{8}(a-b)^2 J^A(\omega_{34})$ [17]. The quantity $S_{0,0}^{\text{trans}}\left(\omega_{\text{I}}\right)$ replaces the term $\left[1+\left(a^2-b^2\right)^2\right] J_{\text{trans}}\left(\omega_{\text{I}}\right)$ in Eq. 7.11.

We shall come back to the role of the electronic relaxation in the last section of this chapter.

One can derive analogous expressions for the case of $P = 1$ (^{14}N). We shall not present in detail the calculations, but it is worth to compare the final expressions for the proton spin–lattice relaxation rate, $R_{1I}^{trans}(\omega_I)$, for the cases of $P = 1/2$ and $P = 1$; for $P = 1$, it yields [10]:

$$R_{1I}^{trans}(\omega_I) = \frac{2}{3}S(S+1)\left(\frac{\mu_0}{4\pi}\gamma_I\gamma_S\hbar\right)^2 \times$$
$$\left\{\begin{array}{l} \frac{1}{2}\left[2+\left(a^2-b^2\right)^2+\left(c^2-d^2\right)^2+\left(e^2-f^2\right)^2+\left(g^2-h^2\right)^2\right]J^{trans}(\omega_I) \\ \left[(ac-bd)^2 J^{trans}(\omega_{23})+(eg-fh)^2 J^{trans}(\omega_{45})\right] \\ \frac{7}{3}\left[b^2 J^{trans}(\omega_{12})+d^2 J^{trans}(\omega_{13})+e^2 J^{trans}(\omega_{46})+g^2 J^{trans}(\omega_{56})\right]+ \\ \frac{7}{3}\left[(af)^2 J^{trans}(\omega_{24})+(ah)^2 J^{trans}(\omega_{25})+(cf)^2 J^{trans}(\omega_{35})\right] \end{array}\right\}$$
$$= \{(\mathrm{I})+(\mathrm{II})+(\mathrm{III})+(\mathrm{IV})\} \tag{7.16}$$

where the coefficients a, b, c, d, e, f, g, and h have been given in Chapter 6. Figure 7.4 shows the total relaxation rate, $R_{1I}^{trans}(\omega_I)$ for the case of $P = 1$ and the individual relaxation contributions, according to Eq. 7.16, for the same set of parameters as used in Fig. 7.2 (the hyperfine coupling constant scales with the gyromagnetic factor of the nucleus, $A = 43$ MHz for ^{14}N corresponds to $A = 63$ MHz for ^{15}N).

For $A = 0$ the first term converges to $\{3J_{trans}(\omega_I)\}$, while the remaining terms converge to $\{\frac{7}{3}[J_{trans}(\omega_{13}) + J_{trans}(\omega_{46}) + J_{trans}(\omega_{25})] = 7J_{trans}(\omega_S)\}$. Again, in order to include the electron spin relaxation into the spectral densities $J_{trans}(\omega_{\alpha\beta})$, it is sufficient to apply Eq. 7.12, that is, to transform $J_{trans}(\omega_{\alpha\beta}, \tau_{trans})$ into $J_{trans}(\omega_{\alpha\beta}, \tau_{\alpha\beta})$, with the relaxation rates $R_{\alpha\beta\alpha\beta}$ (some of them are given in Chapter 6, Eq. 6.41) [17]. To calculate the generalized spectral density, $S_{0,0}^{trans}(\omega_I)$ one has to express the operator S_0 in the Liouville eigenbasis $\{|\psi_\alpha\rangle\langle\psi_\beta|\}$ defined by Eq. 6.27; this representation was already necessary for deriving the expression for $R_{1I}(\omega_I)$ (Eq. 7.16).

As in the present case, $S_0 = |1\rangle\langle 1| + |2\rangle\langle 2| + |3\rangle\langle 3| - |4\rangle\langle 4| - |5\rangle\langle 5| - |6\rangle\langle 6|$, the required representation takes the form:

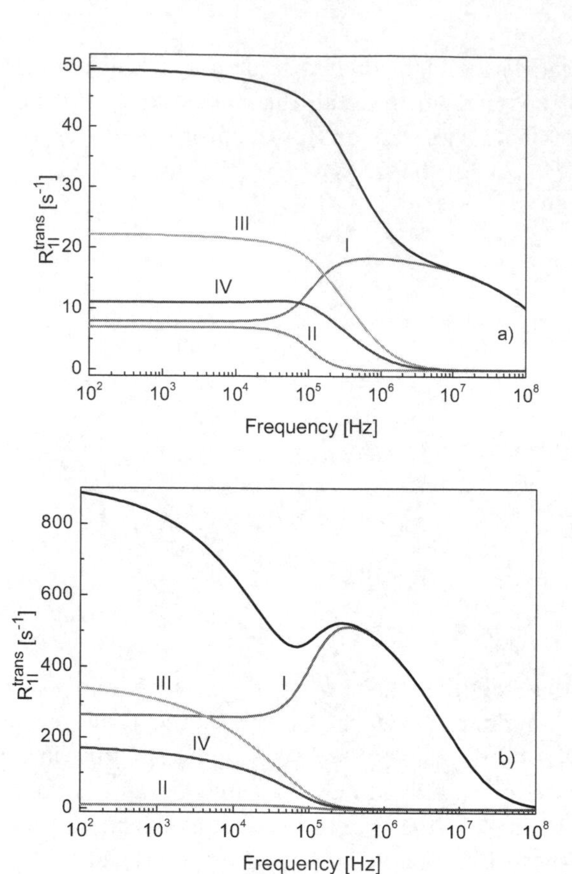

Figure 7.4 Total relaxation rate, $R_{1I}(\omega_I)$ (black line) and individual contributions (I), (II), (III), and (IV) according to Eq. 7.16; $A = 43$ MHz, $d =$ Å, $\tau_{trans} = 5 \times 10^{-8}$ s. 10 mMol concentration of the paramagnetic molecules is assumed.

$$\begin{aligned} S_0 = & \left|\psi_1\right\rangle\left\langle\psi_1\right| + \left(a^2 - b^2\right)\left|\psi_2\right\rangle\left\langle\psi_2\right| + \left(c^2 - d^2\right)\left|\psi_3\right\rangle\left\langle\psi_3\right| \\ & + \left(e^2 - f^2\right)\left|\psi_4\right\rangle\left\langle\psi_4\right| \left(g^2 - h^2\right)\left|\psi_5\right\rangle\left\langle\psi_5\right| + \left|\psi_6\right\rangle\left\langle\psi_6\right| \\ & (ac - bd)\left(\left|\psi_2\right\rangle\left\langle\psi_3\right| + \left|\psi_3\right\rangle\left\langle\psi_2\right|\right) \\ & + (eg - fh)\left(\left|\psi_4\right\rangle\left\langle\psi_5\right| + \left|\psi_5\right\rangle\left\langle\psi_4\right|\right) \end{aligned} \tag{7.17}$$

where the last line gives rise to the terms, $J_{trans}(\omega_{23})$ and $J_{trans}(\omega_{45})$ in Eq. 7.16. In consequence, the $S_{0,0}^{trans}(\omega_I)$ quantity is given as (in full analogy to Eq. 7.15 for the case of ^{15}N) [17]:

$$S_{0,0}^{\text{trans}}(\omega_I) = \frac{1}{2}\text{Re}\left\{[c]^+\begin{bmatrix} \Lambda_{11} & R_{1122} & R_{1133} & R_{1144} & R_{1155} & R_{1166} \\ R_{1122} & \Lambda_{22} & R_{2233} & R_{2244} & R_{2255} & R_{2266} \\ R_{1133} & R_{2233} & \Lambda_{33} & R_{3344} & R_{3355} & R_{3366} \\ R_{1144} & R_{2244} & R_{3344} & \Lambda_{44} & R_{4455} & R_{4466} \\ R_{1155} & R_{2255} & R_{3355} & R_{3355} & \Lambda_{55} & R_{5566} \\ R_{1166} & R_{2266} & R_{3366} & R_{3366} & R_{5566} & \Lambda_{66} \end{bmatrix}^{-1}[c]\right\}$$

(7.18)

where again $\Lambda_{\alpha\alpha} = i\omega + u^2/\tau_{\text{trans}} + R_{\alpha\alpha\alpha\alpha}$. The vector $[c]^+$ contains the elements of the S_0 representation associated with the coherences $\{|\psi_\alpha\rangle\langle\psi_\alpha|\}$, $[c]^+ = [1, a^2 - b^2, c^2 - d^2, e^2 - f^2, g^2 - h^2, 1]$. The reader can find the relaxation rates $R_{\alpha\alpha\beta\beta}$ in Ref. [17]). Now, the quantity $2S_{0,0}^{\text{trans}}(\omega_I)$ replaces the term $\{\lfloor 2 + (a^2 - b^2) + (c^2 - d^2) + (e^2 - f^2) + (g^2 - h^2)\rfloor J_{\text{trans}}(\omega_I)\}$ of Eq. 7.16 and converges to it when the electron spin relaxation is negligible. At higher frequencies (magnetic fields), there is no difference between the ^{14}N and ^{15}N cases as then the electron spin–nitrogen spin hyperfine coupling is negligible ($\omega \ll \omega_S$), and at low frequencies, the ^{1}H relaxation is somewhat faster for the case of $P = 1/2$ (^{15}N) than for $P = 1$ (^{14}N). This effect is confirmed by the expressions, $R_{1I}^{\text{trans}}(\omega_I)$ for the ^{15}N system converges in the low frequency range to $R_{1I}^{\text{trans}}(0) \cong \frac{1}{2}\left(\frac{\mu_0}{4\pi}\gamma_I\gamma_S\hbar\right)^2[5J_{\text{trans}}(0) + 5J_{\text{trans}}(A)]$ and for ^{14}N system to $R_{1I}^{\text{trans}} \cong \frac{1}{2}\left(\frac{\mu_0}{4\pi}\gamma_I\gamma_S\hbar\right)^2\left[\frac{110}{27}J_{\text{trans}}(0) + \frac{160}{27}J_{\text{trans}}\left(\frac{3A}{2}\right)\right]$, which indeed gives the experimentally obtained relationship [10].

Figures 7.5 and 7.6 show examples of ^{1}H spin–lattice relaxation dispersion for 4-oxo-TEMPO-d$_{16}$$^{14(15)}$N dissolved in glicerol and propanol, respectively. In the last case, the analysis in terms of Eq. 7.16 has led to the following values of relative diffusion coefficients of propanol and radical molecules, 3.5×10^{-10} m^2/s (291 K), 5.4×10^{-10} m^2/s (279 K), 5.0×10^{-10} m^2/s (273 K), and 8.2×10^{-10} m^2/s (253 K). Further, examples including the electron spin relaxation effects on ^{1}H spin–lattice relaxation dispersion are shown in Section 7.3.

Eventually, it is worth to stress again the validity range of the presented theory. As explained in Chapter 3, the perturbation relaxation theory is valid only when the Redfield condition [1, 15, 16] is fulfilled, which in this case means that $\omega_{\text{aniso}}\tau_{\text{rot}} \ll 1$. As A

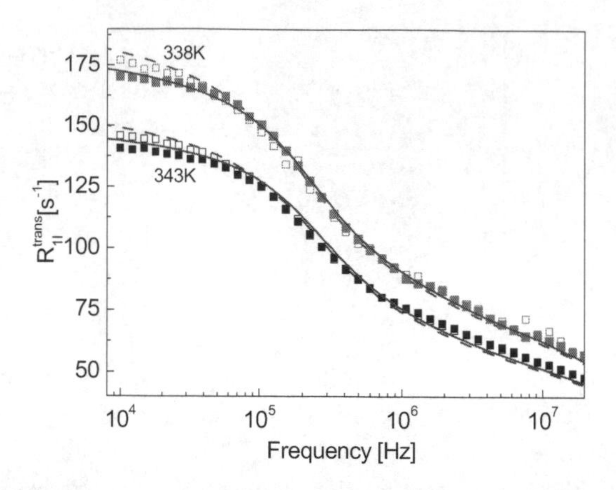

Figure 7.5 ^{1}H spin–lattice relaxation rate, $R_{1I}^{\text{trans}}(\omega_I)$, for glycerol solution of 4-oxo-TEMPO-d_{16}^{15}N (open symbols) and 4-oxo-TEMPO-d_{16}^{14}N (solid symbols); the concentration of the paramagnetic molecules was 10 mMol. Dashed and solid lines: Theoretical curves obtained by means of Eq. 7.11 (^{15}N)-dashed lines and Eq. 7.16 (^{14}N)-solid line; electron spin relaxation not included. Reprinted with permission from Ref. [10]. Copyright 2012, AIP Publishing LLC.

is of the order of 50 MHz (Chapter 6), the limit for the rotational correlation time, τ_{rot}, is about 5×10^{-9} s, for slower rotation this description breaks down.

7.2 Translational Dynamics in Paramagnetic Liquids

It has been shown in Chapter 2 that for diamagnetic liquids (for instance, glycerol), the spin–lattice relaxation rate, $R_1(\omega_I)$ plotted versus $\sqrt{\omega_I}$, that is, $R_1\left(\sqrt{\omega_I}\right)$, is linear in the low frequency range (in which $\omega_I \tau_{\text{trans}} \ll 1$) and from the slope of this dependence the translational diffusion coefficient can be determined. As the linearity stems from mathematical features of the translational spectral density (Eq. 2.53), one can expect a similar behavior of $R_{1I}^{\text{trans}}(\omega_I)$ for paramagnetic liquids. To examine this issue in Fig. 7.7 the relaxation data already shown in Fig. 7.6 are plotted versus $\sqrt{\omega_I}$; the very low frequency range, $\omega_S < A$, has been marked. In this figure one clearly sees two ranges of linearity. For glycerol solution of 4-oxo-TEMPO-d_{16}^{14}N, one can also see a linear dependence of $R_1\left(\sqrt{\omega_I}\right)$

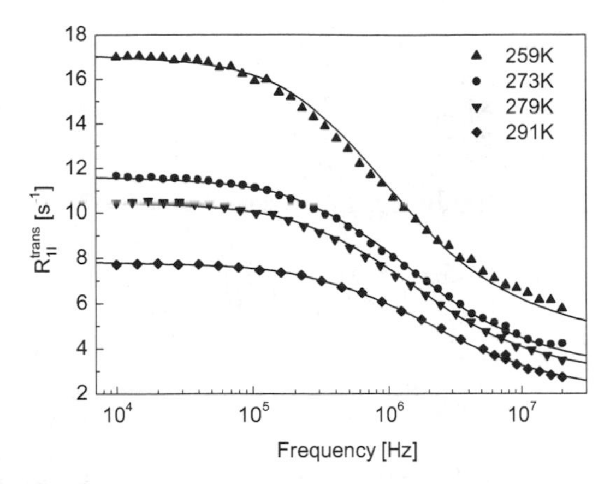

Figure 7.6 ^{1}H spin–lattice relaxation rate, $R_{1I}^{\mathrm{trans}}(\omega_I)$, for propanol solution of 4-4-oxo-TEMPO-d$_{16}$ ^{14}N; the concentration of the paramagnetic molecules was 10 mMol. Solid lines: Theoretical curves obtained by means of Eq. 7.16.

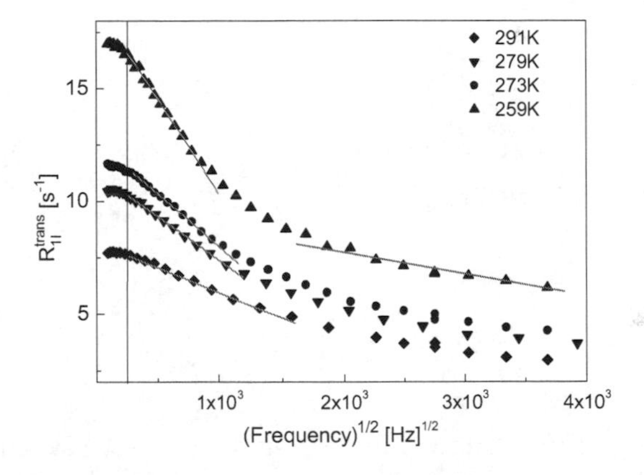

Figure 7.7 ^{1}H spin–lattice relaxation rate for propanol solution of 4-oxo-TEMPO-d$_{16}$ ^{14}N versus $\sqrt{\upsilon_I}\,\omega_I = 2\pi\,\nu_I$. Solid lines indicate ranges of linearity.

in the low frequency range ($\omega_S < A$) as demonstrated in Fig. 7.8. Thus, for solutions of nitroxide radicals one can distinguish the low, intermediate, and high frequency ranges of linearity labeled as I, II, and III, respectively [11, 18].

The low frequency range for the data presented in Fig. 7.8 has been enlarged in Fig. 7.9 to show that the slope of R_{1I}^{trans}

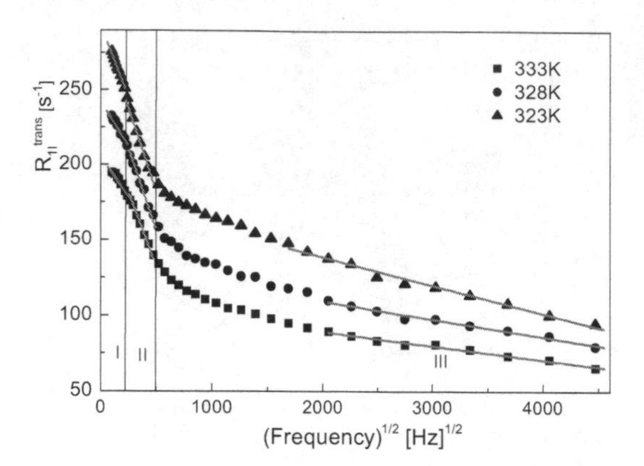

Figure 7.8 ^{1}H spin–lattice relaxation rate for glycerol solution of 4-oxo-TEMPO-d$_{16}$^{14}N versus $\sqrt{\upsilon_I}$. Solid lines: three ranges of linearity. Reprinted with permission from Ref. [18]. Copyright 2013, AIP Publishing LLC.

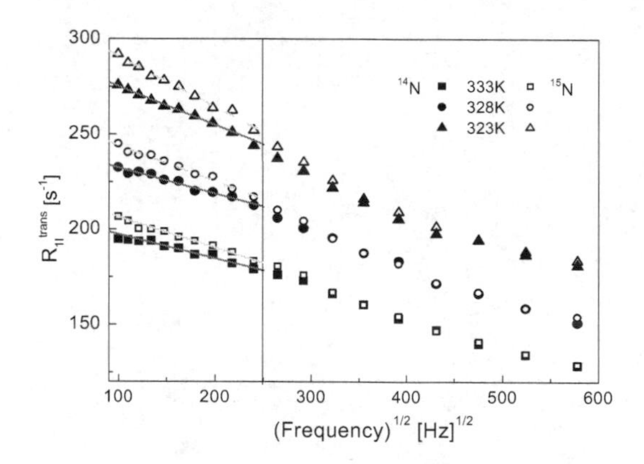

Figure 7.9 ^{1}H spin–lattice relaxation rate for glycerol solution of 4-oxo-TEMPO-d$_{16}$^{14}N (solid symbols) and 4-oxo-TEMPO-d$_{16}$^{15}N (open symbols) versus $\sqrt{\upsilon_I}$, in the low frequency range. Reprinted with permission from Ref. [18]. Copyright 2013, AIP Publishing LLC.

plotted versus $\sqrt{\upsilon_I}$ is (in this range) different for the ^{14}N and ^{15}N systems. However, as the slope is supposed to provide the value of the coefficient of relative translation diffusion of the solvent and solute molecules, applying Eq. 2.54 for this purpose, one would get different values from the ^{1}H relaxation data for ^{14}N and ^{15}N radicals, which does not make much sense. This issue should be clarified.

The range of the lowest frequencies is defined by the conditions $\omega_S < A$ and $\omega_S\tau_{trans} \ll 1$ (if the last condition does not hold in this range, the linearity is not observed). The hyperfine coupling for nitroxide radicals is of the order of 50 MHz and the condition $\nu_s < 50$ MHz implies that approximately $\nu_I < 50$ kHz, that is, $\sqrt{\nu_I} < 250\sqrt{\text{Hz}}$ (as indicated in Figs. 7.7–7.9). One can determine from this part of the relaxation dispersion translational correlation times $\tau_{trans} < 3 \times 10^{-9}$ s (the limit results from the condition $\omega_S\tau_{trans} \ll 1$). It has been shown that for the case of $P = 1/2$ (^{15}N), the coefficient B_1 in the expansion $R_1(\nu_I) \cong A_1 - B_1\sqrt{\nu_I}$ is given as [11, 18]:

$$B_1 = \frac{1}{15}\pi^{3/2}\left(1 + \frac{7}{3\sqrt{2}}\sqrt{\frac{\gamma_S}{\gamma I}}\right)\left(\frac{\mu_0}{4\pi}\gamma_I\gamma_S\hbar\right)^2 \frac{N_S}{D_{12}^{3/2}} \tag{7.19}$$

As one can expect, analogous derivations for the case of $P = 1$ (^{14}N) are more cumbersome. The starting point is Eq. 7.16. As we are in the low field range we shall use the limiting values (for $\omega_S \ll A$) of the coefficients, $a = b = f = g = 2\sqrt{3}$ and $c = e = -b = -h = \sqrt{1/3}$ (this is already an approximation, but one should be aware that this way of determining translational diffusion coefficients is approximated anyway). Thus, Eq. 7.16 takes at low frequencies the form [18]:

$$R_{1I}^{trans}(\omega_I) = \frac{1}{18}\left(\frac{\mu_0}{4\pi}\gamma_I\gamma_S\hbar\right)^2 \times \left\{\begin{matrix} 11J(\omega_I) + 8J(\omega_{23}) + 8J(\omega_{45}) + 7J(\omega_{12}) + 14J(\omega_{13}) \\ +7J(\omega_{46}) + 14J(\omega_{56}) \\ +\frac{28}{3}J(\omega_{24}) + \frac{14}{3}J(\omega_{25}) + \frac{14}{3}J(\omega_{34}) + \frac{7}{3}J(\omega_{35}) \end{matrix}\right\} \tag{7.20}$$

where it has been set for simplicity, $J_{trans}(\omega) = J(\omega)$; we shall use this notation further. In the next step, as $\omega_S < A$ one can neglect the terms of the order of $(\omega_S < A)^2$ in the transition frequencies, $\omega_{\alpha\beta}$, $\sqrt{\frac{9A^2}{4} \pm \omega_s A + \omega_S^2} - \sqrt{\left(\frac{3A}{2} + \frac{\omega_S}{3}\right)^2 \frac{8}{9}A^2} \simeq \frac{3A}{2} + \frac{\omega_S}{3}$. This leads to the further simplification [18]:

$$R_{1\mathrm{I}}^{\mathrm{trans}}(\omega_\mathrm{I}) = \frac{1}{18}\left(\frac{\mu_0}{4\pi}\gamma_\mathrm{I}\gamma_\mathrm{S}\hbar\right)^2 \left\{ \begin{array}{l} 11J(\omega_\mathrm{I}) + \frac{28}{3}J\left(\frac{3A}{2}\right) + 8J\left(\frac{3A}{2}+\frac{\omega_\mathrm{S}}{3}\right) + 8J\left(\frac{3A}{2}-\frac{\omega_\mathrm{S}}{3}\right) \\ +14J\left(\frac{3A}{2}+\frac{2\omega_\mathrm{S}}{3}\right) + 14J\left(\frac{3A}{2}-\frac{2\omega_\mathrm{S}}{3}\right) + \frac{77}{3}J\left(\frac{\omega_\mathrm{S}}{3}\right) \end{array} \right\} \tag{7.21}$$

As $\left[J\left(\frac{3A}{2}+\zeta\omega_\mathrm{S}\right)+J\left(\frac{3A}{2}-\zeta\omega_\mathrm{S}\right)\right] \cong 2J\left(\frac{3A}{2}\right)$, Eq. 7.24 can be reduced to the form:

$$R_{1\mathrm{I}}^{\mathrm{trans}}(\omega_\mathrm{I}) = \frac{1}{18}\left(\frac{\mu_0}{4\pi}\gamma_\mathrm{I}\gamma_\mathrm{S}\hbar\right)^2 \left[11J(\omega_\mathrm{I}) + \frac{77}{3}J\left(\frac{\omega_\mathrm{S}}{3}\right) + \frac{160}{3}J\left(\frac{3A}{2}\right)\right] \tag{7.22}$$

Applying the expansion $J_{\mathrm{trans}}(\omega) \cong a - \frac{2^{3/2}\pi}{45 D_{12}^{3/2}}\sqrt{\omega}$ to the frequency dependent terms of Eq. 7.22 one obtains, $R_1(\nu_\mathrm{I}) \cong \tilde{A}_1 - \tilde{B}_1\sqrt{\upsilon_\mathrm{I}}$, where [18]:

$$\tilde{B}_1 = \frac{22}{405}\pi^{3/2}\left(1+\frac{7}{3\sqrt{3}}\sqrt{\frac{\gamma_\mathrm{S}}{\gamma_\mathrm{I}}}\right)\left(\frac{\mu_0}{4\pi}\gamma_\mathrm{I}\gamma_\mathrm{S}\hbar\right)^2 \frac{N_\mathrm{S}}{D_{12}^{3/2}} \tag{7.23}$$

Comparing Eqs. 7.23 and 7.19 one obtains the relationship, $\tilde{B}_1/B_1 \approx 0.75$ [18].

At higher frequencies there is no difference between the relaxation slopes for ^{14}N and ^{15}N systems (one can see that the experimental data overlap). In the intermediate frequency range it is required that $\omega_\mathrm{S} < A$, but, at the same time the condition $\omega_S\tau_{\mathrm{trans}} \ll 1$ is fulfilled. This implies that when the linearity range ends at about $\nu_\mathrm{I} \cong 2$ MHz, that is, $\nu_S \cong 2$ MHz, the translational correlation time which can be determined from the slop has to fulfill the condition $\tau_{\mathrm{trans}} < 10^{-10}$ s. For $\omega_S < A$ one can set $a = d = e = h = 1$ and $b = c = f = g = 0$, which gives:

$$R_{1\mathrm{I}}^{\mathrm{trans}}(\omega_\mathrm{I}) = \frac{1}{2}\left(\frac{\mu_0}{4\pi}\gamma_\mathrm{I}\gamma_\mathrm{S}\hbar\right)^2 \left[3J(\omega_\mathrm{I}) + \frac{7}{3}J(\omega_{13}) + \frac{7}{3}J(\omega_{46}) + \frac{7}{3}J(\omega_{25})\right] \tag{7.24}$$

Now, one can neglect in the transition frequencies the terms of the order of $(A/\omega_S)^2$, this leads to $\omega_{13} = \omega_\mathrm{S} + A$, $\omega_{46} = \omega_\mathrm{S} - A$, and $\omega_{25} = \omega_\mathrm{S}$. As $[J(\omega_\mathrm{S}+A) + J(\omega_\mathrm{S}-A)] \cong 2J(\omega_\mathrm{S})$, one gets from Eq. 7.27 the well-known form of the high field relaxation formula (compare

Eq. 7.2):

$$R_{1I}^{\text{trans}}(\omega_I) = \frac{1}{2}\left(\frac{\mu_0}{4\pi}\gamma_I\gamma_S\hbar\right)^2 [3J_{\text{trans}}(\omega_I) + 7J_{\text{trans}}(\omega_S)] \tag{7.25}$$

in which we come back to the original notation for the translational spectral density. Equation 7. 25 leads to the relation $R_{1I}^{\text{trans}}(\upsilon_I) = \tilde{A}_2 - \tilde{B}_2\sqrt{\upsilon_I}$, with [11]:

$$\tilde{B}_2 = B_2 = \frac{2}{15}\pi^{3/2}\left(1 + \frac{7}{3}\sqrt{\frac{\gamma_S}{\gamma_I}}\right)\left(\frac{\mu_0}{4\pi}\gamma_I\gamma_S\hbar\right)^2 \frac{N_S}{D^{3/2}} \tag{7.26}$$

In the third, high frequency range, it is assumed that $\omega_S \gg A$, $\omega_S\tau_{\text{trans}} \gg 1$, and $\omega_I\tau_{\text{trans}} < 1$. This implies that one can neglect in Eq. 7.25 the term $7J_{\text{trans}}(\omega_S)$ as this contribution to the overall relaxation already disappeared, while the spectral density $J_{\text{trans}}(\omega_I)$ can be expanded into the power series ($\omega_I\tau_{\text{trans}} < 1$). This gives $R_{1I}^{\text{trans}}(\upsilon_I) = \tilde{A}_3 - \tilde{B}_3\sqrt{\upsilon_I}$, where [11]:

$$\tilde{B}_3 = B_3 = \frac{2}{15}\pi^{3/2}\left(\frac{\mu_0}{4\pi}\gamma_I\gamma_S\hbar\right)^2 \frac{N_S}{D^{3/2}} \tag{7.27}$$

This is an important expression, but does not take into account the electron spin relaxation. This subject should be discussed in more details.

7.3 Effects of Electron Spin Relaxation

The electron spin relaxation can be of importance for the nuclear spin relaxation when the electron spin relaxation rates $R_{\alpha\beta\alpha\beta}$ and $R_{\alpha\alpha\beta\beta}$, $R_{\alpha\alpha\alpha\alpha}$ become larger or at least comparable with the factor $u^2\tau_{\text{trans}}^{-1}$ (see Eqs. 7.13, 7.15, and 7.18). One could argue that this relation can always be fulfilled for small u (which ranges from zero to infinity). However, after closer inspection of the translational spectral densities including the electron relaxation rates (Eq. 7.12), one sees that the functions, $g = \left\{\frac{\tau_{\alpha\beta,u}}{1+\left(\omega_{\alpha\beta}\tau_{\alpha\beta,u}\right)^2}\right\}$ are weighted by the factors, $f(u) = \frac{u^2}{81+9u^2-2u^4+u^6}$, which for small u are also small. This means that although for a given u, the relationship $u^2\tau_{\text{trans}}^{-1} < R_e$ (R_e is the representative electron spin relaxation rate) holds, the contribution of the g term to the overall relaxation rate can be insignificant, as shown in Fig. 7.10.

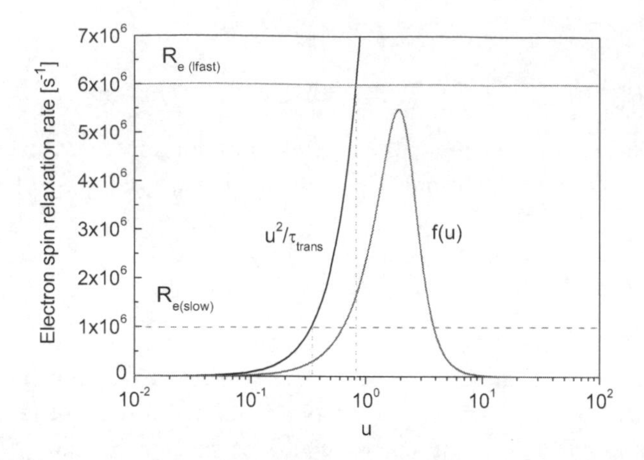

Figure 7.10 Comparison of the influence of the electron spin relaxation and the translational correlation time on the spectral density (discussion in the text). $f(u)$ reaches maximum for $u \cong 2$. Reprinted with permission from Ref. [17]. Copyright 2013, AIP Publishing LLC.

In this figure the quantity $u^2\tau_{\text{trans}}^{-1}$ is compared with two electron spin relaxation rates. For a given u_0 value one gets $u_0^2\tau_{\text{trans}}^{-1} = R_e$, then for $u < u_0$, the electron spin relaxation prevails. However, in the case of slower electron spin relaxation ($R_{\text{e(slow)}} \cong 10^6\ \text{s}^{-1}$) the factor $f(u)$ is close to zero for $u < u_0$ and when $f(u)$ becomes considerably large, the translational dynamics already prevails. The situation is different for $R_{\text{e(fast)}} \cong 5 \times 10^6\ \text{s}^{-1}$; then one can expect a visible effect of the electron spin relaxation on the ^{1}H spin–lattice relaxation.

This discussion brings us to the next question about how fast, in fact, the electron spin relaxation is. The electronic relaxation is multi-exponential and depends on the magnetic field (frequency), the rotational correlation time and the anisotropy of the hyperfine interaction. In Fig. 7.11, examples of calculated electron spin relaxation rates, $R_{\alpha\alpha\alpha\alpha}$ for the case of ^{15}N are plotted.

For the case of ^{15}N the relaxation rates, $R_{\alpha\alpha\alpha\alpha}$ converge in the high frequency range to $R_{\alpha\alpha\alpha\alpha} \to \frac{1}{8}J^A(0)$. The terms containing $J^A(\omega_S)$ vanish, and as we have neglected the nitrogen–Zeeman interaction the remaining terms which are, in fact, $J^A(\omega_P)$ (ω_P is the nitrogen resonance frequency) behave like $J^A(0)$. It is of interest to compare the electron spin relaxation scenario for ^{15}N radicals with those for

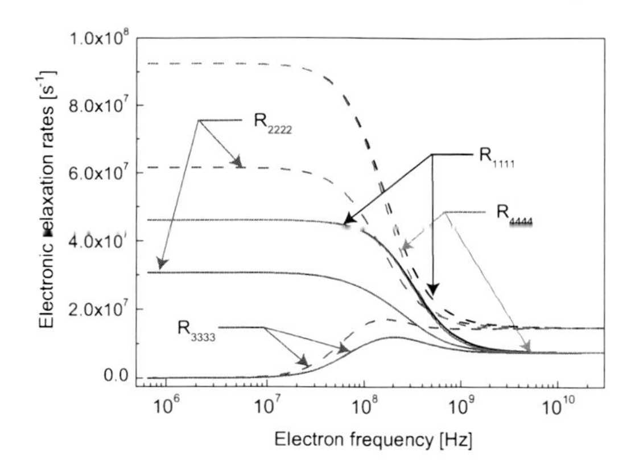

Figure 7.11 Electron spin relaxation rates for ^{15}N radicals versus electron frequency, ν_S; $A_{xx} = 22$ MHz, $A_{yy} = 23$ MHz, $A_{zz} = 142$ MHz; $\tau_{rot} = 5 \times 10^{-10}$ s (solid lines); $\tau_{rot} = 1 \times 10^{-9}$ s (dashed lines).

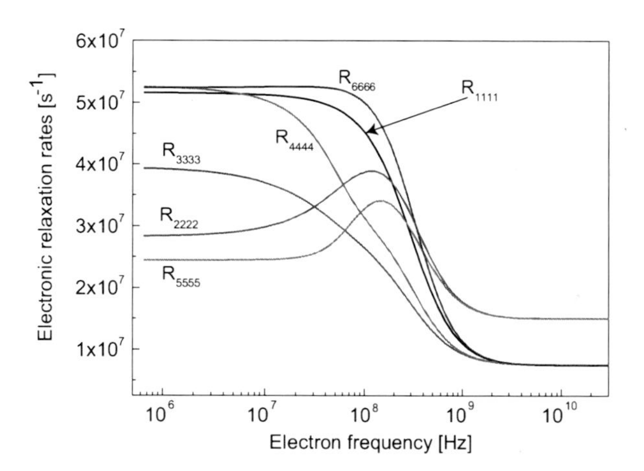

Figure 7.12 $R_{\alpha\alpha\alpha\alpha}$ electron spin relaxation rates for ^{14}N radicals versus electron frequency ν_S, when $A_{xx} = 15.4$ MHz, $A_{yy} = 16.1$ MHz, $A_{zz} = 99.5$ MHz, and $\tau_{rot} = 5 \times 10^{-10}$ s.

the case of ^{14}N. The electron spin relaxation rates for $P = 1$ (^{14}N) are shown in Fig. 7.12. The relaxation coefficients $R_{\alpha\alpha\alpha\alpha}$ for $\alpha = 1, 3, 4, 6$ converge to $\frac{9}{8}J^{\text{A}}(0)$, while R_{2222} and R_{5555} converge to $\frac{1}{2}J^{\text{A}}(0)$. This shows that the specific electron spin relaxation features for the ^{14}N ($P = 1$) case are different and more complex than for the

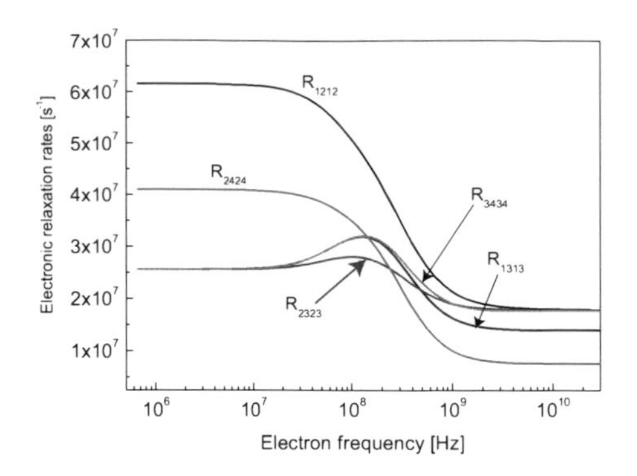

Figure 7.13 $R_{\alpha\beta\alpha\beta}$ electron spin relaxation rates for ^{15}N radicals versus electron frequency, ν_S, when $A_{xx} = 22$ MHz, $A_{yy} = 23$ MHz, $A_{zz} = 142$ MHz, and $\tau_{rot} = 5 \times 10^{-10}$ s.

^{15}N ($P = 1/2$) system. One should realize that the difference stems from the nitrogen spin quantum number. The different electron spin relaxation features depending on the nitrogen spin quantum number can also be seen for other relaxation rates associated with single- and double-quantum electron spin transitions (Figs. 7.13 and 7.14). One should also notice the different frequency dependencies of the individual electron spin relaxation rates.

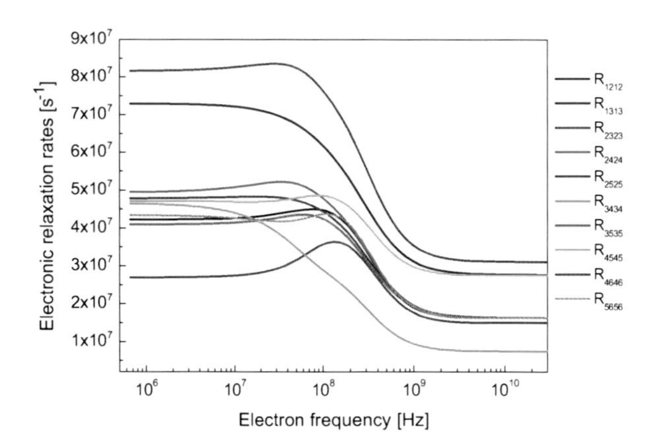

Figure 7.14 $R_{\alpha\beta\alpha\beta}$ electron spin relaxation rates for ^{14}N radicals versus electron frequency, ν_S, when $A_{xx} = 15.4$ MHz, $A_{yy} = 16.1$ MHz, $A_{zz} = 99.5$ MHz, and $\tau_{rot} = 5 \times 10^{-10}$ s.

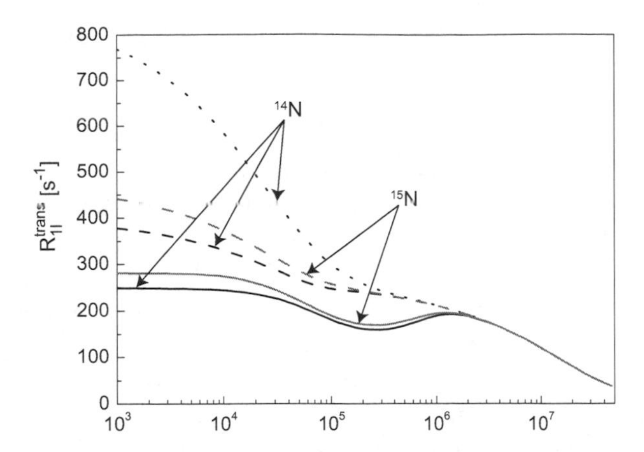

Figure 7.15 ^{1}H spin–lattice relaxation rate R_{1I}^{trans} versus ν_I for ^{14}N and ^{15}N systems. Dotted line: Predictions of Eq. 7.2; Dashed lines: Effects of isotropic hyperfine coupling; Solid lines: Isotropic hyperfine coupling and electron spin relaxation effects; $d = 4$ Å and $\tau_{\text{rot}} = 2 \times 10^{-8}$ s. The concentration of the paramagnetic molecules is assumed to be 10 mMol.

To understand how the electron spin relaxation affects the nuclear (^{1}H) spin–lattice relaxation, a comparison of the relaxation rates for systems with ^{15}N and ^{14}N radicals is shown in Fig. 7.15. The effects are gradually stepped: first the predictions of Eq. 7.2 (neglecting the hyperfine coupling) are presented, then the isotropic part of the electron spin–nitrogen spin hyperfine coupling is included, and eventually one can see the electron spin relaxation effects. According to the theoretical predictions shown in Fig. 7.14, one can expect that the ^{1}H spin–lattice relaxation dispersion profiles exhibit a local relaxation maximum at $\nu_I \cong (1\text{–}2)$ MHz. The effect is indeed observed experimentally, as one can see in Fig. 7.15 [17]. The relaxation maximum stems from the $J_{\text{trans}}(\omega_I)$ term (Eq. 7.11 and Eq. 7.16 for the ^{15}N and ^{14}N case, respectively) [17]. This is a result of competition between the spectral density $J_{\text{trans}}(\omega_I)$ which decreases with increase in frequency and the pre-factor $\left\lfloor 1 + (a^2 - b^2)^2 \right\rfloor$ (Eq. 7.11) for ^{15}N) which ranges between 1 (at low frequencies) and 2 (at high frequencies); that is it increases with increasing frequency. When the electron spin relaxation is included the effective correlation time describing the modulations of the electron spin–proton spin dipole–dipole coupling becomes shorter and the maximum shifts to higher frequencies. Analogous explanations can be given for the ^{14}N system [17].

7.4 Hilbert Space and Spin Relaxation

For the complex systems as discussed in this chapter, it might be convenient to choose a somewhat different way of calculating the ^{1}H spin–lattice relaxation rate. The method depends on whether we are interested in a complete description of the relaxation effects, including all transitions and coherences, or we just want to get only the expression for the spin–lattice relaxation rate of the spin I, neglecting the electron spin relaxation. In such a case one can apply the formula [1, 13, 14]:

$$R_1(\omega_I) = 2\int_0^\infty \tilde{C}_{+-}(t)\exp(i\omega_I t)\,dt + 2\int_0^\infty \tilde{C}_{-+}(t)\exp(-i\omega_I t)\,dt \quad (7.28)$$

where the dipolar correlation functions $\tilde{C}_{+-}(t)$ and $\tilde{C}_{-+}(t)$ are defined as [11]:

$$\tilde{C}_{+-}(t) = \overline{\left\langle \frac{1}{2}, m_S \middle| H_{DD}(I,S)(t) \middle| -\frac{1}{2}, m_S \right\rangle \left\langle -\frac{1}{2}, m_S \middle| H_{DD}(I,S)(0) \middle| \frac{1}{2}, m_S \right\rangle} \quad (7.29a)$$

$$\tilde{C}_{-+}(t) = \overline{\left\langle -\frac{1}{2}, m_S \middle| H_{DD}(I,S)(t) \middle| \frac{1}{2}, m'_S \right\rangle \left\langle \frac{1}{2}, m'_S \middle| H_{DD}(I,S)(0) \middle| -\frac{1}{2}, m_S \right\rangle} \quad (7.29b)$$

In Eq. 7.29 we consider the pairs of states, $|m_S, m_I\rangle$, with different m_I values ($1/2 \to -1/2$ for $\tilde{C}_{+-}(t)$ and $-1/2 \to 1/2$ for $\tilde{C}_{-+}(t)$) including all possible combinations of the m_S quantum numbers. This means that Eq. 7.29 includes pairs of states, $|m_S, m_I\rangle$, which are associated with different tensor components of the dipole–dipole Hamiltonian. For $\tilde{C}_{+-}(t)$, the considered pairs of states are: $\{|1/2\rangle, |-1/2\rangle\}$, $\{|1/2\rangle, |-1/2\rangle\}$, and $\{|1/2\rangle, |-1/2\rangle\}$. The first pair is connected by the $T_1^2(I,S) = \frac{1}{2}[I_Z S_+ + I_+ S_Z]$ tensor component, the second one by $T_2^2(I,S) = \frac{1}{2} I_+ S_+$, while the last one by $T_0^2(I,S) =$

$\frac{1}{\sqrt{6}}\left[2I_Z S_Z - \frac{1}{2}(I_+ S_- + I_- S_+)\right]$. Analogously, for $\tilde{C}_{-+}(t)$ the pairs of states are $\{|-1/2\rangle, |1/2\rangle\}$, $\{|-1/2\rangle, |1/2\rangle\}$, and $\{|-1/2\rangle, |1/2\rangle\}$; they are connected by $T^2_{-1}(I,S) = -\frac{1}{2}[I_Z S_- + I_- S_Z]$, $T^2_{-2}(I,S) = \frac{1}{2}I_- S_-$, and $T^2_0(I,S)$, respectively. The "bar" in Eq. 7.29 denotes assemble averaging according to the definition of a correlation function (Eqs. 2.30 and 2.31). Let us explicitly calculate the contributions to the correlation function $\tilde{C}_{+-}(t)$ from the individual tensor components:

$$\overline{\left\langle \frac{1}{2}, \frac{1}{2} \right| T^2_1(t) \left| -\frac{1}{2}, \frac{1}{2} \right\rangle \left\langle -\frac{1}{2}, \frac{1}{2} \right| T^2_{-1}(t) \left| \frac{1}{2}, \frac{1}{2} \right\rangle}$$
$$= \frac{1}{4}\overline{S_Z(t)\,S_Z(0)} = \frac{1}{4}\langle S_Z(t)\,S_Z(0)\rangle \tag{7.30a}$$

$$\overline{\left\langle \frac{1}{2}, \frac{1}{2} \right| T^2_2(t) \left| -\frac{1}{2}, -\frac{1}{2} \right\rangle \left\langle -\frac{1}{2}, -\frac{1}{2} \right| T^2_{-2}(t) \left| \frac{1}{2}, \frac{1}{2} \right\rangle}$$
$$= \frac{1}{4}\overline{S_+(t)\,S_-(0)} = \frac{1}{4}\langle S_+(t)\,S_-(0)\rangle \tag{7.30b}$$

$$\overline{\left\langle -\frac{1}{2}, \frac{1}{2} \right| T^2_0(t) \left| \frac{1}{2}, -\frac{1}{2} \right\rangle \left\langle \frac{1}{2}, -\frac{1}{2} \right| T^2_{-0}(t) \left| -\frac{1}{2}, \frac{1}{2} \right\rangle}$$
$$= \frac{1}{24}\overline{S_-(t)\,S_+(0)} = \frac{1}{24}\langle S_-(t)\,S_+(0)\rangle \tag{7.30c}$$

In the last equality the "bar" has been replaced by the "< >" averaging symbol. The task of calculating two-spin correlation functions (Eq. 7.29) has been reduced to the set of correlation functions for the electron spin only (Eq. 7.30). Summing up the individual contributions, one gets for the correlation functions:

$$\tilde{C}_{+-}(t) = \frac{a^2_{DD}}{4}\left\{ \langle S_+(t)\,S_-(0)\rangle + \langle S_Z(t)\,S_Z(0)\rangle + \frac{1}{6}\langle S_-(t)\,S_+(0)\rangle \right\} C(t) \tag{7.31a}$$

$$\tilde{C}_{-+}(t) = \frac{a^2_{DD}}{4}\left\{ \langle S_-(t)\,S_+(0)\rangle + \langle S_Z(t)\,S_Z(0)\rangle + \frac{1}{6}\langle S_+(t)\,S_-(0)\rangle \right\} C(t) \tag{7.31b}$$

where $\frac{a_{\mathrm{DD}}^2}{4} = \frac{3}{2}\left(\frac{\mu_0}{4\pi}\gamma_I\gamma_S\hbar\right)^2$. The correlation functions for the electron spin can be obtained from the formula:

$$\langle S_\alpha(t)\, S_\beta(0)\rangle = \frac{1}{6}\sum_{p,q=1}^{6} \exp(i\omega_p t)\,\langle p\,|S_\alpha|\,q\rangle \exp(-i\omega_q t)\,\langle q\,|S_\beta|\,p\rangle \tag{7.32}$$

Calculating these correlation functions, we need to use the obtained energy level structure resulted from the superposition of the electron Zeeman and hyperfine interactions (Eq. 6.4 and Eq. 6.27). The individual electron spin correlation functions calculated in this way have the form:

$$\langle S_Z(t)\, S_Z(0)\rangle = \frac{1}{12}\left[\begin{array}{l} 1 + \frac{\left(a^2-b^2\right)^2}{2} + \frac{\left(c^2-d^2\right)^2}{2} + \frac{\left(e^2-f^2\right)^2}{2} + \frac{\left(g^2-h^2\right)^2}{2} + \\ \frac{(ac-bd)^2}{2}\left(\exp(i\omega_{23}t) + \exp(-i\omega_{23}t)\right) + \\ \frac{(eg-fh)^2}{2}\left(\exp(i\omega_{45}t) + \exp(-i\omega_{54}t)\right) \end{array}\right] \tag{7.33}$$

$$\langle S_\pm(t)\, S_\pm(0)\rangle = \frac{1}{6}\left[\begin{array}{l} b^2\exp(\pm i\omega_{12}t) + d^2\exp(\pm i\omega_{13}t) \\ +e^2\exp(\pm i\omega_{46}t) + g^2\exp(\pm i\omega_{56}t) \\ +(af)^2\exp(\pm i\omega_{24}t) + (ah)^2\exp(\pm i\omega_{25}t) \\ +(cf)^2\exp(\pm i\omega_{34}t) + (ch)^2\exp(\pm i\omega_{35}t) \end{array}\right] \tag{7.34}$$

Substituting these expressions to the initial formula for $R_1(\omega_I)$ (Eq. 7.28) and taking into account that $J(\omega) = J(-\omega)$, one obtains the expression of Eq. 7.16.

References

1. Abragam, A. (1961). *The Principles of Nuclear Magnetism* (New York: Oxford University Press).
2. Bloembergen, N. (1957). Proton relaxation times in paramagnetic solutions. *J. Chem. Phys.*, **27**, pp. 572–573.
3. Bloembergen, N. and Morgan, L. O. (1961). Proton relaxation times in paramagnetic solutions: Effects of electron spin relaxation. *J. Chem. Phys.*, **34**, pp. 842–850.

4. Kowalewski, J., Kruk, D. and Parigi, G. (2005). NMR relaxation in solution of paramagnetic complexes: Recent theoretical progress for S $\geq$ 1. *Adv. Inorg. Chem.*, **57**, pp. 41–104.
5. Kruk, D. (2007). *Theory of Evolution and Relaxation of Multi-Spin Systems* (Bury St Edmunds: Arima).
6. Solomon, I. (1955). Relaxation processes in a system of two spins. *Phys. Rev.*, **99**, pp. 559–565.
7. Solomon, I. and Bloembergen, N. (1956). Nuclear magnetic interactions in the HF molecule. *J. Chem. Phys.*, **25**, pp. 261–266.
8. Kruk, D., Kubica, A., Masierak, W., Privalov, A. F., Wojciechowski, M. and Medycki, W. (2011). Quadrupole relaxation enhancement: Application to molecular crystals. *Solid State NMR*, **40**, pp. 114–120.
9. Nilsson, T. and Kowalewski, J. (2000). Slow-motion theory of nuclear spin relaxation in paramagnetic low-symmetry complexes: A generalization to high electron spin. *J. Magn. Reson.*, **146**, pp. 345–358.
10. Kruk, D., Korpala, A., Kowalewski, J., Rossler, E. A., and Moscicki, J. (2012). ^{1}Hrelaxation dispersion in solutions of nitroxide radicals: Effects of hyperfine interactions with ^{14}N and ^{15}N nuclei. *J. Chem. Phys.*, **137**, pp. 044512–044524.
11. Belorizky, E., Gilies, D. G., Gorecki, W., Lang, K., Noack, F., Roux, C., Struppe, J., Suteliffe, L. H., Travers, J. P. and Wu, X. (1998). Translational diffusion constants and intermolecular relaxation in paramagnetic solutions with hyperfine coupling on the electronic site. *J. Phys. Chem. A*, **102**, pp. 3674–3680.
12. Barci, L., Bertini, I. and Luchinat, C. (1991). *Nuclear and Electronic Relaxation* (Weinheim: Wiley-VCH).
13. Nilsson, T. and Kowalewski, J. (2000). Slow-motion theory of nuclear spin relaxation in paramagnetic low-symmetry complexes: A generalization to high electron spin. *J. Magn. Reson.*, **146**, pp. 345–358.
14. Mayne, C. L. and Smith, S. A. (1996). *Encyclopedia of Nuclear Magnetic Resonance, Relaxation processes in Coupled-Spin Systems* (Chichester: Wiley), pp. 4053–4071.
15. Redfield, A. G. (1996). *Encyclopedia of Nuclear Magnetic Resonance, Relaxation Theory: Density Matrix Formulation* (Chichester: Wiley & Sons Ltd.), pp. 4085–4092.
16. Slichter, C. P. (1990). *Principles of Magnetic Resonance* (Berlin: Springer-Verlag).

17. Kruk, D., Korpala, A., Kubica, A., Kowalewski, J., Rossler, E. A. and Moscicki, J. (2013). ^{1}H relaxation dispersion in solutions of nitroxide radicals: Influence of electron spin relaxation. *J. Chem. Phys.*, **138**, pp. 124506–124521.

18. Kruk, D., Korpaa, A., Kubica, A., Meier, R., Rossler, E. A. and Moscicki, J. (2013). Translational diffusion in paramagnetic liquids by ^{1}H NMR relaxometry: Nitroxide radicals in solution. *J. Chem. Phys.*, **138**, pp. 024506–024523.

Chapter 8

Spin Resonance Beyond Perturbation Range

In previous chapters we have been concerned with limiting cases, discussing very fast or very slow dynamics (rigid systems). The examples of nuclear magnetic relaxation dispersion (NMRD) and electron spin resonance (ESR) lineshape analysis presented until now were based on the perturbation theory. The spectral shape was related to the molecular dynamics by spin relaxation rates which are given as linear combinations of spectral densities. This implies that in all cases it was required that the condition $\omega_{\mathrm{rel}}\tau_{\mathrm{c}} \ll 1$ is fulfilled (ω_{rel} denotes the amplitude of the interaction causing the relaxation). In this chapter we shall give answer to the question how to describe spin resonance spectra when this condition is not fulfilled.

8.1 Intermediate Spin Resonance Spectra

The condition $\omega_{\mathrm{rel}}\tau_{\mathrm{c}} \ll 1$ considerably limits the applicability of the presented treatment as far as the time scale of the dynamical processes is concerned. Moreover, when the dynamics is fast

Understanding Spin Dynamics
Danuta Kruk

ISBN 978-981-4463-49-2 (Hardcover), 978-981-4463-50-8 (eBook)
www.panstanford.com

enough to fulfill the perturbation condition, the spectral shapes are rather uncharacteristic (they are described as a superposition of Lorentzian functions) and, therefore, they can be equally well reproduced in terms of different motional models. In other words, in this motional limit, the spectral analysis can hardly be treated as an unambiguous source of information on motional mechanisms; the information is mostly limited to the time scale of the dynamics. Thus, the examples give us the opportunity to understand the quantum-mechanical formalism, but they do not fulfill our expectations regarding the potential of NMR relaxometry and spectroscopy and ESR spectroscopy as a source of information about mechanisms of motion in condensed matter.

Nevertheless, ^{2}H NMR spectroscopy is one of the key methods providing information on the nature of dynamical processes. ^{2}H NMR spectral shapes (as well ^{2}H NMR spin relaxation) are dominated by strong quadrupolar interactions, which are one-spin interactions (in contrary to dipole–dipole couplings dominating ^{1}H NMR methods). This implies that by ^{2}H spectroscopy and relaxometry one directly probes dynamics of single molecules (molecular unit). Combining this great advantage of ^{2}H NMR studies with the isotope labeling technique (selective substitution of ^{1}H by ^{2}H) one gets a powerful tool for studying dynamical processes of complex molecules as detailed information can be obtained on a site-specific basis. The information is, however, encoded into ^{2}H spectral shapes reflecting dynamical processes on a time scale between the fast motion limit $\omega_{\mathrm{rel}}\tau_c \ll 1$ (Lorentzian lines) and rigid limit $\omega_{\mathrm{rel}}\tau_c \ll 1$ (Pake doublet), so called intermediate spectra. This is the range of correlation times in which the perturbation theory breaks down.

Examples of ^{2}H spectral shapes ranging between the fast motion and rigid limits are presented in Ref. [1]. Characteristic shapes of the intermediate spectra reflect not only the time scale of the motion, but also the dynamical mechanism, which will be presented soon after. To extract the dynamical information, one needs, however, theoretical models which link the lineshape function with parameters describing the molecular motion independently of its time scale. In other words, a tool which allows calculating ^{2}H

NMR spectra beginning with a Loretzian line and ending with Pake doublet is required.

8.2 Stochastic Liouville Formalism

Such a general theory of spin resonance spectra can be formulated in terms of the stochastic Liouville formalism [2–16]. As this sounds complicated and does not tell us much at this stage, we shall explain the issue by examples.

We know that the lineshape function, $L(\omega)$, is given by the expression (Eq. 4.59):

$$L(\omega) \propto [I_{\pm}]^{+} [M]^{-1} \left[I_{\pm}^{1}\right] \tag{8.1}$$

To set the matrix representations of the spin operators and interactions, a basis has to be specified. In the examples which we have discussed so far, the basis was provided by Zeeman spin functions or by their linear combinations. The basis has always been limited to the spin states as the motional degrees of freedom are included into the spectral densities determining the relaxation rates according to the perturbation theory of relaxation. However, when the requirements of the perturbation treatment are not fulfilled, spin relaxation rates cannot be explicitly defined. In consequence, the basis has to include both, the spin and the motional degrees of freedom. This means that for a nucleus of spin quantum number I placed in a molecule which undergoes rotational motion, the basis has to be formed as a combination (outer product) of rotational and spin functions [2, 3, 8–11, 17–19]:

$$|O_i) = |LKM) \otimes |\Sigma\sigma) \tag{8.2}$$

The rotational function $|LKM) = |LKM\rangle\langle LKM|$ are determined via Wigner matrices, $D^{L}_{K,M}(\Omega)$, associated with the molecular tumbling [2, 3, 8–11, 17–19]:

$$|LKM) = |LKM\rangle\langle LKM| = \sqrt{\frac{2L+1}{8\pi^2}} D^{L}_{KM}(\Omega) \tag{8.3}$$

where Ω denotes the molecular orientation, while the spin functions are defined in terms of the $|I, m_{\mathrm{I}}\rangle\left\langle I, m_{\mathrm{I}}'\right|$ quantities. To construct them it is convenient to use spin operators defined as [2, 3, 5, 7–11, 14, 15, 17–20]:

$$Q_\sigma^\Sigma \equiv |\Sigma\sigma) = \sum_m (-1)^{I-m-\sigma}\sqrt{2\Sigma+1} \times \begin{pmatrix} I & I & \Sigma \\ m+\sigma & -m & -\sigma \end{pmatrix} |I, m+\sigma\rangle\langle I, m| \tag{8.4}$$

where Σ ranges from 1 to $2I$, while $\begin{pmatrix} I & I & \Sigma \\ m+\sigma & -m & -\sigma \end{pmatrix}$ are $3j$ coefficients [21–23]. Although this definition looks rather complicated, the operators $Q_\delta^\Sigma \equiv |\Sigma\sigma)$ are related to the standard tensor components, T_σ^Σ, used in previous chapters: $T_0^1(I) = I_z$, $T_{+1}^1(I) = \pm I_+/\sqrt{2}$; $T_0^1(I) = \frac{1}{\sqrt{6}}\left[3I_z^2 - I(I+1)\right]$, $T_{\pm1}^2(I) = \pm\frac{1}{2}(I_z I_\pm + I_\pm I_z)$, $T_{\pm2}^2 = \frac{1}{2}I_\pm I_\pm$ only by a constant:

$$Q_\sigma^1 = |1,\sigma) = \sqrt{\frac{3}{(2I+1)(I+1)I}}T_\sigma^1(I) \tag{8.5}$$

$$Q_\sigma^1 = |2,\sigma) = \sqrt{\frac{30}{(2I+3)(2I+1)(I+1)(2I-1)I}}T_\sigma^1(I) \tag{8.6}$$

This renormalization is needed for the basic vectors to satisfy the condition $(O_i|O_j) = \delta_{ij}$ (they are orthonormal).

From the form of the basis vectors $|LKM\rangle\langle LKM|$ and $|I, m+\sigma\rangle\langle I, m|$, it is apparent that we are preparing ourselves for calculations in the Liouville space. In fact, we have used the concept of Liouville space in previous chapters, but we have not stressed this explicitly. The first step is to express the quadrupolar Hamiltonian, $H_Q^{(L)}(I)$ (Eq. 4.10) in the basis $\{|O_i)\}$. This is straightforward when one uses the relationship of Eq. 8.6:

$$\begin{aligned} H_Q^{(L)}(I) &= \frac{1}{2}\sqrt{\frac{3}{2}}\frac{a_Q}{I(2I-1)}\sum_{m=-2}^{2}(-1)^m\left(\sum_{k=-2}^{2}V_k^{2(P)}D_{k,-m}^2(\Omega)\right)T_m^2(I) \\ &= \sqrt{\frac{(2I+3)(2I+1)(I+1)(2I-1)}{30}}\frac{1}{2}\sqrt{\frac{3}{2}}\frac{a_Q}{I(2I-1)} \\ &\quad\times\sum_{m=-2}^{2}(-1)^m\left(\sum_{k=-2}^{2}V_k^{2(P)}D_{k,-m}^2(\Omega)\right)Q_m^2(I) \end{aligned} \tag{8.7}$$

One should stress here that we need the form of the quadrupole Hamiltonian in the laboratory frame; nothing changed in this matter compared to the perturbation treatment. In Eq. 8.7 we have explicitly included the transformation between the principal axis system of the electric field gradient tensor (P) and the laboratory frame (L) (Eq. 4.9, $\Omega_{PL} = \Omega$):

$$V_{-m}^{2(\mathrm{L})}(\Omega) = \sum_{k=-2}^{2} V_k^{2(\mathrm{P})} D_{k,-m}^2(\Omega) \text{ where } V_0^{2(\mathrm{P})} = 1,$$
$$V_{\pm 1}^{2(\mathrm{P})} = 0,\ V_{\pm 2}^{2(\mathrm{P})} = \eta/\sqrt{6}.$$

After adapting the Hamiltonian to our basis, the next step is to find its matrix representation. In the previous (perturbation) cases the basis was limited to spin functions and the matrix representation could be set up by just using the well-known rules of how the operators I_z and $I_\pm$ act on the Zeeman states, $|I, m_\mathrm{I}\rangle$. Now, the basis consists of spin and rotational elements. As in the Hamiltonian (Eq. 8.6), the spin and rotational degrees of freedom are tangled, to find the matrix representation of the quadrupole interaction Wigner–Eckart theorem is needed [21–23]. According to this theorem one can separate, in the evaluations of the matrix elements of the Hamiltonian, the spin and rotational variables [10, 11, 18]:

$$\begin{aligned}
&\left(L'K'M' \right| \Sigma'\sigma' \left| L_\mathrm{Q}(I) \right| \Sigma\sigma \left| LKM \right) \\
&= \sqrt{\frac{(2I+3)(2I+1)(I+1)(2I-1)}{30}} \frac{1}{2} \sqrt{\frac{3}{2}} \frac{a_\mathrm{Q}}{I(2I-1)} \\
&\times \sum_{m=-2}^{2} (-1)^m Tr\left\{ Q_{\sigma'}^{\Sigma'+} \left[Q_{-m}^2, Q_\sigma^\Sigma \right] \right\} \\
&\times \left\{ \sum_{k=-2}^{2} V_k^{2(\mathrm{P})} \int \psi_{K'M'}^{L'*}(\Omega) D_{k,m}^2(\Omega) \psi_{KM}^L(\Omega)\, d\Omega \right\} \qquad (8.8)
\end{aligned}$$

where $\psi_{KM}^L(\Omega) = |LKM\rangle\langle LKM| = \sqrt{\frac{2L+1}{8\pi^2}} D_{KM}^L(\Omega)$ (Eq. 8.3), while the Liouville operator $L_\mathrm{Q}(I)$ is associated with the quadrupole Hamiltonian, $L_\mathrm{Q}(I) = \left[H_\mathrm{Q}^{(\mathrm{L})}(I), \ldots \right]$ (Eq. 3.60) and Tr denotes trace operation. After the separation, the spin and rotational terms in

Eq. 8.8 can be explicitly calculated using the relations [10, 11, 18]:

$$Tr\left\{Q_{\alpha}^{A+}\left\lfloor Q_{\beta}^{B}, Q_{\gamma}^{C}\right\rfloor\right\} = \sqrt{(2A+1)(2B+1)(2C+1)}\,(-1)^{-\alpha} \times \begin{pmatrix} A & B & C \\ -\alpha & \beta & \gamma \end{pmatrix} \begin{Bmatrix} A & B & C \\ S & S & S \end{Bmatrix} \left[(-1)^{A+B+C} - 1\right] \tag{8.9}$$

$$\int \psi_{K'M'}^{L'*}(\Omega) D_{m,n}^{2} \psi_{KM}^{L}(\Omega)\, d\Omega = \sqrt{(2L'+1)(2L+1)}\,(-1)^{K'-M'} \times \begin{pmatrix} L' & 2 & L \\ -K' & m & K \end{pmatrix} \begin{Bmatrix} L' & 2 & L \\ -M' & n & M \end{Bmatrix} \tag{8.10}$$

This leads to the following expression for the matrix representation of the quadrupole interaction [18, 19]:

$$\left(L'K'M'\right|\Sigma'\sigma'\left|L_{Q}\right|\Sigma\sigma\left|LKM\right) = \sqrt{\frac{(2I+3)(2I+1)(I+1)(2I-1)}{30}}\,\frac{1}{2}\sqrt{\frac{2}{2}}\,\frac{a_{Q}}{I(2I-1)} \times \sqrt{5}\sqrt{(2L'+1)(2L+1)(2\Sigma+1)(2\Sigma'+1)} \begin{Bmatrix} \Sigma' & 2 & \Sigma \\ S & S & S \end{Bmatrix} \times \sum_{m,k=-2}^{2} (-1)^{m-\sigma'+K'-M'} \left[(-1)^{\Sigma+\Sigma'} - 1\right] V_{k}^{2(P)} \times \begin{pmatrix} \Sigma' & 2 & \Sigma \\ -\sigma' & -m & \sigma \end{pmatrix} \begin{pmatrix} L' & 2 & L \\ -K' & k & K \end{pmatrix} \begin{pmatrix} L' & 2 & L \\ -m' & k & M \end{pmatrix} \tag{8.11}$$

This expression can be further simplified employing some properties of the $3j$ coefficients. Eventually, it takes the form [19]:

$$\left(Q_{i}\left|L_{Q}\right|O_{j}\right) = \frac{1}{6}\frac{a_{Q}}{I(2I-1)}(-1)^{\sigma'} V_{|K-K'|}^{2(P)} \left[(-1)^{\Sigma'+\Sigma} - 1\right] \times \sqrt{(2L+3)(2L+1)(I+1)I(2I-1)(2L'+1)(2L+1)(2\Sigma'+1)(2\Sigma+1)} \times \begin{pmatrix} L' & 2 & L \\ -K' & K'-K & K \end{pmatrix} \begin{pmatrix} L' & 2 & L \\ -M' & M'-M & M \end{pmatrix} \begin{pmatrix} \Sigma' & 2 & \Sigma \\ -\sigma' & M'-M & \sigma \end{pmatrix} \begin{pmatrix} \Sigma' & 2 & \Sigma \\ I' & I & I \end{pmatrix} \tag{8.12}$$

where {} denotes $6j$ coefficients [21–23]. The calculations are quite involving, but worth undertaking as the final result can be rather straightforwardly adapted to other spin interactions. The representation of the Zeeman coupling in the basis $\{|Q_i)\}$ is very simple [2–4, 11–15, 17–19, 23]:

$$\left(L'K'M'\right|\Sigma'\sigma'|L_Z|LKM)|\sigma) = \delta_{LL'}\delta_{KK'}\delta_{MM'}\delta_{\Sigma\Sigma'}\delta_{\sigma\sigma'}\omega_I\sigma \quad (8.13)$$

The rotational dynamic is represented by the diffusion operator, $L_{\text{rot}} = -i D_{\text{rot}}\nabla_{\Omega}^2$, where D_{rot} is the rotational diffusion coefficient related to the rotational correlation time as $\tau_{\text{rot}} = 1/6D_{\text{rot}}$. In analogy to the Zeeman interaction, the rotational diffusion operator also contributes only to the diagonal elements of the matrix $[M]$ [2–4, 10–15, 17–19, 23]:

$$\left(L'K'M'\right|\left(\Sigma'\sigma'\,|L_{\text{rot}}|\,|LKM\right)|\sigma)$$

$$= \delta_{LL'}\delta_{KK'}\delta_{MM'}\delta_{\Sigma\Sigma'}\delta_{\sigma\sigma'}i D_{\text{rot}}L(L+1) \quad (8.14)$$

The matrix $[M]$ contains also on the diagonal the frequency ω (compare, for instance, Eq. 6.8 or Eq. 6.37). Due the rotational dynamics the dimension of the matrix is, in principle, infinite. As we intend to describe a classical stochastic motion (Brownian diffusion) we should achieve a continuity of the rotational levels (as expected for classical mechanics). However, it has been checked for a broad range of dynamical parameters that the results converge for $L = 12$ (for faster dynamics it is enough to set $L = 8$).

The representation of the operators $I_{\pm}$ in the basis $\{|Q_i)\}$ consists of only one element $|LKM)|\Sigma\sigma) = |000)|1 \pm 1)$. This implies that the lineshape function $L(\omega)$ is completely described by only one element of the inverted matrix $[M]^{-1}$ [18, 19]. This approach allows calculating quadrupolar spectra not only for arbitrary motional conditions (from very fast motion to the rigid limit), but also for an arbitrary spin quantum number. In Fig. 8.1 examples of ^{17}O spectra ($I = 5/2$) are shown. To remind, the quadrupolar coupling constant is defined by Eq. 4.5, where Q is a quadrupolar moment of the nucleus, while q is the zz component of the electric field gradient tensor.

The spectral shape for high spin quantum numbers is not Lorentzian even for fast dynamics. This effect is seen in Fig. 8.2

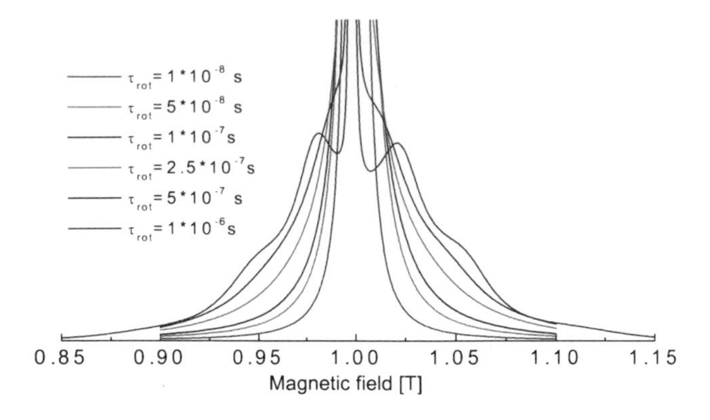

Figure 8.1 ^{17}O spectra at 1T, $a_Q = 3$ MHz, $\eta = 0$.

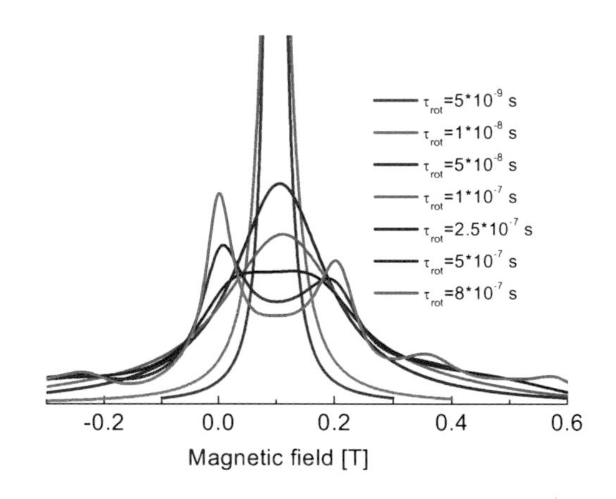

Figure 8.2 ^{139}La spectra at 0.1T, $a_Q = 15$ MHz, $\eta = 0$.

in which simulated quadrupole spectra for ^{139}La $I = 7/2$ are presented.

It is also of interest to inquire into how the anisotropy parameter influences the quadrupolar lineshape; some examples for the spin quantum number $I = 1$ at low magnetic fields are shown in Fig. 8.3a,b.

One can clearly see that anisotropy of the quadrupole interaction changes the lineshape very considerably. This is also observed for

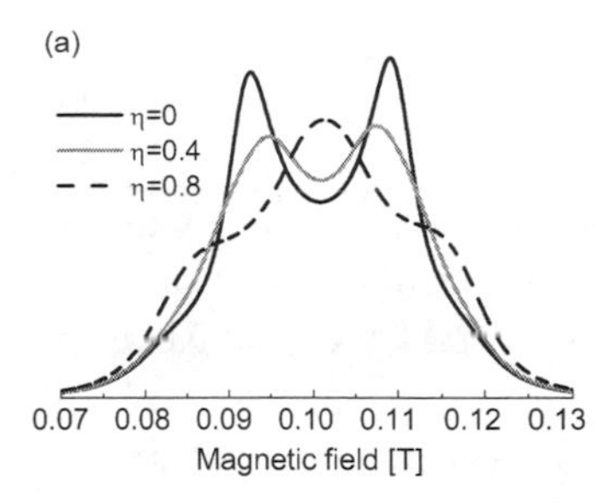

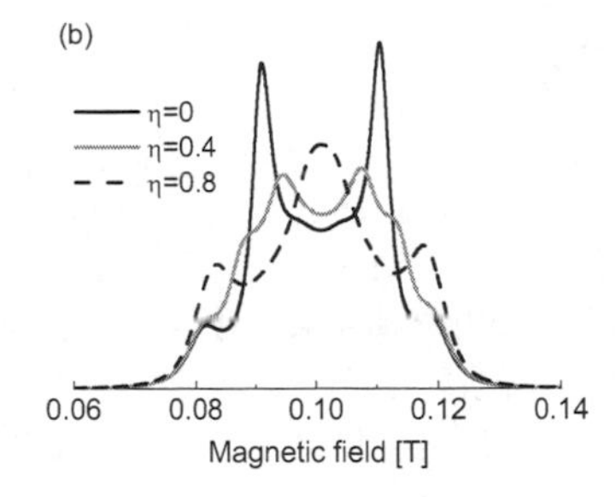

Figure 8.3 ^{2}H lineshapes for different asymmetry parameters of the quadrupole interaction at 0.1 T, $a_Q = 220$ kHz; (a) $\tau_{rot} = 5$ μs, (b) $\tau_{rot} = 10$ μs.

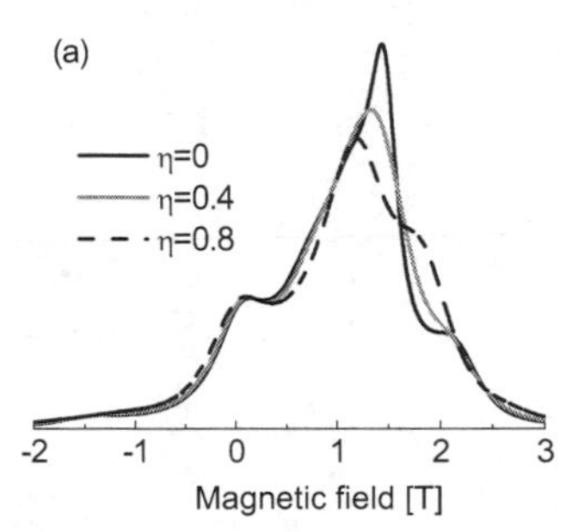

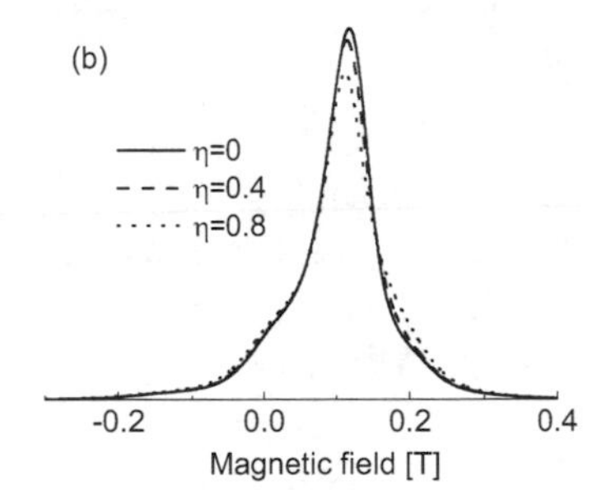

Figure 8.4 (a) ^{79}Br spectra at 1 T, $a_Q = 30$ MHz, $\tau_{rot} = 50$ ns; (b) ^{17}O spectra at 0.1 T, $a_Q = 3$ MHz, $\tau_{rot} = 50$ ns.

high integer spin quantum numbers as demonstrated in Fig. 8.4a,b for ^{79}Br ($I = 3/2$) and for ^{17}O ($I = 5/2$), respectively.

8.3 ^{2}H NMR Spectroscopy and Motional Heterogeneity

^{2}H NMR spectroscopy is widely used to probe dynamics of glass forming liquids. This is a perfect example when ^{2}H spectral shapes cover a broad range of correlation times, from fast dynamics (Lorentzian lines) to the rigid spectrum. It is believed that dynamics of such systems is heterogeneous. The heterogeneity is especially pronounced for binary systems (i.e., systems containing two kinds of molecular species). This feature is reflected by ^{2}H NMR spectral shapes which look like a superposition of a liquid-like

(Lorentzian) and solid-like (Pake doublet) lines. This is a result of a broad distribution of correlation times, covering several orders of magnitude. The distribution implies that to reproduce such spectra, a theoretical tool which is valid for an arbitrary time scale of the molecular motion is necessary. As such, a general approach is quite demanding (as we could see), in most cases the lineshapes are discussed only qualitatively. They are approximated by a sum of liquid-like and solid-like lines and their relative contributions versus temperature are discussed [24]. Such analysis does not fully exploit the potential of ^{2}H spectroscopy.

For a distribution of correlation times described by a probability density function, $g(\tau_c)$, the spectrum is given as an integral:

$$L(\omega) = \int L(\omega, \tau_c) g(\tau_c)\, d\tau_c \tag{8.15}$$

Gaussian and Cole–Davidson distributions of correlation times are most often assumed. The distributions are described by $g_G(\tau_c)$ and $g_{CD}(\tau_c)$ functions [25, 26], respectively. The Gaussian distribution yields:

$$g_G(\tau_c) \equiv g_G(\tau_c, \tau_0, \alpha) \propto \alpha \exp\left[-\left(\alpha \ln \frac{\tau_c}{\tau_0}\right)^2\right] \tag{8.16}$$

while the Cole–Davidson distribution is defined by Eq. 2.49, which we repeat below for convenience, replacing the rotational correlation time, τ_{rot}, by a "generalized" correlation time, τ_c:

$$g_{CD}(\tau_c) \equiv g_{CD}(\tau_c, \tau_0, \beta) \propto \sum_{k=0}^{\infty} (-1)^k \frac{\sin(\pi\beta k)\,(\beta k + 1)}{k!} \left(\frac{\tau_c}{\tau_0}\right)^{\beta k} \tag{8.17}$$

τ_0 denotes here a characteristic time constant; $g_G(\tau_c, \tau_0, \alpha)$ reaches maximum for $\tau_c = \tau_0$, while for $g_{CD}(\tau_c, \tau_0, \beta)$ the maximum is reached for $\tau_c = \beta\tau_0$; α and β are parameters determining the distribution width. In Fig. 8.5, illustrative ^{2}H NMR spectra for Gaussian distribution of correlation times are shown. The spectral shapes resulted from the weighted (according to the Gaussian distribution) superposition of the individual spectra (for a given τ_c) show features which cannot be obtained for a single correlation time. Depending on τ_0, the spectra are of more Lorentzian-like or more Pake-like forms.

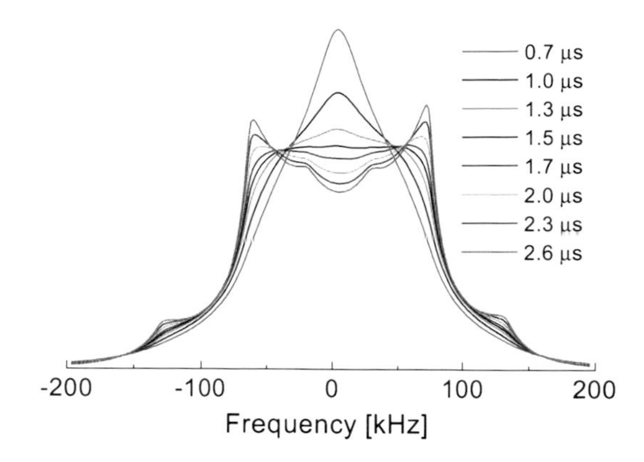

Figure 8.5 ^{2}H NMR spectra for Gaussian distribution of correlation times. Magnetic field =1 T, $a_Q = 220$ kHz, $\alpha = 0.83$, and τ_0 values are indicated in the figure.

This description of the spectral shape allows for a quantitative analysis of ^{2}H spectroscopy results for high viscosity liquids and binary systems. An example of such analysis is shown in Fig. 8.7. It has been demonstrated that from the ^{2}H spectral shapes one can determine the kind of distribution (for the system shown in Fig. 8.6, a Gaussian distribution leads to a better agreement with the experimental data than a Cole–Davidson distribution) and its characteristic parameters [27].

8.4 ^{2}H NMR Spectroscopy and Mechanisms of Motion

Discussing the spectral shapes we have always assumed that the motion can be described by an exponential correlation function, $C(t) = \exp(-t/\tau_c)$, where the correlation time, τ_c, has been, in most cases, attributed to the rotational correlation time, $\tau_c = \tau_{rot}$ (i.e., isotropic molecular tumbling has been assumed). This is a highly idealized picture of molecular dynamics. In this section we shall demonstrate how different motional models influence ^{2}H NMR spectral shapes.

Let us consider two examples, free diffusion and jump diffusion [3, 4, 16, 27, 29]. Free diffusion means that the molecule reorients freely with a correlation time τ_{rot} for a given time τ and then jumps to a new orientation (for instance due to collision with neighboring

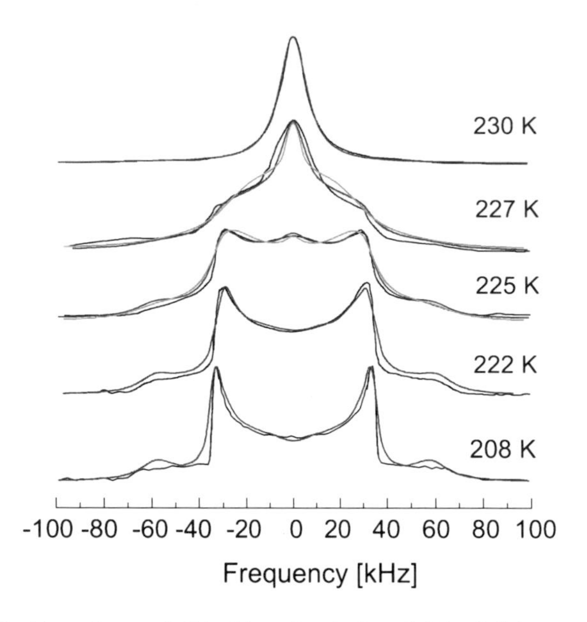

Figure 8.6 Experimental (black) and calculated (blue) ^{2}H spectra for of benzene-d_6 in tricresyl phosphate. Gaussian distribution (blue lines), Cole–Davidson distribution (green lines). The experimental data are taken from [28] and the figure is reprinted with permission from Ref. [27]. Copyright 2012, AIP Publishing LLC.

molecules) and then again reorients freely. By jump diffusion one understands a process when the molecule has a fixed average direction for a time τ around which it moves with correlation time τ_{rot} and then jumps to a new (random) direction. The correlation functions for free diffusion, $C_{\mathrm{FD}}(t)$, and jump diffusion, $C_{\mathrm{JD}}(t)$, respectively, yield the forms [3, 4, 16, 27, 29]:

$$C_{\mathrm{FD}}(t) = \exp\left\{ L(L+1)\frac{\tau}{6\tau_{\mathrm{rot}}}\left[1 - \left(\left(\frac{t}{\tau}\right)^2 + 1\right)^{1/2}\right]\right\} \quad (8.18)$$

$$C_{\mathrm{JD}}(t) = \exp\left\{ \frac{L(L+1)}{6\tau_{\mathrm{rot}}}\left[1 + L(L+1)\frac{\tau}{6\tau_{\mathrm{rot}}}\right]^{-1} t\right\} \quad (8.19)$$

For $\tau \ll \tau_{\mathrm{rot}}$ Eqs. 8.18 and 8.19 converge to $C(t) = \exp\left\{-\frac{L(L+1)}{6\tau_{\mathrm{rot}}}t\right\}$ (which is equal to $C(t) = \exp(-t/\tau_{\mathrm{rot}})$ for $L = 2$). In Fig. 8.7, the correlation function $C_{\mathrm{FD}}(t)$ and $C_{\mathrm{JD}}(t)$, are compared

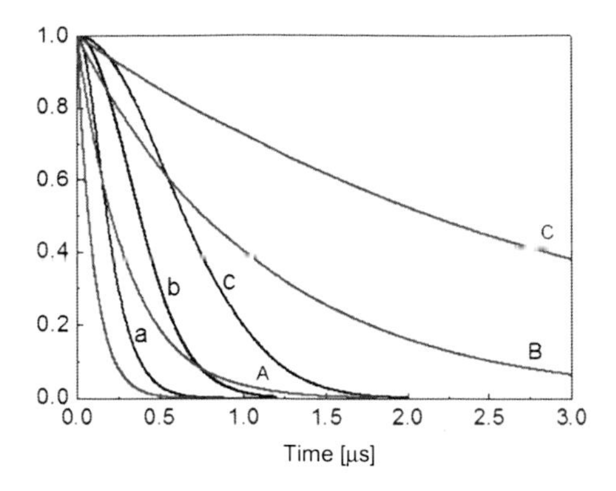

Figure 8.7 Comparison of $C_{\mathrm{FD}}(t)$, $C_{\mathrm{JD}}(t)$, and $C(t)$ for $L = 2$, $\tau = 10^{-7}$ s. $C(t)$ (the shortest one); $C_{\mathrm{FD}}(t)$ ($\tau = 2 \times 10^{-7}$ s (a), $\tau = 10^{-6}$ s (b), $\tau = 3 \times 10^{-6}$ s (c)); $C_{\mathrm{JD}}(t)$ ($\tau = 2 \times 10^{-7}$ s (A), $\tau = 10^{-6}$ s (B), $\tau = 3 \times 10^{-6}$ s (C)).

with the exponential correlation function, $C(t)$, for $L = 2$ as the quadrupolar coupling is a rank-two interaction.

To calculate the shape of ^{2}H spectra for free and jump diffusion one has to replace Eq. 8.14 by the expression [3, 4, 16, 19]:

$$\left(L'K'M' \middle| \Sigma'\sigma' \left| L_{\mathrm{rot}} \right| \Sigma\sigma \right) \left| LKM \right) = \delta_{LL'}\delta_{KK'}\delta_{MM'}\delta_{\Sigma\Sigma'}\delta_{\sigma\sigma'} \frac{iL(L+1)(6\tau_{\mathrm{rot}})^{-1}}{\left\lfloor 1 + L(L+1)(6\tau_{\mathrm{rot}})^{-1}\tau \right\rfloor^{p}} \tag{8.20}$$

with $p = 1/2$ for free diffusion and $p = 1$ for jump diffusion.

The different decays of the correlation functions are reflected (in the frequency range) by the ^{2}H spectral shapes. In Fig. 8.8, ^{2}H spectra for the free-diffusion and jump-diffusion models are compared with a spectrum calculated for isotropic molecular tumbling. Closer inspection of Fig. 8.8 leads to two observations. First, the spectral shape is more sensitive to the increasing value of the lifetime τ for jump diffusion than for free diffusion. This could be anticipated from Fig. 8.7, for the same τ value the correlation function, $C_{\mathrm{JD}}(t)$, decays considerably slower than $C_{\mathrm{FD}}(t)$. The second observation is that for jump diffusion the minimum becomes less pronounced with increase in τ; this is not observed for free diffusion.

At this stage the question arises to what extent the spectral features associated with the different motional models are unique,

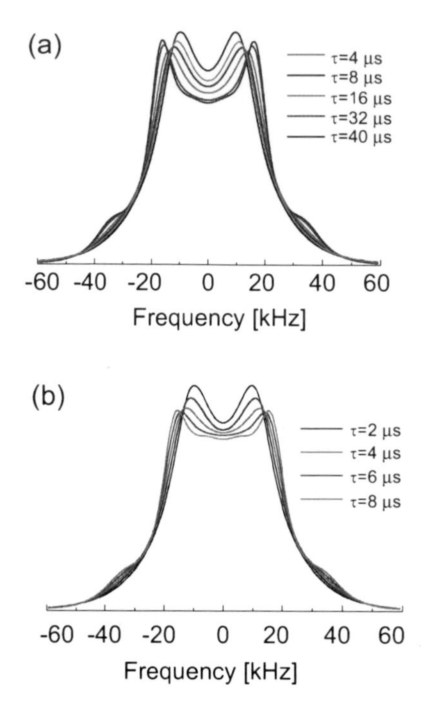

Figure 8.8 ^{2}H spectra (1 T, $a_Q = 220$ kHz) for (a) free diffusion and (b) jump diffusion compared with ^{2}H spectrum for isotropic rotation (black lines), $\tau_{rot} = 8$ ns.

that is, whether by varying parameters one can obtain the same ^{2}H spectral shapes assuming different models of motion.

This question can be, at least partially, answered by the example shown in Fig. 8.9. One can see that the illustrative spectra for the free-diffusion and jump-diffusion models (which are of quite peculiar shapes) cannot be reproduced assuming isotropic molecular tumbling.

Eventually, finishing this section it is of interest to show that the free- and jump-diffusion models well describe ^{2}H spectral shapes of highly viscous liquids. Figure 8.10 shows a ^{2}H spectrum for partially deuterated glycerol [19, 30]. The free-diffusion and jump-diffusion models give a good agreement with the experimental results, while the assumption of isotropic molecular tumbling falls (which is not surprising for such systems) [19].

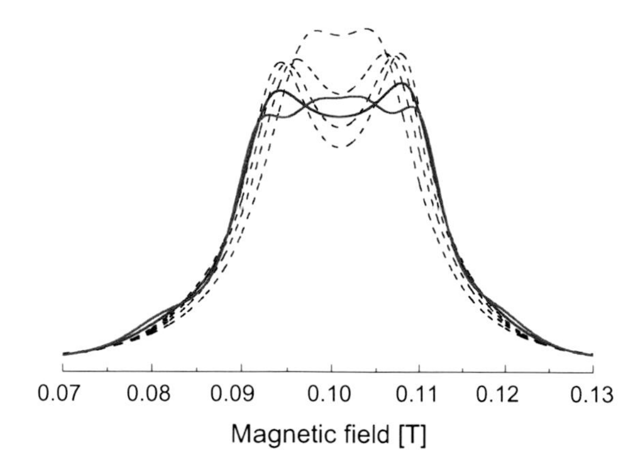

Figure 8.9 ^{2}H spectrum at 0.1 T, $a_Q = 220$ kHz, $\tau_c = 0.5$ μs, $\tau = 2$ μs. Free diffusion (red line), jump diffusion (blue line), isotropic molecular tumbling (dashed black lines); τ_{rot} is in the range of 1.5–4 μs. Reprinted with permission from Ref. [19]. Copyright 2011, AIP Publishing LLC.

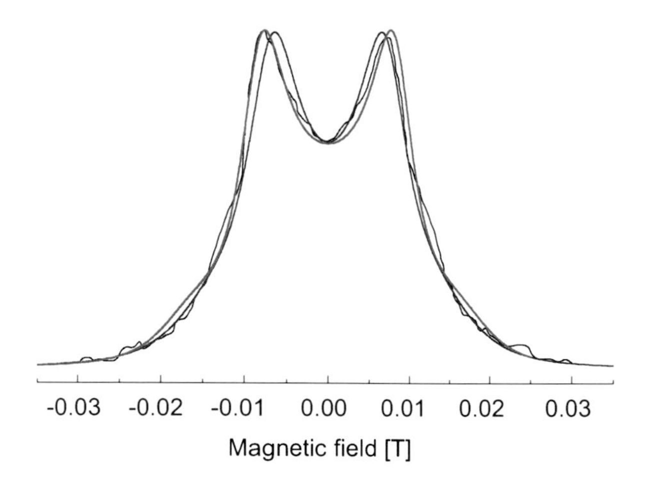

Figure 8.10 Experimental ^{2}H NMR spectrum for glycerol-h$_5$ (black line) [30]; calculations by means of isotropic diffusion (red line), jump diffusion, $\tau_{rot} = 4.6$ μs, $\tau = 0.5$ μs (green line). Reprinted with permission from Ref. [19]. Copyright 2011, AIP Publishing LLC.

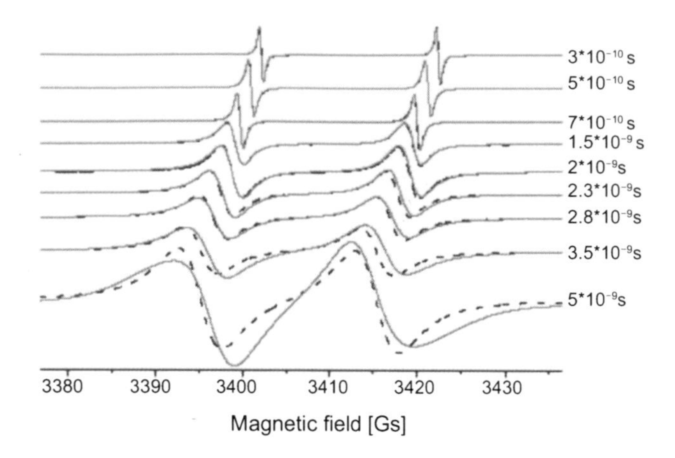

Figure 8.11 Simulated ESR spectra for a ^{15}N radical at 0.34T (3,400 Gs) for different rotational correlation times, τ_{rot} in terms of the perturbation approach described in Chapter 6 (dashed black lines) and a general theory based on the Stochastic Liouville equation (solid gray lines). (The simulations in terms of the general theory have been performed using the software package [5, 6, 15] downloaded from http://www.acert.cornell.edu/index_files/acert_resources.php; the same software has been used for calculating the spectrum for the longest correlation time in Fig. 6.4).

8.5 Deviations from Perturbation Approach

In Chapter 6 the perturbation approach to ESR lineshape analysis for nitroxide radicals, we have pointed out that the theory breaks down when the rotational correlation time of the paramagnetic molecule becomes about 2–3 ns (as then the condition $\omega_{aniso}\tau_{rot} < 1$ does not apply any more).

Thus, for longer correlation times the perturbation approach has to be replaced by a general theory based on the stochastic Liouville equation. The general framework of the theory remains unchanged; one has to set up a matrix form of all relevant spin interactions and operators representing the molecular motion (using Wigner–Eckart theorem) in a basis constructed from the spin and motional degrees of freedom of the system. For the specific details of the ESR theory for nitroxide radicals ($S = 1/2$) the reader is referred to [2–5, 8, 9, 14, 15], while the general theory of ESR lineshape for species of $S \geq 1$ has been described in detail in [10, 11, 18].

To illustrate the validity range of the perturbation approach, in Fig. 8.11 the predictions of the theory presented in Chapter 6 have been compared with ESR spectral shapes for ^{15}N radicals calculated by means of a general theory [2–5, 8, 9, 14, 15]. One sees that indeed, as anticipated, the perturbation approach breaks down when the time scale of the molecular tumbling exceeds few nanoseconds.

References

1. Kruk, D., Privalov, A., Medycki, W., Uniszkiewicz, C., Masierak, W. and Jakubas, R. (2012). NMR studies of solid-state dynamics. *Annu. Rep. NMR Spectrosc.*, **76**, pp. 67–138.
2. Freed, J. H., Bruno, G. V. and Polnaszek, C. F. (1971). Electron spin resonance line shapes and saturation in the slow motional region. *J. Phys. Chem.*, **75**, pp. 3385–3392.
3. Freed, J. H. (1976). *"Theory of Slow Tumbling ESR Spectra for Nitroxides," in Spin labeling: Theory and applications*, ed. Berliner, L. J. (New York: Academic Press).
4. Goldman, S. A., Bruno, G, V., Polnaszek, C. F. and Freed, J. H. (1972). An ESR study of anisotropic rotational reorientation and slow tumbling in liquid and frozen media. *J. Chem. Phys.*, **56**, pp. 716–736.
5. Liang, Z. and Freed, J. H. (1999). An assessment of the applicability of multifrequency ESR to study the complex dynamics of biomolecules. *J. Phys. Chem. B*, **103**, pp. 6384–6396.
6. Moro, G. and Freed, J. H. (1981). Calculation of ESR spectra and related Fokker-Planck forms by the use of the Lanczos algorithm. *J. Chem. Phys.* **74**, pp. 3757–3773.
7. Lynden-Bell, R. M. (1971). A density matrix formulation of the theory of magnetic resonance spectra in slowly reorienting systems. *Mol. Phys.*, **22**, pp. 837–851.
8. Moro, G. and Freed, J. H. (1981). Calculation of ESR spectra and related Fokker–Planck forms by the use of the Lanczos algorithm. *J. Chem. Phys.*, **74**, pp. 3757–3774.
9. Nevzorov, A. and Freed, J. H. (2000). Spin relaxation by dipolar coupling: From motional narrowing to the rigid limit. *J. Chem. Phys.*, **112**, pp.1413–1444.
10. Nilsson, T., Svoboda, J. Westlund, P.-O. and Kowalewski, J. (1998). Slow-motion theory of nuclear spin relaxation in paramagnetic complexes (S = 1) of arbitrary symmetry. *J. Chem. Phys.*, **109**, pp. 6364–6375.

11. Nilsson, T. and Kowalewski, J. (2000). Slow-motion theory of nuclear spin relaxation in paramagnetic low-symmetry complexes: A generalization to high electron spin. *J. Magn. Reson.*, **146**, pp. 345–358.
12. Sanktuary, B. C. (1976). Multipole operators for an arbitrary number of spins. *J. Chem. Phys.*, **64**, 4352–4361.
13. Sanktuary, B. C. (1983). Multipole NMR. 3. Multiplet spin theory. *Mol. Phys.*, **48**, pp. 1155–1176.
14. Schneider, D. J. and Freed, J. H. (1989). Spin relaxation and molecular dynamics. *Adv. Chem. Phys.*, **73**, pp. 387–528.
15. Schneider, D. J. and Freed, J. H. (1989). Calculating slow motional magnetic resonance spectra. *Biol. Magn. Reson.*, **8**, pp. 1–76.
16. Vasavada, K. V. and Kaplan, J. I. (1985). NMR lineshape under the conditions where the exchange time approaches correlation time. *J. Magn. Reson.*, **64**, pp. 32–37.
17. Westlund, P.-O. (1995). *Dynamics of Solutions and Fluid Mixtures by NMR*, ed. J. J. Delpuech (Chichester: Wiley).
18. Kruk, D. (2007). *Theory of Evolution and Relaxation of Multi-Spin Systems* (Bury St Edmunds: Arima).
19. Kruk, D., Earle, K. A., Mielczarek, A., Kubica, A., Milewska, A. and Moscicki, J. (2011). Nuclear quadrupole resonance lineshape analysis for different motional models: Stochastic Liouville approach. *J. Chem. Phys.*, **135**, pp. 224511–224520.
20. Bowden, G. J. and Hutchison, W. D. (1986). Tensor operator formalism for multiple-quantum NMR. 1. Spin-1 nuclei. *J. Magn. Reson.*, **67**, pp. 403–414.
21. Brink, D. M. and Satchler, G. R. (1979) *Angular Momentum* (Oxford: Clarendon Press).
22. Edmunds, A. R. (1974). *Angular Momentum in Quantum Mechanics* (Princeton: Princeton University Press).
23. Varshalovich, D. A., Moskalev, A. N. and Khersonkii, V. K. (1988). *Quantum Theory of Angular Momentum* (Singapore: World Scientific).
24. Bohmer, R., Diezemann, G., Hinze, G. and Rossler, E. (2001). Dynamics of supercooled liquids and glassy solids. *Prog. Nucl. Magn. Reson. Spectrosc.*, **39**, pp. 191–267.
25. Böttcher, C. J. F. and Bordewijk, P. (1973). *Theory of Electric Polarization*, Vol. 2. (Amsterdam: Elsevier).
26. Lindsey, C. P. and Patterson, G. D. (1980). Detailed comparison of the Williams–Watts and Cole–Davidson functions. *J. Chem. Phys.*, **73**, pp. 3348–3358.

27. Kruk, D., Mielczarek, A., Korpala, A., Kozlowski, A., Earle, K. A, Moscicki, J. (2012). Sensitivity of H-2 NMR spectroscopy to motional models: Proteins and highly viscous liquids as examples. *J. Chem. Phys.*, **136**, pp. 244509–244516.
28. Blochowicz, Th., Karle, C., Kudlik, A., Medick, P., Roggatz, I., Tschirwitz, Ch., Wolber, J., Senker, J. and Rossler, E. (1999). Molecular dynamics in binary organic glass formers. *Phys. Chem. B*, **103**, pp. 4032–4044.
29. Egelstaff, P. A. (1970). Cooperative rotation of spherical molecules. *J. Chem. Phys.*, **53**, pp. 2590–2599.
30. Vogel, M. and Rossler, E. (2000). On the nature of slow β-process in simple glass formers:? A ^{2}H NMR study. *J. Phys. Chem. B*, **104**, pp. 4285–4287.

Chapter 9

Dipolar Relaxation and Quadrupolar Interactions

In Chapter 8, a theory of quadrupolar lineshapes valid for an arbitrary time-scale of molecular motion has been presented. The theory uses the stochastic Liouville formalism. It has been stressed that when the condition $\omega_Q \tau_c < 1$ (ω_Q is the amplitude of the quadrupole coupling, while τ_c denotes the correlation time characterizing stochastic fluctuations of the quadrupole interaction) is not fulfilled, one cannot explicitly define quadrupole relaxation rates, although one can fully describe NMR spectra, linking their shapes to the correlation time, τ_c. In this chapter we shall inquire into how relaxation processes in systems consisting of spins $I = 1/2$ and $P \geq 1$ coupled by dipole–dipole interactions are influenced by quadrupolar interactions of the P spin.

9.1 Quadrupole Relaxation Enhancement

Consider a nucleus of spin quantum number $P \geq 1$ possessing a quadrupole moment and interacting by a dipole–dipole coupling

---

Understanding Spin Dynamics
Danuta Kruk

ISBN 978-981-4463-49-2 (Hardcover), 978-981-4463-50-8 (eBook)
www.panstanford.com

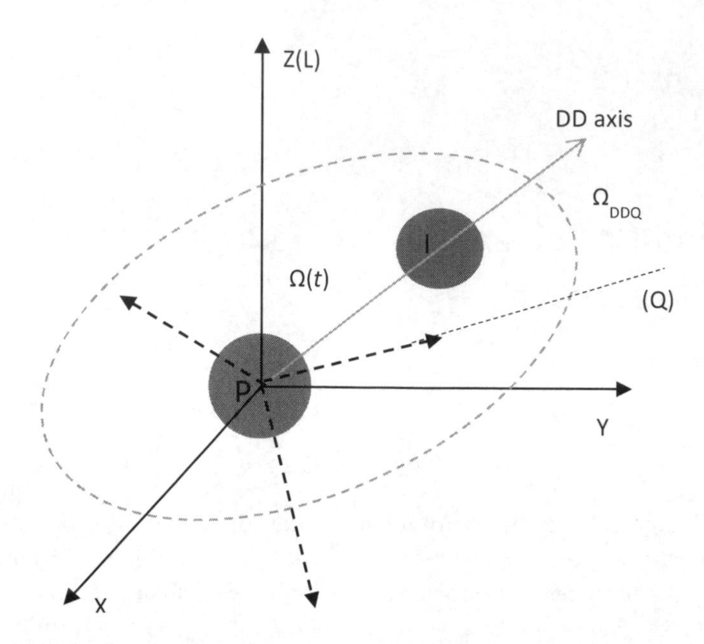

Figure 9.1 A pair of nuclear spins $I = 1/2$ and $P \geq 1$ interacting by a dipole–dipole (DD) coupling. The orientation of the DD axis with respect to the laboratory frame (L) is described by the angle $\Omega(t)$ which fluctuates in time as a result of the overall molecular tumbling. The principal axis system of the electric field gradient at the position of the P spin nucleus (Q frame) is oriented with respect to the DD axis by angle Ω_{DDQ} which depends on the molecular geometry, but does not fluctuate in time (as a special case one can set $\Omega_{\mathrm{DDQ}} = 0$). Thus, the relative orientation of the (Q) and (L) frames are also modulated by the molecular rotation.

with a nucleus of spin quantum number $I = 1/2$. When both nuclei belong to the same molecule (for instance ^{1}H and ^{14}N) the quadrupolar as well as dipolar interactions are modulated by the molecular tumbling, as schematically illustrated in Fig. 9.1. As the I–P dipole–dipole coupling is usually considerably weaker than the quadrupolar interaction of the spin P, it is likely that the condition $\omega_{\mathrm{DD}}\tau_{\mathrm{rot}} < 1$ is fulfilled, while at the same time the relationship $\omega_{\mathrm{Q}}\tau_{\mathrm{rot}} < 1$ does not hold.

One can expect that the I spin relaxation rate, $R_{1\mathrm{I}}$ is given as a linear combination of spectral densities, in analogy to the Solomon–Bloembergen–Morgan (SBM) formula for a nuclear spin–electron spin system, discussed in Section 5.5 (Eq. 5.58) [1, 2, 4, 17]:

$$R_{1I}(\omega_I) = \frac{2}{15}\left(\frac{\mu_0}{4\pi}\frac{\gamma_I\gamma_P\hbar}{r^3}\right)^2 P(P+1)\left[\frac{\tau_{c2}}{1+(\omega_I-\omega_P)^2\tau_{c2}^2} + \frac{3\tau_{c1}}{1+\omega_I^2\tau_{c1}^2} + \frac{6\tau_{c2}}{1+(\omega_I+\omega_P)^2\tau_{c2}^2}\right] \quad (9.1)$$

with $\tau_{ci}^{-1} = \tau_{rot}^{-1} + R_{iP}$, where R_{1P} and R_{2P} denote the quadrupole spin–lattice and spin–spin relaxation rates of the P spin, respectively. Equation 9.1 requires, however, that the quadrupole spin relaxation rates are explicitly defined, and this is possible only when $\omega_Q\tau_{rot} < 1$. This limitation can be avoided by using the generalized expression for $R_{1I}(\omega_I)$, given by Eq. 5.2, which we remind below for convenience:

$$R_{1I}(\omega_I) = \frac{2}{15}\left(\frac{\mu_0}{4\pi}\frac{\gamma_I\gamma_P\hbar}{r^3}\right)^2 P(P+1)\left[S_{1,1}(\omega_I) + S_{0,0}(\omega_I) + S_{-1,-1}(\omega_I)\right] \quad (9.2)$$

Nevertheless, so far (for instance, Chapter 7) the generalized spectral densities, $S_{0,0}(\omega)$, have been expressed in terms of the relaxation rates $R_{\alpha\alpha'\beta\beta'}$ of the spin P. We shall see that this is not necessary.

The relaxation rate $R_{1I}(\omega_I)$ of Eq. 9.2 can be expressed as a real part of a "composite lattice" spectral density $K_{1,1}^{DD}(\omega_I)$ [4, 5, 8, 13, 19, 20]:

$$R_{1I}(\omega_I) = 2\mathrm{Re}\left\{K_{1,1}^{DD}(\omega_I)\right\} \quad (9.3)$$

defined as:

$$K_{1,1}^{DD}(\omega_I) = \frac{1}{2P+1}\int_0^\infty Tr_{\mathrm{Lattice}}\left\{T_1^{1(DD)+} \times \left[\exp(-iL\tau)\,T_1^{1(DD)}\right]\right\}\exp(-i\omega_I\tau)\,d\tau \quad (9.4)$$

The concept of a "composite lattice" and the meaning of the symbols used in Eq. 9.4 require explanations. We shall not present here the full mathematical treatment leading to Eq. 9.3 and Eq. 9.4 as this has been done in Ref. [8, 13]; but we shall explain in detail their physical meaning.

Let us begin with the dipolar tensor operator $T_1^{1(\mathrm{DD})}$. The dipole–dipole Hamiltonian, $H_{\mathrm{DD}}^{(L)}(I, P) = a_{\mathrm{DD}} \sum_{m=-2}^{2} (-1)^m F_{-m}^{2(L)} T_m^2 (I, P)$ (Eq. 2.12) can be rewritten in the form ((L) refers here to the laboratory frame):

$$H_{\mathrm{DD}}^{(L)}(I, P) = \sum_{m=-1}^{1} (-1)^m T_m^1 (I)\, T_{-m}^{1(\mathrm{DD})} (P) \tag{9.5}$$

where $T_0^1(I) = I_z$, $T_{\pm 1}^1(I) = \mp I_\pm/\sqrt{2}$. The idea of the representation of Eq. 9.5 is to separate the I and P spin operators. The tensor operators $T_m^{1(\mathrm{DD})}$ contain, besides the P spin operators, $T_0^1(P)$, $T_{\pm 1}^1(P)$, the rotational variable $\Omega(t)$ included into the functions, $F_{-m}^{2(L)}(t) = D_{0,-m}^2(\Omega(t))$. The operator $T_1^{1(\mathrm{DD})}(P)$ is defined as [4, 5, 8, 13, 15, 16, 19, 20]:

$$T_1^{1(\mathrm{DD})} = a_{\mathrm{DD}} \sqrt{\frac{5(2P+1)(P+1)P}{3}} \times \sum_{q=-1}^{1} \begin{pmatrix} 2 & 1 & 1 \\ 1-q & q & -1 \end{pmatrix} |1, \sigma)\, D_{0,1-q}^2 (\Omega) \tag{9.6}$$

where the $3j$ symbols $\begin{pmatrix} 1 & 2 & 1 \\ -1 & 2 & -1 \end{pmatrix}$, $\begin{pmatrix} 1 & 2 & 1 \\ 0 & 1 & -1 \end{pmatrix}$ and $\begin{pmatrix} 1 & 2 & 1 \\ 1 & 0 & -1 \end{pmatrix}$ for $q = 2,1,0$ respectively, are in the ratio, 6:3:1 (they are equal to 1/5, 1/10, and 1/30, respectively) [3, 4, 19]. This explains the origin of the pre-factors in Eq. 9.2, which, in fact, stems from Eq. 9.3 [9, 11]. The quantity $|1, \sigma)$ has already been defined in Chapter 8 (Eq. 8.5), now it reads:

$$|1, \sigma) = Q_\sigma^1 = \sqrt{\frac{3}{(2P+1)(P+1)P}}\, T_\sigma^1 (P) \tag{9.7}$$

The Liouville operator, L has also been introduced in Chapter 8. It is referred to as lattice Liouville operator as it contains all degrees of freedom (including spin and dynamical variables) which are relevant for the I spin relaxation. The collection of all degrees of freedom is referred to as "composite lattice". In the present case the operator is defined as a sum of the quadrupolar interaction, Zeeman coupling of the P spin and rotational operators, $L = L_{\mathrm{Q}} + L_{\mathrm{Z}}(P) + L_{\mathrm{rot}}$, all defined in Chapter 8. The symbol Tr_{Lattice} denotes a trace (averaging) over the lattice degrees of freedom.

In analogy to ESR lineshape analysis, Eq. 9.4 can be rewritten in a matrix form [9, 11, 12, 20, 21]:

$$K_{1,1}^{(DD)}(\omega_I) = \frac{1}{2P+1}\left[T_1^{1(DD)}\right]^+ [M]^{-1}\left[T_1^{1(DD)}\right] \tag{9.8}$$

where $\left[T_1^{1(DD)}\right]$ and $[M]$ are matrix representations of the $T_1^{1(DD)}$ and M operators in the basis $\{|O_i) = |LKM) \otimes |\sum \sigma)\}$ introduced in Chapter 8 (Eq. 8.1) as an outer product of the rotational and spin (P) states. The operator M is now defined as, $M = -iL\text{-}1\omega$. Thus, one can see that the only difference between the matrix $[M]$ used for the analysis of the quadrupolar lineshape of the spin P and for calculating the spin–lattice relaxation rate R_{1I} resulting from I-P dipole–dipole interactions lies in the frequency contributing to the diagonal elements of the matrix $[M]$ (ω for the lineshape analysis and ω_I for R_{1I}). However, while the quadrupolar lineshape is given as only one element of the inverted matrix $[M]^{-1}$, in the case of R_{1I} the situation is more complex.

The representation of the dipolar tensor $\left[T_1^{1(DD)}\right]$ in the basis $|O_i)$ contains three non-zero elements corresponding to the states $|202)\,|1-1)$, $|201)\,|10)$, and $|200)\,|11)$. In consequence, the spin–lattice relaxation rate, R_{1I}, is given by the 3×3 block of the inverted matrix $[M]^{-1}$ corresponding to the non-zero elements. [9, 11, 12, 20, 21]. In Figs. 9.2a,b illustrative proton relaxation spin–lattice dispersion curves, $R_{1I}(\omega_I)$, for a ^{1}H–^{2}H spin system are shown. For fast molecular tumbling (i.e., when $\omega_Q\tau_{rot} \ll 1$ is fulfilled) the shape of the proton relaxation dispersion is typical; the relaxation rates just decrease with frequency. In fact, $R_{1I}(\omega_I)$, does not significantly differ from the predictions of Eq. 9.1, in which the quadrupolar relaxation is neglected. The quadrupolar relaxation (represented in Eq. 9.1 by the relaxation rates R_{1P} and R_{2P}) is caused by the rotational dynamics. The condition $\omega_Q\tau_{rot} < 1$ implies that $R_{iP} \ll \tau_{rot}^{-1}$ (the relaxation process is much slower than the motion causing it). In consequence, the contribution of the quadrupolar relaxation to the modulations of the $I - P$ dipole–dipole coupling is negligible, that is, $\tau_{ci}^{-1} = \tau_{rot}^{-1} + R_{iP} \cong \tau_{rot}^{-1}$. For slower dynamics quite peculiar features of the ^{1}H relaxation dispersion emerge. One observes local relaxation maxima which positions (in terms of the ^{1}H frequency, ω_I, depend on the quadrupolar parameters (the

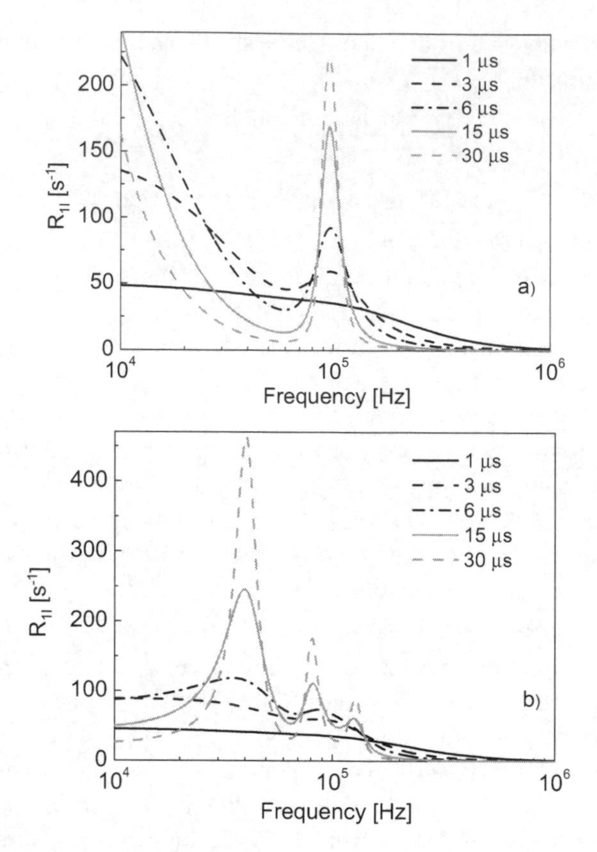

Figure 9.2 ^{1}H spin–lattice relaxation rate, R_{11} versus frequency for a ^{1}H–^{2}H system. $a_{\mathrm{Q}} = 150$ kHz, $r = 300$ pm; (a) $\eta = 0$, (b) $\eta = 0.9$.

quadrupole coupling constant, a_{Q}, and the asymmetry parameter, η). To understand this effect let us imagine that the nuclei of $P \geq 1$ are fixed at their positions in a crystal lattice (let us take solids as an example). There are numerous solids containing mobile $I = 1/2$ nuclei (^{1}H, ^{19}F) and high spin ($P \geq 1$) nuclei fixed at their lattice sites (for instance $\mathrm{LaF_3}$ [7, 8, 13, 15]). In such a case the energy level structure of the P spin is determined by a Hamiltonian H_0 which is a superposition of the Zeeman and quadrupole interaction [7, 8, 10, 12, 13, 15]:

$$H_0\left(P\right)\left(\Omega_{\mathrm{QL}}\right) = H_{\mathrm{Z}}\left(P\right) + H_{\mathrm{Q}}^{(L)}\left(P\right)\left(\Omega_{\mathrm{QL}}\right) \tag{9.9}$$

As the quadrupolar interaction is expressed in the laboratory frame, the Hamiltonian is orientation dependent. To make the consideration simpler let us limit ourselves at this stage to the case when the principal axis system of the electric field gradient coincides with the laboratory axis.

The relaxation maxima correspond to the ν_{I} (resonance frequency of the spin I) values at which some of the energy levels of the P and I spins cross [7–12, 15]. For $P = 1$ and $\eta \neq 0$ there are three quadrupolar transitions $\frac{3}{4}a_{\mathrm{Q}}(1 + \eta/3)$, $\frac{3}{4}a_{\mathrm{Q}}(1 - \eta/3)$, and $a_{\mathrm{Q}}\eta/2$; for $\eta = 0$ they converge to only one quadrupolar transition frequency, $3a_{\mathrm{Q}}/4$. These frequencies are modified by the Zeeman interaction, $H_Z(P)$. This mainly concerns the transition frequency, $a_{\mathrm{Q}}\eta/2$ as it is often comparable with the Zeeman splitting of the P spin, ν_P. When the nucleus carrying the P spin undergoes rotational dynamics one can hardly talk about its well-defined energy level structure. This is actually the reason why the perturbation theory of relaxation does not apply; the quadrupolar coupling is neither the origin of the quadrupolar relaxation (as the condition $\omega_{\mathrm{Q}}\tau_{\mathrm{rot}} \ll 1$ does not hold) nor contributes to the energy level structure (as the opposite condition $\omega_{\mathrm{Q}}\tau_{\mathrm{rot}} \gg 1$ is not fulfilled, neither). Therefore, to provide a relatively simple explanation of the observed effects we have referred to a system in which the P spin nuclei are immobile (i.e., the quadrupolar interaction does not fluctuate in time). Staying for a while with this concept one can say that when the energy levels of the I and P spins cross (their transition frequencies are equal) the spins can easily "exchange magnetization". The energy which is released due to the transition of the I spin between its energy levels can directly be used by the P spin for its transition. One can say that the two processes enhance each other leading, from the viewpoint of the spin I, to a faster decay of its magnetization, that is, to a locally faster relaxation; this phenomenon is referred to as quadrupole relaxation enhancement (QRE) [11, 12]. This scenario should not be directly transferred to every case when one sees such relaxation maxima (we shall comment on this soon); however, it helps to intuitively understand the underlying quantum-mechanical effects.

Coming back to the shape of the relaxation dispersion, $R_{1\mathrm{I}}(\omega_{\mathrm{I}})$, one can see in Fig. 9.3 further examples, this time for a ^{1}H–^{14}N spin

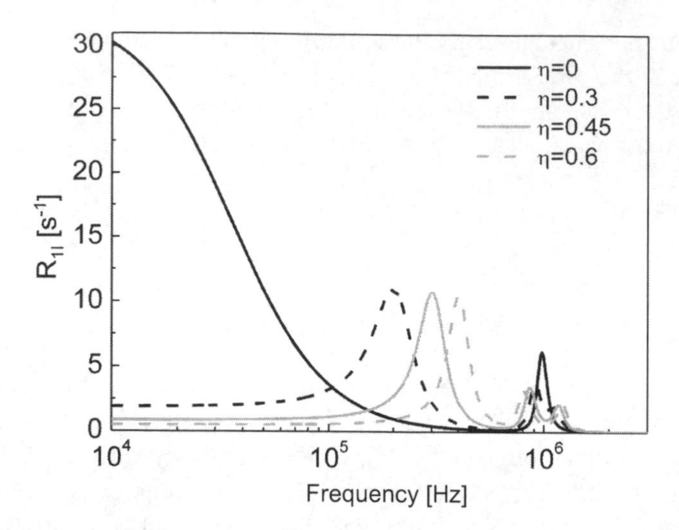

Figure 9.3 ^{1}H spin–lattice relaxation rate R_{1I} versus frequency for a ^{1}H–^{14}N system; $a_Q = 1.5$ MHz, $r = 300$ pm, $\tau_{rot} = 3$ μs.

system. One should note that the asymmetry of the quadrupolar interaction leads to a significant reduction of the relaxation rate $R_{1I}(\omega_I)$ in the low frequency range [11]. It is of interest to inspect more closely the QRE pattern for higher spin quantum numbers, P. For this purpose in Fig. 9.4 ^{1}H spin–lattice relaxation dispersion, $R_{1I}(\omega_I)$, for ^{1}H–^{35}Cl and ^{1}H–^{79}Br spin systems ($P = 3/2$) are compared for an axially symmetric electric field gradient tensor.

The energy levels of the P spin for $\eta = 0$ yield $\omega_{3/2} = a_Q/4 + 3\nu_P/2$, $\omega_{1/2} = a_Q/4 + \omega_P/2$, and $\omega_{-3/2} = a_Q/4 - 3\omega_P/2$. This implies the following transition frequencies $\omega_{3/2\to1/2} = a_Q/2+\omega_P$, $\omega_{1/2\to-1/2} = \omega_P$, and $\omega_{3/2\to-1/2} = a_Q/2+2\omega_P$. Two of the P spin transitions, $\omega_{3/2\to1/2}$ and $\omega_{3/2\to-1/2}$, can be matched by the I spin (^{1}H in this case) transition frequency, ω_I. In consequence one observes two relaxation maxima (Fig. 9.4) separated by the frequency ω_P. The transition $3/2\to1/2$ is a single-quantum transition as $|\Delta m_P| = 1$, while the transition $3/2\to -1/2$ implies $|\Delta m_P| = 2$, that is, this is a double-quantum spin transition. One should, however, remember that this is a simplified picture assuming that the quadrupolar interaction is time-independent. We consider a situation when the quadrupolar coupling fluctuates in time but on a time-scale which

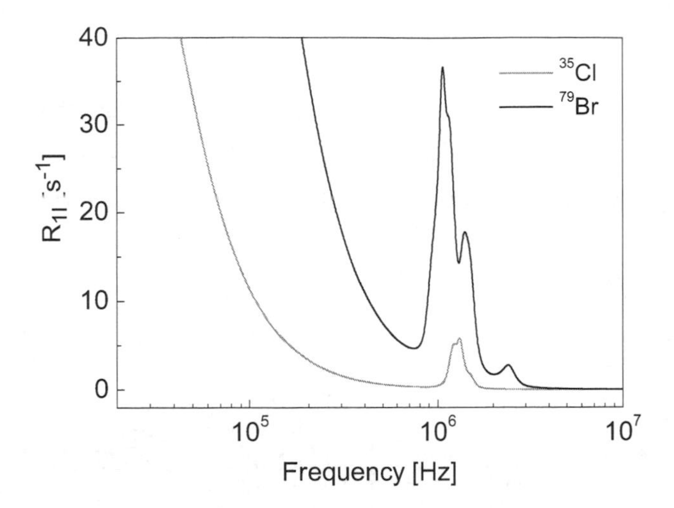

Figure 9.4 ^{1}H spin–lattice relaxation rate R_{1I} versus frequency for ^{1}H–^{35}Cl and ^{1}H–^{79}Br system; $a_Q = 3.0$ MHz, $\eta = 0$, $r = 300$ pm, $\tau_{rot} = 5$ μs.

makes the perturbation approach invalid. This leads to a somewhat more complex QRE pattern as shown in Fig. 9.4. The shape of the QRE pattern is affected by anisotropy of the quadrupolar interaction (examples for $\eta \neq 0$ are shown in Fig. 9.5).

In complex molecular systems there are several relaxation pathways. Typical examples for which the presented above description is highly relevant are compounds containing ^{1}H and ^{14}N nuclei. For such systems the measured ^{1}H spin–lattice relaxation rate, $R_1(\omega_I)$, is a sum of two contributions, $R_1^{H-H}(\omega_I)$ and $R_1^{H-N}(\omega_I) = R_{1I}(\omega_I)$, originating from ^{1}H–^{1}H and ^{1}H–^{14}N dipole-dipole interactions, respectively:

$$R_1(\omega_I) = R_1^{H-H}(\omega_I) + R_{1I}(\omega_I) \tag{9.10}$$

The first contribution, $R_1^{H-H}(\omega_I)$ [18], is given by Eq. 2.37, which now explicitly yields:

$$R_1^{H-H}(\omega_I) = \frac{3}{2}\left(\frac{\mu_0}{4\pi}\frac{\gamma_I^2 \hbar}{r_{HH}^3}\right)^2 [J(\omega_I) + 4J(2\omega_I)] \tag{9.11}$$

where r_{HH} is the inter-proton distance. The second contribution, $R_{1I}(\omega_I)$, can be evaluated as described above.

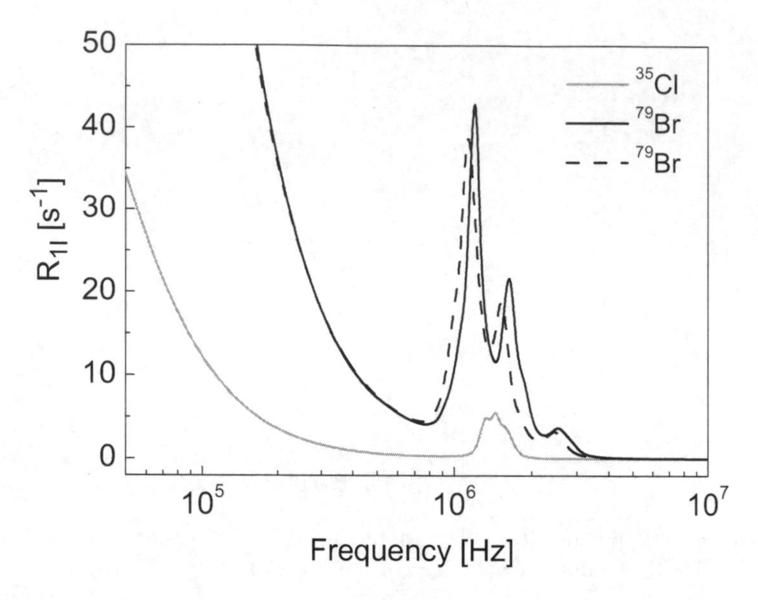

Figure 9.5 ^{1}H spin–lattice relaxation rate R_{1I} versus frequency for a ^{1}H–^{35}Cl (gray line) and ^{1}H–^{79}Br (black line) system; $a_Q = 3.0$ MHz, $r = 300$ pm, $\tau_{rot} = 5$ μs. $\eta = 0.75$ (solid line), $\eta = 0.45$ (dashed line).

Examples of the overall relaxation rates, $R_1(\omega_I)$, for a hypothetic ^{1}H–^{2}H spin system, for which the ^{1}H–^{1}H and ^{1}H–^{2}H inter-spin distances are equal ($r_{HH} = r_{HD} = r$, $D = {}^2$H) are shown in Fig. 9.6. As the proton gyromagnetic factor is much larger than for deuteron, the second contribution is negligible except of the frequencies at which the QRE effect is observed.

Let us now examine this phenomenon in a somewhat different way looking at the magnetization decay which can be then described as:

$$\frac{M_{ZI}(t)}{M_{ZI}(0)} = \exp(-R_{1I}t) \tag{9.12}$$

In Fig. 9.7 the fraction of the magnetization $M_{ZI}(t)/M_{ZI}(0)$ which remains after a time t is plotted versus ω_I for different t values. When the time t is much shorter than the relaxation time, $T_1 = R_1^{-1}$, the magnetization remains almost unchanged ($M_{ZI}(t)/M_{ZI}(0) \cong 1$) in the whole frequency range. Then, for longer times, t, one can

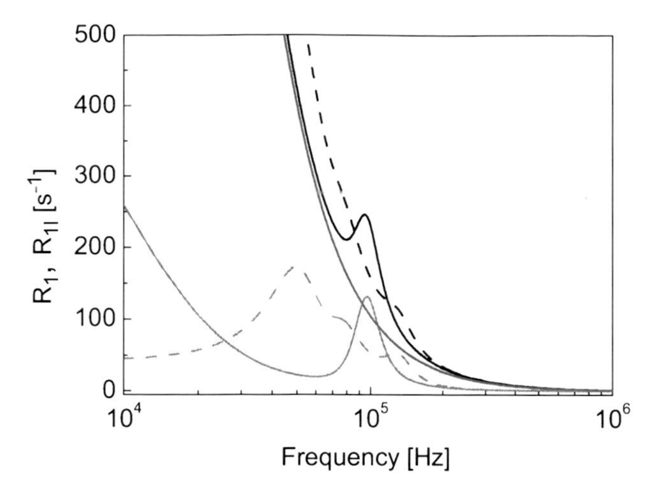

Figure 9.6 ^{1}H spin–lattice relaxation rates R_1 (Eq. 9.10) versus frequency for a ^{1}H–^{2}H spin system decomposed into R_1^{H-H} and R_{1I} contributions; R_1 for $a_Q = 150$ kHz, $r = 300$ pm, $\tau_{rot} = 10$ μs, $\eta = 0$ (black solid line), $\eta = 0.75$ (black dashed line), R_{1I} for $a_Q = 150$ kHz, $r = 300$ pm, $\tau_{rot} = 10$ μs, $\eta = 0$ (gray solid line), $\eta = 0.75$ (gray dashed line), R_1^{H-H} (green line).

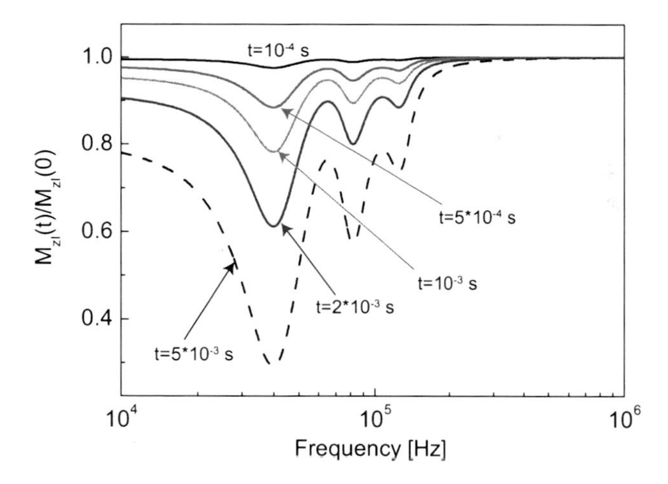

Figure 9.7 Relative ^{1}H magnetization, $M_{ZI}(t)/M_{ZI}(0)$ versus frequency for a ^{1}H–^{2}H spin system; $a_Q = 150$ kHz, $r = 300$ pm, $\eta = 0.75$, $\tau_{rot} = 15$ μs.

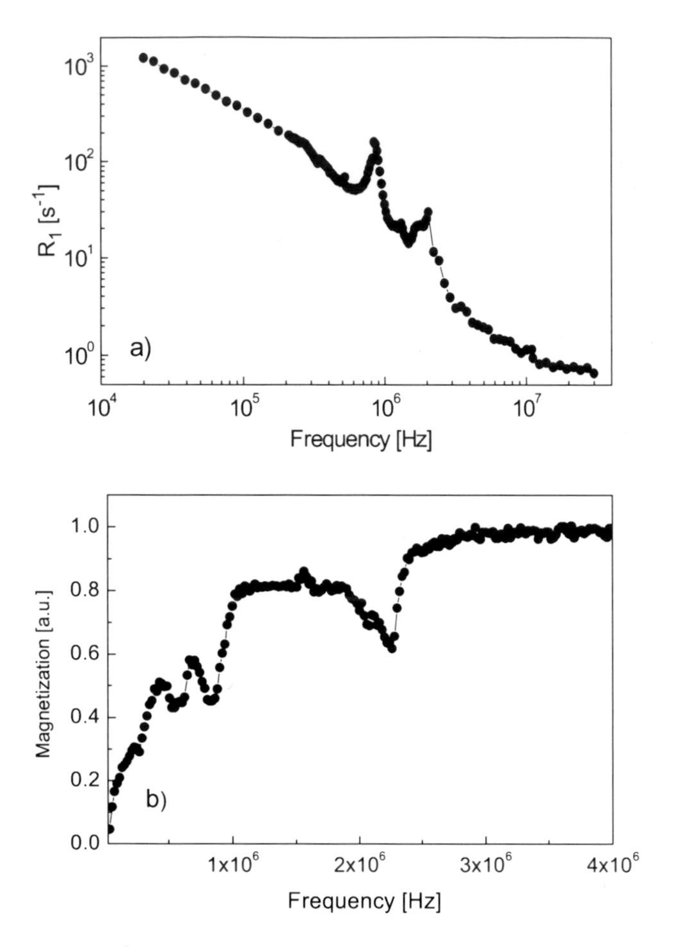

Figure 9.8 (a) ^1H spin–lattice relaxation rate R_1 versus frequency for $Pr_3Sb_2Br_9$ (Pr denotes pyridinium cation) at 350 K; (b) ^1H magnetization versus frequency recorded after $t = 15$ μs.

see a progressive formation of local magnetization minima (at frequencies at which quadrupolar relaxation enhancement effects are observed), referred to as magnetization dips. The positions of the magnetization dips correspond to the frequencies at which the maxima of R_1 (ω_I) are present as they reflect the same effect; a faster relaxation at some frequencies implies that at these frequencies a larger part of the magnetization is gone after a given time, t.

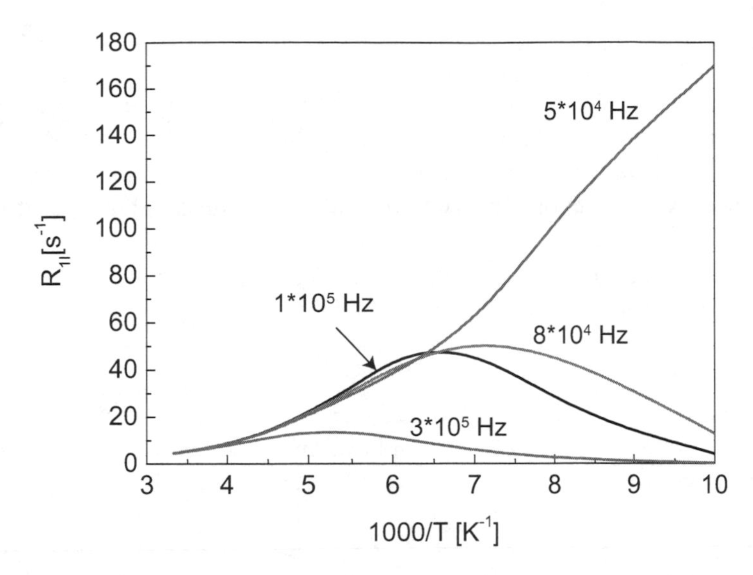

Figure 9.9 ^{1}H spin–lattice relaxation rate R_{1I} versus $1000/T$ for a ^{1}H–^{2}H system; $a_Q = 150$ kHz, $r = 300$ pm, $\eta = 0$, $\tau_0 = 4.0 \times 10^{-9}$ s, $E_A = 0.95$ kJ

Quadrupole relaxation enhancement effects have been observed experimentally for many systems. In Fig. 9.8a, ^{1}H spin–lattice relaxation dispersion results for some organic solids are shown, while Fig. 9.8b presents corresponding ^{1}H magnetization curves ($M_{ZI}(t)/M_{ZI}(0)$ versus frequency ν_I).

It is interesting to inspect the influence of QRE on temperature dependencies of the $R_{1I}(\omega_I)$. For this purpose in Fig. 9.9, $R_{1I}(\omega_I)$ for a ^{1}H–^{14}N spin system is plotted versus reciprocal temperature assuming Arrhenius dependence of the correlation time, τ_{rot} on temperature $\tau_{rot} = \tau_0 \exp\left(\frac{E_A}{RT}\right)$ (E_A denotes activation energy) and neglecting the quadrupolar interaction. This dependence is compared with the relaxation rate $R_{1I}(\omega_I)$ including the QRE effects.

9.2 Perturbation Approach to QRE

The description of QRE effects outlined above concerns the most complex situation which cannot be handled in terms of a perturbation approach. The reason for that is the quadrupolar interaction which fluctuates in time on an intermediate time scale (see Chapter 8); the fluctuations are too fast to be treated as a time-independent contribution to the energy level structure of the P spin and too slow to be treated as a relaxation mechanism. When the quadrupolar interaction is static (as in the example given in the previous section, referring to solid state) one can describe QRE effects by means of a perturbation approach, that is, restrict the basis $\{|O_i)\}$ to only spin variables. As already explained in the previous section, the assumption that the quadrupolar interaction does not fluctuate in time implies that the orientation of the electric field gradient tensor (Q) remains unchanged with respect to the laboratory frame (L). Thus, the energy levels of the spin system are given as eigenvalues of the Hamiltonian of Eq. 9.9 (for $P = 1$, its matrix representation is expressed by Eq. 4.12 in which ω_{I} should be replaced by ω_{P}).

In the general case of $P \geq 1$ the system is characterized by $(2P + 1)$ orientation dependent energy levels, $E_\alpha(\Omega_{\mathrm{QL}})$, and corresponding eigenvectors $|\psi_\alpha(\Omega_{\mathrm{QL}})\rangle$ given as linear combinations of the Zeeman states of the P spin $|n\rangle = |m_{\mathrm{P}}\rangle$; $|\psi_\alpha(\Omega_{\mathrm{QL}})\rangle = \sum_{n=1}^{2P+1} a_{\alpha n}(\Omega_{\mathrm{QL}})|n\rangle$ with orientation dependent coefficients $a_\alpha(\Omega_{\mathrm{QL}})$ [11]. Then, the generalized spectral densities $S_{m,m}(\omega_{\mathrm{I}}, \Omega_{\mathrm{QL}})$ (they are now orientation dependent) can be obtained (Section 5.1) as a matrix product (Eq. 5.3):

$$S_{m,m}(\omega_{\mathrm{I}}, \Omega_{\mathrm{QL}}) = \frac{1}{2P+1} Re\left\{\left[P_m^1(\Omega_{\mathrm{QL}})\right]^+ \times \left[M(\Omega_{\mathrm{QL}}) - i\omega 1\right]^{-1}\left[P_m^1(\Omega_{\mathrm{QL}})\right]\right\} \quad (9.13)$$

The matrices entering Eq. 9.13 are of the dimension $(2P + 1)^2$; they are set up in the Liouville basis $\{|\psi_\alpha\rangle\langle\psi_\beta|(\Omega_{\mathrm{QL}})\}$ constructed from the eigenvectors $|\psi_\alpha(\Omega_{\mathrm{QL}})\rangle$. The matrix $\lfloor M(\Omega_{\mathrm{QL}}) - i\omega_{\mathrm{I}}1\rfloor$ is diagonal with the elements [11]:

$$\left[M(\Omega_{\mathrm{QL}}) - i\omega_{\mathrm{I}}1\right]_{\alpha\beta,\alpha\beta} = i\left(\omega_{\alpha\beta}(\Omega_{\mathrm{QL}}) - \omega_{\mathrm{I}}\right) + \tau_c^{-1} \quad (9.14)$$

where $E_{\alpha\beta}(\Omega_{\mathrm{QL}}) = E_\beta(\Omega_{\mathrm{QL}}) - E_\alpha(\Omega_{\mathrm{QL}})$. The correlation time in Eq. 9.14 is denoted as τ_c. The correlation time characterizes the dynamics of the nuclei currying the I spin—this does not need to be rotation—in solids one can think for instance about jumps between available lattice sites (like, for instance, for fluorine ions in LaF_3 [7, 8, 11, 12]). To obtain the representation of the operators

$$P_0 = \sum_{m_P>0} \left(|m_P\rangle \langle m_P| - |-m_P\rangle \langle -m_P|\right),\ P_1 = \sum_{m_P} \left(|m_P\rangle \langle m_P - 1|\right),$$

and

$$P_{-1} = -\sum_{m_P} |m_P - 1\rangle \langle m_P|,$$

the inverse relationship between the Zeeman basis and the eigenbasis,

$$|n\rangle = \sum_{\alpha=1}^{2P+1} a_{\alpha n} \left(\Omega_{QL}\right) |\psi_\alpha\rangle$$

can be used. It is worth to note that as the coefficient matrix $[a_{n\alpha}]$ is hermitian the inverse relationship can be obtained from the transposed matrix $[a_{n\alpha}]$. Eventually, as a result of the matrix product of Eq. 9.13, the relaxation rate $R_{1I}(\omega_I)$ is given as a linear combination of functions of the type

$$f\left(\omega_{\alpha\beta}, \omega_I, \tau_c\right) = \frac{\tau_c}{1 + \left(\omega_{\alpha\beta} - \omega_I\right)^2 \tau_c^2}.$$

Such a function reaches its maximum when $\omega_{\alpha\beta} = \omega_I$, that is, when the transition frequency of the P spin, $\omega_{\alpha\beta}$, matches the transition frequency of the I spin, ω_I. This describes the condition for the quadrupole relaxation enhancement. The enhancement is seen when the correlation time τ_c is relatively long. For short, correlation times (fast dynamics) when the condition $\left|\left(\omega_{\alpha\beta} - \omega_I\right)\right| \tau_c \ll 1$ is fulfilled) the function $f\left(\omega_{\alpha\beta}, \omega_I, \tau_c\right)$ is not sensitive to the changes in the frequency (such a situation is referred to as the extreme narrowing condition). On the other hand, one should be aware that to describe the relaxation rate, $R_{1I}(\omega_I)$ in terms of spectral densities, the condition $\omega_{DD}^{IP}\tau_c < 1$ must hold (ω_{DD}^{IP} denotes the amplitude of the I–P dipole–dipole coupling).

Finishing this discussion, it should be stressed that for polycrystals, averaging over all orientations (Ω_{QL} angles) should be performed to obtain $R_{1I}(\omega_I)$.

9.3 Polarization Transfer

The condition $\omega_{DD}^{IP}\tau_c < 1$ limiting the applicability of the described model of QRE gives rise to the question what happens when it breaks down. Before we shall answer to this question it is worth to summarize the dynamical scenarios discussed so far. Originally, we have considered a pair of spins $I-P$ belonging to a molecule which undergoes overall rotational dynamics. In consequence, the fluctuations of the quadrupolar interaction of the P spin and the $I-P$ dipole–dipole coupling occur on the same time scale as they are caused by the same process; the molecular rotation. Assuming that $\omega_{DD}^{IP}\tau_{rot} < 1$ (i.e., the condition of the perturbation theory of relaxation is fulfilled for the I spin) the spin–lattice relaxation, R_{1I}, can be described by Eq. 9.3 and Eq. 9.4 independently of the value of the product $\omega_Q\tau_{rot}$ (but QRE effects are observed only for rather slow rotation). Then, we have allowed for different origins of the modulations of the quadrupolar and dipole–dipole interactions, assuming that the motion modulating the quadrupole coupling is very slow (the interaction is, in fact, time independent). It has been explained that then the relaxation R_{1I} can be treated by means of a perturbation approach applied to both I and P spins. Local magnetization dips (like those shown in Fig. 9.8b) can be observed also for rigid systems when neither the quadrupolar interaction nor the dipolar one fluctuates in time. Such systems do not relax—the magnetization of the I spin remains unchanged for a long time (i.e., polarizing the spins to create the magnetization also takes long time) except of the specific frequencies at which it drops pretty fast—this effect is referred to as polarization transfer [8, 9, 15].

For a rotating molecule carrying the I and P spins, a rigid system means that $\omega_{DD}^{IP}\tau_{rot} \gg 1$ and $\omega_Q\tau_{rot} \gg 1$. The time independent $I-P$ dipole–dipole coupling leads to a mixing of the energy levels of the participating spins; the spins cannot be considered separately anymore. The total Hamiltonian of the spin system is given now as:

$$H(I, P) = H_Z(I) + H_Z(P) + H_Q(P) + H_{DD}(I, P) \qquad (9.15)$$

For $P = 1$ its matrix representation in the Zeeman basis $\{|n\rangle = |m_1, m_P\rangle\}$, labeled as $|1\rangle = |1/2, 1\rangle$, $|2\rangle = |1/2, 0\rangle$, $|3\rangle = |1/2, -1\rangle$, $|4\rangle = |-1/2, -1\rangle$, $|5\rangle = |-1/2, 0\rangle$, $|6\rangle = |-1/2, -1\rangle$

has the form,

$$H\,(I,\,P) = H_{\mathrm{Z}}\,(I) + H_{\mathrm{Z}}\,(P) + H_{\mathrm{Q}}\,(P) + H_{\mathrm{DD}}\,(I,\,P),$$

where [9]:

$$[H_{\mathrm{Z}}\,(I) + H_{\mathrm{Z}}\,(P) + H_{\mathrm{Q}}\,(P)] =$$

$$\begin{bmatrix}
\frac{1}{2}\omega_{\mathrm{I}} + \omega_{\mathrm{P}} + \frac{1}{\sqrt{6}} a_{\mathrm{Q}} V_0^2 & \frac{1}{\sqrt{2}} a_{\mathrm{Q}} V_{-1}^2 & a_{\mathrm{Q}} V_{-1}^2 & 0 & 0 & 0 \\
-\frac{1}{\sqrt{2}} a_{\mathrm{Q}} V_1^2 & \frac{1}{2}\omega_{\mathrm{I}} - \frac{2}{\sqrt{6}} a_{\mathrm{Q}} V_0^2 & \frac{1}{\sqrt{2}} a_{\mathrm{Q}} V_{-1}^2 & 0 & 0 & 0 \\
a_{\mathrm{Q}} V_2^2 & -\frac{1}{\sqrt{2}} a_{\mathrm{Q}} V_{-1}^2 & \frac{1}{2}\omega_{\mathrm{I}} - \omega_{\mathrm{P}} + \frac{1}{\sqrt{6}} a_{\mathrm{Q}} V_0^2 & 0 & 0 & 0 \\
0 & 0 & 0 & \frac{1}{2}\omega_{\mathrm{I}} - \omega_{\mathrm{P}} + \frac{1}{\sqrt{6}} a_{\mathrm{Q}} V_0^2 & \frac{1}{\sqrt{2}} a_{\mathrm{Q}} V_{-1}^2 & a_{\mathrm{Q}} V_2^2 \\
0 & 0 & 0 & -\frac{1}{\sqrt{2}} a_{\mathrm{Q}} V_1^2 & -\frac{1}{2}\omega_{\mathrm{I}} - \frac{1}{\sqrt{6}} a_{\mathrm{Q}} V_0^2 & -\frac{1}{\sqrt{2}} a_{\mathrm{Q}} V_{-1}^2 \\
0 & 0 & 0 & a_{\mathrm{Q}} V_2^2 & \frac{1}{\sqrt{2}} a_{\mathrm{Q}} V_1^2 & -\frac{1}{2}\omega_{\mathrm{I}} - \omega_{\mathrm{P}} + \frac{1}{\sqrt{6}} a_{\mathrm{Q}} V_0^2
\end{bmatrix} \quad (9.16a)$$

and

$$\lfloor H_{\mathrm{DD}}\,(I,\,P) \rfloor = a_{\mathrm{DD}}$$

$$\begin{bmatrix}
\frac{1}{\sqrt{6}} F_0^2 & \frac{1}{2\sqrt{2}} F_{-1}^2 & 0 & \frac{1}{2} F_{-1}^2 & \frac{1}{\sqrt{2}} F_{-2}^2 & 0 \\
-\frac{1}{2\sqrt{2}} & 0 & \frac{1}{2\sqrt{2}} F_{-1}^2 & -\frac{1}{2\sqrt{3}} F_0^2 & 0 & \frac{1}{\sqrt{2}} F_{-2}^2 \\
0 & -\frac{1}{2\sqrt{2}} F_1^2 & -\frac{1}{\sqrt{6}} F_0^2 & 0 & -\frac{1}{2\sqrt{3}} F_0^2 & -\frac{1}{2} F_{-1}^2 \\
-\frac{1}{\sqrt{2}} F_1^2 & -\frac{1}{2\sqrt{3}} F_1^2 & 0 & -\frac{1}{\sqrt{6}} F_0^2 & \frac{1}{2\sqrt{2}} F_{-1}^2 & 0 \\
\frac{1}{\sqrt{2}} F_2^2 & 0 & -\frac{1}{2\sqrt{3}} F_0^2 & -\frac{1}{2\sqrt{2}} F_1^2 & 0 & \frac{1}{2\sqrt{2}} F_{-1}^2 \\
0 & \frac{1}{\sqrt{2}} F_2^2 & \frac{1}{2} F_1^2 & 0 & -\frac{1}{2\sqrt{2}} F_1^2 & \frac{1}{\sqrt{6}} F_0^2
\end{bmatrix} \quad (9.16b)$$

The structure of the Hamiltonian matrix reflects the physical mechanism of polarization transfer. The energy levels of the I and P spins obtained from the matrix $\lfloor H_Z(I) + H_Z(P) + H_Q(P) \rfloor$ are separated; there is no coupling between spin states characterized by different m_I quantum numbers. A connection between them is only introduced by the dipole–dipole interaction. The coupling between the states of $m_I = 1/2$ and $m_I = -1/2$ is necessary for creating the transfer of polarization between I and P spins as it links their transitions. They cannot occur independently. When the transition frequency of the P spin matches the transition frequency ν_I of the I spin the energy released by the transition of the I spin can be used by the P spin, provided there is a coupling between these spins. This is a somewhat simplified statement as one should not think about the spins separately; nevertheless it helps to understand the physical picture of the polarization transfer effect.

A rigorous quantum-mechanical description of polarization transfer can be given in terms of the time-independent Liouville equation, which yields:

$$\frac{d\rho(t)}{dt} = i\,[H, \rho(t)] \tag{9.17}$$

where ρ is the density operator of the entire I–P spin system, while H is the total (time independent) Hamiltonian. The elements $\rho_{\alpha\beta}$ of the density operator are equivalent to the $|\psi_\alpha\rangle\langle\psi_\beta|$ quantity, where $\{|\psi_\alpha\rangle\}$ are eigenvectors of the I–P system, which can be obtained by diagonalizing the Hamiltonian matrix $[H]$. Equation 9.17 gives an explicit, straightforward expression for the time evolution of the individual density matrix elements [8–10]:

$$\rho_{\alpha\beta}(t) = \rho_{\alpha\beta}(0) \exp\{-i\omega_{\alpha\beta}t\} \tag{9.18}$$

where $\omega_{\alpha\beta}$ denotes the transition frequency between the eigenstates described by the functions $|\psi_\alpha\rangle$ and $|\psi_\beta\rangle$. Assuming that initially both spins have been polarized in a strong magnetic field, the initial density operator is given as $\rho(0) \propto \gamma_I I_Z + \gamma_P P_Z$. When $\gamma_P \ll \gamma_I$, one can set $\rho(0) \propto I_Z$. The elements of the initial density operator can be obtained, directly from the definition $\rho_{\alpha\beta}(t) = \langle\psi_\alpha|\rho(t)|\psi_\beta\rangle$. The I spin magnetization is represented by expectation value of the I_Z operator, $\langle I_Z\rangle$. Expectation value of an operator O can also be calculated just from its definition $\langle O\rangle(t) = \mathrm{Tr}\{O\rho(t)\}$, which is convenient to evaluate in the matrix representation as:

$$\langle 0 \rangle (t) = \sum_{\alpha,\beta=1}^{N} O_{\alpha\beta}\rho_{\alpha\beta}(t) \tag{9.19}$$

where N denotes the number of spin states. Applying Eq. 9.19 to the present case one obtains:

$$\langle I_Z \rangle (t) = \sum_{\alpha\beta=1}^{(2I+1)(2P+1)} \langle \psi_\alpha | I_Z | \psi_\beta \rangle^{?} \exp(-i\omega_{\alpha\beta}t) \tag{9.20}$$

In this way one can follow how the magnetization M_{ZI} evolves in time at a frequency ω_I entering the Hamiltonian of Eq. 9.15. The Hamiltonian is orientation dependent (we have not stressed this explicitly), so the result of Eq. 9.20 has to be averaged over molecular orientations. The description can be straightforwardly generalized to a system containing more spins. For a detailed theory of polarization transfer phenomena the reader is referred to [8–10].

Polarization transfer requires that there is a coupling between the I and P spins which is not averaged out due to molecular motion. For this purpose the system does not need to be rigid. When the motion modulating the $I-P$ dipole–dipole interaction is anisotropic, the averaging of the dipolar interaction is not complete (even if the dynamics is fast). The averaged, non-zero part of the dipolar coupling can provide a pathway for the polarization transfer. In consequence, one observes a combined effect of relaxation (caused by the fluctuating part of the dipolar interaction) and polarization transfer (caused by the non-zero average). It may happen that one observes then two-step magnetization decay, as schematically shown in Fig. 9.10.

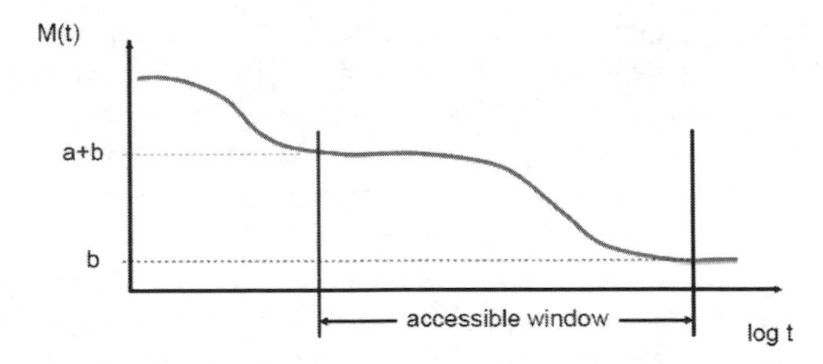

Figure 9.10 Two-step magnetization decay caused first by polarization transfer and later by relaxation.

9.4 QRE and Internal Dynamics of Molecules

So far it has been assumed that the quadrupole interaction fluctuates in time due to overall molecular tumbling. The motional scenario can be, however, more complex due to internal dynamics of the molecule. The internal motion can be of different origin, for instance, local vibrations or restricted motion of molecular units (CD_3 groups). Then, there are relatively fast fluctuations of the quadrupolar interaction inducing relaxation of the P spin and leading to a partial averaging of the interaction (the situation is analogous to the dipolar coupling discussed in Section 9.3). Then, the remaining part of the quadrupolar coupling fluctuates on a much longer time scale due to slow overall tumbling.

To account for the internal dynamics of the molecule one has to extend the basis $\{|O_i) = |LKM) \otimes |\Sigma\sigma)\}$ by including the internal degrees of freedom. The generalization depends on the mechanism of the internal dynamics. Here, we shall limit the discussion to internal rotation. Examples of other mechanisms of motion, like classical and quantum vibrations can be found in [9] in the context of paramagnetic relaxation enhancement (instead of a nuclear spin $P \geq 1$ possessing a quadrupole moment, an electron spin $S \geq 1$ experiencing zero-field splitting interactions is considered). The motional scenario can be described as follows. The total quadrupole interaction, $H_Q(t)$, characterized by the quadrupole parameters a_Q and η is expressed as:

$$H_Q(t) = (1 - f)\, H_Q^{\text{local}}(t) + f\, H_Q^{\text{global}}(t) \tag{9.21}$$

According to the internal (local) rotation, the quadrupolar coupling $H_Q(t)$ gets reduced by a factor, $(1 - f)$ and then, the remaining part, $f H_Q(t)$ fluctuates in time due to the overall (global) tumbling (Eq. 9.21). In the simplest case, the "reduction" of the quadrupolar interaction means that the quadrupolar coupling associated with the term $(1 - f)\, H_Q^{\text{local}}(t)$ yields $(1 - f)\, a_Q$ while the asymmetry parameter, η, remains unchanged. Analogously, the coupling constant for $f H_Q^{\text{global}}(t)$ is equal to $f a_Q$ (η remains unchanged). Now, the basis $\{|O_i)\}$ contains three components, that is, the P spin states $\left|\sum\sigma\right)$, the rotational states $|LKM)$

representing the overall (global) rotation, and rotational states $|ABC)$ associated with the internal (local) rotation, defined as (Eq. 8.3) [6, 14]:

$$|ABC) = |ABC\rangle\langle ABC| = \sqrt{\frac{2A+1}{8\pi^2}}D^{A}_{BC}(\Omega) \tag{9.22}$$

Thus, the vectors $|O_i)$ are now defined as:

$$|O_i) = |ABC) \otimes |LKM) \otimes |\Sigma\sigma) \tag{9.23}$$

The Liouville operator L of Eq. 9.4 representing the degrees of freedom of the "composite" lattice contains now, besides the previous terms, the operator $L^{\text{local}}_{\text{rot}} = -iD^{\text{local}}_{\text{rot}}\nabla^2_{\Omega}$ describing the internal dynamics, where $D^{\text{local}}_{\text{rot}}$ denotes rotational diffusion coefficient for the internal dynamics. The representation of the individual components of the operator L in the extended basis takes the form given below. For the Zeeman coupling of the P spin one obtains:

$$(A'B'C'|\,(L'K'M'|\,(\Sigma'\sigma'\,|L_Z|\,ABC)\,|LKM)\,|\Sigma\sigma)$$

$$= \delta_{AA'}\delta_{BB'}\delta_{CC'}\delta_{LL'}\delta_{KK'}\delta_{MM'}\delta_{\Sigma\Sigma'}\delta_{\sigma\sigma'}\omega_P\sigma \tag{9.24}$$

The $fH^{\text{global}}_{\text{Q}}$ part of the quadrupolar interaction is represented as (see Eq. 8.12 for comparison):

$$(\Sigma'\sigma'|\left(L'K'M'|\,(A'B'C'\left|fL^{\text{global}}_{\text{Q}}\right|ABC)\,|LKM\right)|\Sigma\sigma)$$

$$= \frac{1}{6}\frac{fa_{\text{Q}}}{P(2P-1)}(-1)^{\sigma}V^{2(P)}_{|K-K'|}\left[(-1)^{\Sigma'+\Sigma}-1\right]$$

$$\times\sqrt{(2P+3)(2P+1)(P+1)P(2P-1)(2L'+1)(2L+1)(2\Sigma'+1)(2\Sigma+1)}$$

$$\times\begin{pmatrix} L' & 2 & L \\ -K' & K'-K & K \end{pmatrix}\begin{pmatrix} L' & 2 & L \\ -M' & M'-M & M \end{pmatrix}\begin{Bmatrix} \Sigma' & 2 & \Sigma \\ P' & P & P \end{Bmatrix}\sigma_{AA'}\sigma_{BB'}\sigma_{CC'}$$

$$\tag{9.25}$$

From Eq. 9.25 one can deduce the representation of the $(1-f)\,H^{\text{local}}_{\text{Q}}$ interaction:

$$\left(\Sigma'\sigma'\right|\left(L'K'M'\right|\left(A'B'C'\right|(1-f)\,L_Q^{\text{local}}\left|ABC\right)\left|LKM\right)\left|\Sigma\sigma\right)$$

$$= \frac{1}{6}\frac{(1-f)\,a_Q}{P\,(2P-1)}(-1)^{\sigma}\,V^{2(P)}_{|K-K'|}\left[(-1)^{\Sigma'+\Sigma}-1\right]$$

$$\times\sqrt{(2P+3)\,(2P+1)\,(P+1)\,P\,(2P-1)\,(2L'+1)\,(2L+1)\,(2\Sigma'+1)\,(2\Sigma+1)}$$

$$\times\begin{pmatrix} A' & 2 & A \\ -B' & B'-B & B \end{pmatrix}\begin{pmatrix} A' & 2 & A \\ -C' & C'-C & C \end{pmatrix}\begin{Bmatrix} \Sigma' & 2 & \Sigma \\ P' & P & P \end{Bmatrix}\sigma_{LL'}\sigma_{KK'}\sigma_{MM'} \tag{9.26}$$

In Equation 9.21 the influence of the overall molecular tumbling on the fluctuations of H_Q^{local} has been neglected. In fact, the fluctuations occur due to a combined effect of the local and global dynamics. When the global motion is much slower than the local, its influence can be omitted. Equation 9.26 is a consequence of this assumption. Eventually, for the overall rotation one obtains:

$$\left(A'B'C'\right|\left(L'K'M'\right|\left(\Sigma'\sigma'\right|L_{\text{rot}}^{\text{global}}\left|ABC\right)\left|LKM\right)\left|\Sigma\sigma\right)$$

$$= \delta_{AA'}\delta_{BB'}\delta_{CC'}\delta_{LL'}\delta_{KK'}\delta_{MM'}\delta_{\Sigma\Sigma'}\delta_{\sigma\sigma'}\,i\,D_{\text{rot}}^{\text{global}}\,L\,(L+1) \tag{9.27}$$

Similarly, the representation of the local motion yields:

$$\left(A'B'C'\right|\left(L'K'M'\right|\left(\Sigma'\sigma'\right|L_{\text{rot}}^{\text{local}}\left|ABC\right)\left|LKM\right)\left|\Sigma\sigma\right)$$

$$= \delta_{AA'}\delta_{BB'}\delta_{CC'}\delta_{LL'}\delta_{KK'}\delta_{MM'}\delta_{\Sigma\Sigma'}\delta_{\sigma\sigma'}\,i\,D_{\text{rot}}^{\text{local}}\,A\,(A+1) \tag{9.28}$$

assuming that the overall and local motions are uncorrelated; when they occur on different time scales this condition is fulfilled. In analogy to the case discussed in Section 9.1, the spin–lattice relaxation rate, R_{1I}, is also now given by the 3×3 block of the inverted matrix $[M]^{-1}$ corresponding, this time, to the states $|000)\ 202)\ |1-1)$, $|000)\ |201)\ |10)$, and $|000)\ |200)\ |11)$.

References

1. Abragam, A. (1961). *The Principles of Nuclear Magnetism* (New York: Oxford University Press).
2. Bloembergen, N. and Morgan, L. O. (1961). Proton relaxation times in paramagnetic solutions effects of electron spin relaxation. *J. Chem. Phys.*, **34**, pp. 842–850.

3. Brink, D. M. and Satchler, G. R. (1979). *Angular Momentum* (Oxford: Clarendon Press).
4. Edmunds, A. R. (1974). *Angular Momentum in Quantum Mechanics* (Princeton: Princeton University Press).
5. Kowalewski, J., Kruk, D. and Parigi, G. (2005). NMR Relaxation in solution of paramagnetic complexes: Recent theoretical progress for $S \geq 1$. *Adv. Inorg. Chem.*, **57**, pp. 41–104.
6. Kruk, D. and Kowalewski, J. (2003). Nuclear spin Relaxation in solution of paramagnetic complexes with large transient zero-field splitting. *Mol. Phys.*, **101**, pp. 2861–2874.
7. Kruk, D. and Lips, O. (2005). Evolution of solid state systems containing mutually coupled dipolar and quadrupole spins: Perturbation treatment. *Solid State Nucl. Magn. Reson.*, **28**, pp. 180–192.
8. Kruk, D., Altmann, J., Fujara, F., Gadke, A., Nolte, M. and Privalov, A. F. (2005). Analysis of ^{1}H-^{14}N polarization transfer experiments in molecular crystals. *J. Phys. Condens. Matter*, **17**, pp. 519–533.
9. Kruk, D. (2007). *Theory of Evolution and Relaxation of Multi-Spin Systems* (Bury St Edmunds: Arima).
10. Kruk, D., Fujara, F., Gumann, G., Medycki, W., Privalov, A. F. and Tacke, Ch. (2009). Field cycling methods as a tool for dynamics investigations in solid state systems: Recent theoretical progress. *Solid State Nucl. Magn. Reson.*, **35**, pp. 152–163.
11. Kruk, D., Kubica, A., Masierak, W., Privalov, A. F., Wojciechowski, M. and Medycki, W. (2011). Quadrupole relaxation enhancement: Application to molecular crystals. *Solid State Nucl. Magn. Reson.*, **40**, 114–120.
12. Kruk, D., Privalov, A., Medycki, W., Uniszkiewicz, C., Masierak, W. and Jakubas, R. (2012). NMR studies of solid-state dynamics. *Ann. Rep. Nucl. Magn. Reson. Spectrosc.*, **76**, pp. 67–138.
13. Lips, O., Kruk, D., Privalov, A. F. and Fujara, F. (2007). Simultaneous effects of relaxation and polarization transfer in LaF3-type crystals as source of dynamic information. *Solid State Nucl. Magn. Reson.*, **31**, pp. 141–152.
14. Nilsson, T. and Kowalewski, J. (2000). Slow-motion theory of nuclear spin relaxation in paramagnetic low-symmetry complexes: A generalization to high electron spin. *J. Magn. Reson.*, **146**, pp. 345–358.
15. Nolte, M., Privalov, A., Altmann, J., Anferov, V. and Fujara, F. (2002). ^{1}H-^{14}N cross-relaxation in trinitrotoluene: A step toward improved landmine detection. *J. Phys. D*, **35**, pp. 939–942.

16. Sanktuary, B. C. (1976). Multipole operators for an arbitrary number of spins. *J. Chem. Phys.*, **64**, pp. 4352–4361.
17. Sanktuary, B. C. (1983). Multipole NMR. 3. Multiplet spin theory. *Mol. Phys.*, **48**, pp. 1155–1176.
18. Solomon, I. (1955). Relaxation processes in a system of 2 spins. *Phys. Rev.*, **99**, pp. 559–565.
19. Varshalovich, D. A., Moskalev, A. N. and Khersonkii, V. K. (1988). *Quantum Theory of Angular Momentum*, (Singapore: World Scientific).
20. Westlund, P.-O. (2009). Quadrupol-enhanced proton spin relaxation for a slow reorienting spin pair: (I)-(S). A stochastic Liouville approach. *Mol. Phys.*, **107**, pp. 2141–2148.
21. Westlund, P.-O. (2010). The quadrupole enhanced ^{1}H spin-lattice relaxation of the amide proton in slow tumbling proteins. *Phys. Chem. Chem. Phys.*, **12**, pp. 3136–3140.

Chapter 10

Effects of Mutual Spin Interactions

This chapter focuses on systems contacting more electron spins (i.e. more paramagnetic centers). Strong dipole–dipole interactions between the electron spins considerably influence the shape of electron spin resonance (ESR) spectra and electronic relaxation. The observed effects are dependent on the time scale of stochastic fluctuations of the spin couplings. Discussing different kinds of spin interactions, the influence of the ^{14}N nitrogen quadrupole coupling on the electron and proton spin relaxation in solutions of nitrogen radicals is examined.

10.1 ESR Spectra for Interacting Paramagnetic Centers

In Chapter 6 a theory of ESR lineshape for ^{15}N and ^{14}N radicals has been discussed. It has been shown that the hyperfine coupling between the electron and nitrogen spins affects the lineshape function, $L(\omega)$ in two ways, that is, the isotropic part of the hyperfine interaction leads to a mixing of the electron spin and nitrogen spin energy levels, while the anisotropic part acts as a source of the electron spin relaxation. Let us now go a step further and inquire

Understanding Spin Dynamics
Danuta Kruk

ISBN 978-981-4463-49-2 (Hardcover), 978-981-4463-50-8 (eBook)
www.panstanford.com

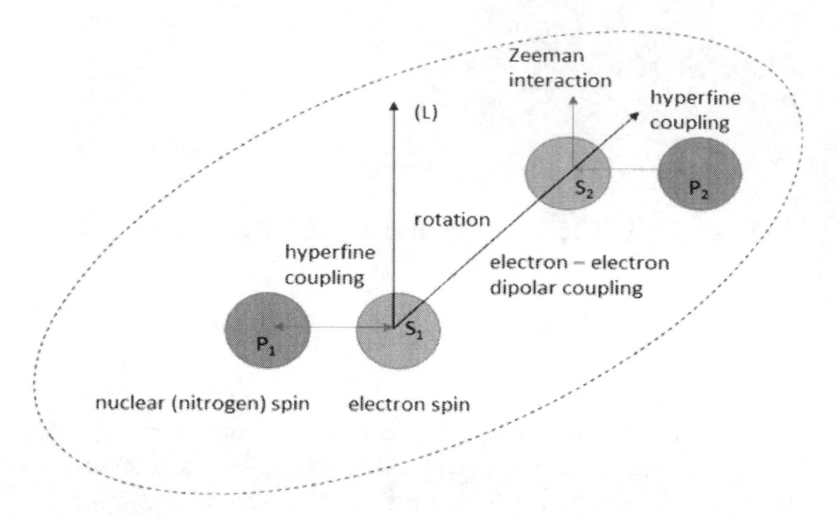

Figure 10.1 Schematic view of a nitroxide biradical which forms a four-spin system.

into how interactions between the electron spins can influence the ESR spectra.

Our spin system (nitroxide biradical) is as follows: there are two paramagnetic centers (electron spins $S_1 = S_2 = 1/2$) coupled by hyperfine interactions with neighboring ^{15}N spins ($P_1 = P_2 = 1/2$). The two P–S spin pairs are not independent as there is a dipole-dipole coupling between the electron spins, S_1–S_2. The system is schematically shown in Fig. 10.1.

To describe it from the quantum-mechanical viewpoint, one has to use four magnetic spin quantum numbers $m_{\mathrm{P}_1} = m_1$, $m_{\mathrm{P}_2} = m_2$, $M_{\mathrm{S}_1} = M_1$, and $M_{\mathrm{S}_2} = M_2$. This implies that the Zeeman basis, $\{|i\rangle\}$, now consists of 16 functions $|i\rangle = |M_1, m_1, M_2, m_2\rangle$. The Hamiltonian H_0 which determines the energy level structure of this spin system contains the following terms:

$$H_0\left(P_1, P_2, S_1, S_2\right) = H_{\mathrm{Z}}\left(S_1\right) + H_{\mathrm{Z}}\left(S_2\right) + H_{\mathrm{SC}}\left(P_1, S_1\right) + H_{\mathrm{SC}}\left(P_2, S_2\right) \tag{10.1}$$

The first two terms, $H_{\mathrm{Z}}(S_1)$ and $H_{\mathrm{Z}}(S_2)$, describe Zeeman interactions of the electron spins (the Zeeman couplings of the nitrogen spins are neglected), while $H_{\mathrm{SC}}(P_1, S_1)$ and $H_{\mathrm{SC}}(P_2, S_2)$ represent the isotropic part of electron spin–nitrogen spin hyperfine couplings (Eq. 6.1):

$$H_{SC}\left(P_{i},\, S_{i}\right) = A\left[P_{iz}S_{iz} + \frac{1}{2}\left(P_{i+}S_{i-} + P_{i-}S_{i+}\right)\right], \quad i = 1, 2 \quad (10.2)$$

For a single pair of spins, P–S, discussed in Chapter 6, the following labeling of the Zeeman functions $|m_S, m_P\rangle = |M, m\rangle$ has been used: $|1\rangle = |1/2, 1/2\rangle$, $|2\rangle = |1/2, -1/2\rangle$, $|3\rangle = |-1/2, 1/2\rangle$, and $|4\rangle = |-1/2, -1/2\rangle$. Following the line of analogy we shall label the four-spin Zeeman functions, $|i\rangle = |M_1, m_1, M_2, m_2\rangle$, of the present system as:

$$|1\rangle = |1/2, 1/2, 1/2, 1/2\rangle,$$

$$|2\rangle = |1/2, 1/2, 1/2, -1/2, 1/2\rangle,$$

$$|3\rangle = |-1/2, 1/2, 1/2, 1/2, 1/2, 1/2\rangle,$$

and

$$|4\rangle = |-1/2, 1/2, 1/2, -1/2, 1/2, 1/2\rangle$$

The magnetic spin quantum number for the first pair of spins, $P_1 - S_1$, are ordered as in Chapter 6, while

$$m_{S_2} = M_2 = 1/2 \text{ and } m_{P_2} = m_2 = 1/2,$$

then the next functions are:

$$|5\rangle = |1/2, 1/2, -1/2, 1/2, 1/2, -1/2\rangle,$$

$$|6\rangle = |1/2, 1/2, -1/2, -1/2, 1/2, -1/2\rangle,$$

$$|7\rangle = |-1/2, 1/2 - 1/2, 1/, 2, 1/2, -1/2\rangle,$$

$$|8\rangle = |-1/2, 1/2, -1/2, -1/2, 1/2, -1/2\rangle$$

$$(m_{S_2} = M_2 = 1/2 \text{ and } m_{P_2} = m_2 = -1/2)$$

$$|9\rangle = |1/2, -1/2, 1/2, 1/2, -1/21/2\rangle,$$

$$|10\rangle = |1/2, -1/2, 1/2, -1/2, -1/2, 1/2\rangle,$$

$$|11\rangle = |-1/2, -1/2, 1/2, 1/2, -1/2, 1/2\rangle,$$

$$|12\rangle = |-1/2, -1/2, 1/2, -1/2, -1/2, 1/2\rangle,$$

$$(m_{S_2} = M_2 = -1/2 \text{ and } m_{P_2} = m_2 = 1/2),$$

and eventually,

$$|13\rangle = |1//2, -1/2, -1/2, /2, -1/2, -1/2\rangle,$$

$$|14\rangle = |1/2, -1/2, -1/2, -1/2, -1/2, -1/2\rangle,$$

$$|15\rangle = |-1//2, -1/2, \;\; 1/2, 1/2, -1/2, -1/2\rangle,$$

$$|16\rangle = |-1/2, -1/2, -1/2, -1/2, -1/2, -1/2\rangle,$$

$$(m_{S_2} = M_2 = -1/2 \text{ and } m_{P_2} = m_2 = -1/2).$$

As the Hamiltonian H_0 does not contain S_1–S_2 interactions (at this stage we consider the case when the S_1–S_2 dipole–dipole coupling is time dependent, so it does not enter the main, static Hamiltonian), its representation in the basis $\{|i\rangle = |M_1, m_1, M_2, m_2\rangle\}$ can be deduced from Eq. 6.3 for a single spin pair, P–S. The Hamiltonian matrix consists of four blocks, labeled as I, II, III, and IV, for the $|1\rangle, \ldots, |4\rangle$; $|5\rangle, \ldots, |8\rangle$; $|9\rangle, \ldots, |12\rangle$; and $|13\rangle, \ldots, |16\rangle$ subsets of the spin functions, respectively, and given by Eq. 10.3:

$$[\mathrm{I}] = \begin{bmatrix} \omega_S + \frac{A}{2} & 0 & 0 & 0 \\ 0 & \omega_S & \frac{A}{2} & 0 \\ 0 & \frac{A}{2} & 0 & 0 \\ 0 & 0 & 0 & \frac{A}{2} \end{bmatrix}, \qquad [\mathrm{II}] = \begin{bmatrix} \omega_S & 0 & 0 & 0 \\ 0 & \omega_S - \frac{A}{2} & \frac{A}{2} & 0 \\ 0 & \frac{A}{2} & -\frac{A}{2} & 0 \\ 0 & 0 & 0 & 0 \end{bmatrix},$$

$$[\mathrm{III}] = \begin{bmatrix} 0 & 0 & 0 & 0 \\ 0 & -\frac{A}{2} & \frac{A}{2} & 0 \\ 0 & \frac{A}{2} & -\omega_S - \frac{A}{2} & 0 \\ 0 & 0 & 0 & -\omega_S \end{bmatrix}, \qquad [\mathrm{IV}] = \begin{bmatrix} \frac{A}{2} & 0 & 0 & 0 \\ 0 & 0 & \frac{A}{2} & 0 \\ 0 & \frac{A}{2} & -\omega_S & 0 \\ 0 & 0 & 0 & -\omega_S + \frac{A}{2} \end{bmatrix} \tag{10.3}$$

The structure of the Hamiltonian matrix is schematically shown in Fig. 10.2.

The first conclusion which can be drawn from the matrix representation of Eq. 10.3 is that some of the energy levels are degenerated. This is not surprising taking into account that the two pairs of spins, P_i–S_i are equivalent. The degeneracy of the energy levels brings us to the conclusion that even neglecting the S_1–S_2 dipole–dipole interaction, the quantum-mechanical picture of the present spin system is different, this is not a duplicated P–S system.

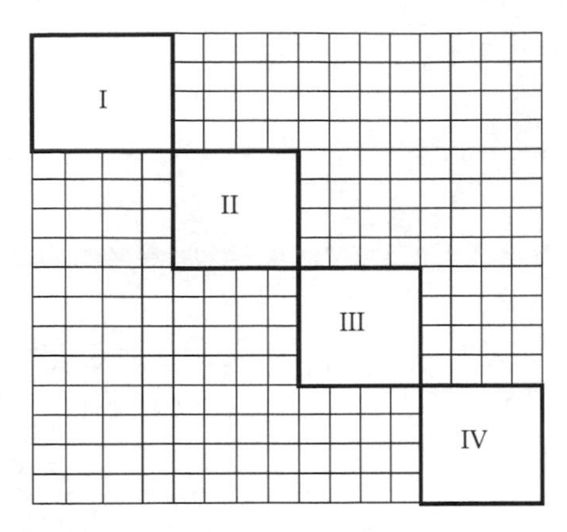

Figure 10.2 Structure of the Hamiltonian matrix $H_0\,(P_1, P_2, S_1, S_2)$ of Eq. 10.1.

The eigenfunctions and eigenvalues (energy levels) of the present system are:

$$|\psi_1\rangle = |1\rangle\,,\ E_1 = \omega_S + \frac{A}{2} \tag{10.4a}$$

$$|\psi_2\rangle = \frac{1}{\sqrt{1+a^2}}\left[a\,|2\rangle + |3\rangle\right],\ E_2 = \frac{\omega_S + \sqrt{\omega_S^2 + A^2}}{2} \tag{10.4b}$$

$$|\psi_3\rangle = \frac{1}{\sqrt{1+b^2}}\left[a\,|2\rangle + |3\rangle\right],\ E_3 = \frac{\omega_S - \sqrt{\omega_S^2 + A^2}}{2} \tag{10.4c}$$

$$|\psi_4\rangle = |4\rangle\,,\ E_4 = \frac{A}{2} \tag{10.4d}$$

$$|\psi_5\rangle = |5\rangle\,,\ E_5 = \omega_S \tag{10.4e}$$

$$|\psi_6\rangle = \frac{1}{\sqrt{1+a^2}}\left[a\,|6\rangle + |7\rangle\right],\ E_6 = \frac{-A + \omega_S + \sqrt{A^2 + \omega_S^2}}{2} \tag{10.4f}$$

$$|\psi_7\rangle = \frac{1}{\sqrt{1+b^2}}\left[a\,|6\rangle + |7\rangle\right],\ E_7 = \frac{-A + \omega_S - \sqrt{A^2 + \omega_S^2}}{2} \tag{10.4g}$$

$$|\psi_8\rangle = |8\rangle\,,\, E_8 = 0 \tag{10.4h}$$

$$|\psi_9\rangle = |9\rangle\,,\, E_9 = 0 \tag{10.4i}$$

$$|\psi_{10}\rangle = \frac{1}{\sqrt{1+a^2}}\left[a\,|10\rangle + |11\rangle\right],\, E_{10} = \frac{-A - \omega_S + \sqrt{A^2+\omega_S^2}}{2} \tag{10.4j}$$

$$|\psi_{11}\rangle = \frac{1}{\sqrt{1+b^2}}\left[b\,|10\rangle + |11\rangle\right],\, E_{11} = \frac{-A - \omega_S - \sqrt{A^2+\omega_S^2}}{2} \tag{10.4k}$$

$$|\psi_{12}\rangle = |12\rangle\,,\, E_{12} = -\omega_S \tag{10.4l}$$

$$|\psi_{13}\rangle = |13\rangle\,,\, E_{13} = \frac{A}{2} \tag{10.4m}$$

$$|\psi_{14}\rangle = \frac{1}{\sqrt{1+a^2}}\left[a\,|14\rangle + |15\rangle\right],\, E_{14} = \frac{-\omega_S + \sqrt{\omega_S^2+A^2}}{2} \tag{10.4n}$$

$$|\psi_{15}\rangle = \frac{1}{\sqrt{1+b^2}}\left[a\,|14\rangle + |15\rangle\right],\, E_{15} = \frac{-\omega_S - \sqrt{\omega_S^2+A^2}}{2} \tag{10.4o}$$

$$|\psi_{16}\rangle = |16\rangle\,,\, E_{16} = -\omega_S + \frac{A}{2} \tag{10.4p}$$

where $a = \frac{\omega_S+\sqrt{A^2+\omega_S^2}}{A}$, $b = \frac{\omega_S-\sqrt{A^2+\omega_S^2}}{A}$. The energy levels are plotted in Fig. 10.3. One can see that there are two pairs of degenerated energy levels $E_8 = E_9 = 0$ and $E_4 = E_{13} = A/2$; moreover, $E_5 = -E_{12}$, $E_2 = -E_{15}$, $E_3 = -E_{14}$.

In the high field approximation ($\omega_S \gg A$), neglecting the off-diagonal elements of the matrices of Eq. 10.3 one obtains $a \to 1$, $b \to 0$, and the eigenvectors $|\psi_i\rangle$ are equal to the corresponding Zeeman functions $|\psi_i\rangle = |i\rangle$. The energy levels are then given as $E_3 = E_8 = E_9 = E_{12} = 0$, $E_2 = E_5 = -E_{12} = -E_{15} = \omega_S$, $E_4 = E_{13} = -E_7 = -E_{10} = A/2$, $E_1 = \omega_S + A/2$, $E_6 = \omega_S - A/2$, $E_{11} = -\omega_S - A/2$, and $E_{16} = \omega_S + A/2$. In this limit one can write a closed mathematical form for the energy levels:

$$E\left(M_1, m_1, M_2, m_2\right) = \left(M_1 = M_2\right)\omega_S + \left(M_1 m_1 + M_2 m_2\right) A/4 \tag{10.5}$$

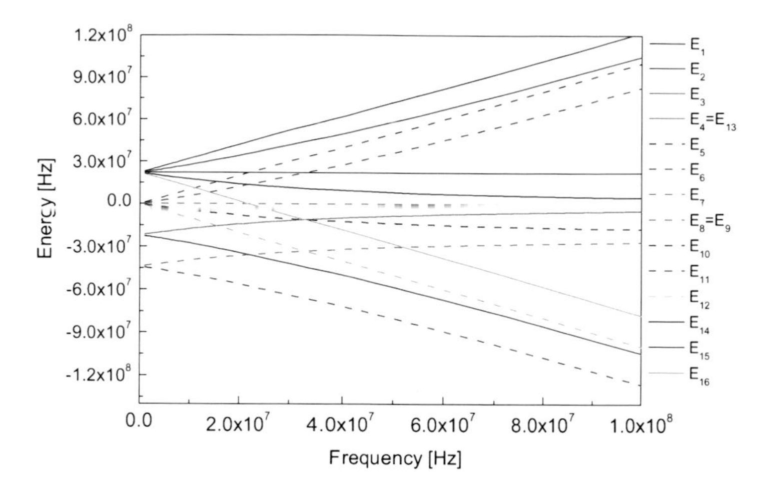

Figure 10.3 Energy level structure of a system consisting of two electron spin and two ^{14}N spins, calculated according to Eq. 10.4.

Comparing Fig. 10.3 with Fig. 10.4 in which the energy level structure given by Eq. 10.4 is plotted, one can see how significant is the role of the off-diagonal elements in the Hamiltonian representation.

To calculate the ESR lineshape function, $L_S(\omega)$, we shall follow the procedure described in Chapter 6. We have seen that it is wise to begin with the representation of the electron spin operator S_+ in the eigenbasis as it determines the spin transitions which are relevant for the lineshape. As both electron spins are equivalent, let us focus on the S_{1+} operator, which is represented by the following combination of the Liouville space elements $|i\rangle\langle i'|$ (we limit ourselves to the high frequency range):

$$S_{1+} \propto |1\rangle\langle 3| + |2\rangle\langle 4| + |5\rangle\langle 7| + |6\rangle\langle 8|$$
$$|9\rangle\langle 11| + |10\rangle\langle 12| + |13\rangle\langle 15| + |14\rangle\langle 16| \quad (10.6)$$

This implies that for the S_1 spin eight transitions is relevant; they are characterized by the frequencies: $\omega_{1,3} = \omega_{5,7} = \omega_{9,11} = \omega_{13,15} = -\omega_S - A/2$ and $\omega_{2,4} = \omega_{6,8} = \omega_{10,12} = \omega_{14,16} = -\omega_S + A/2$. Analogously, one can write down the representation of the

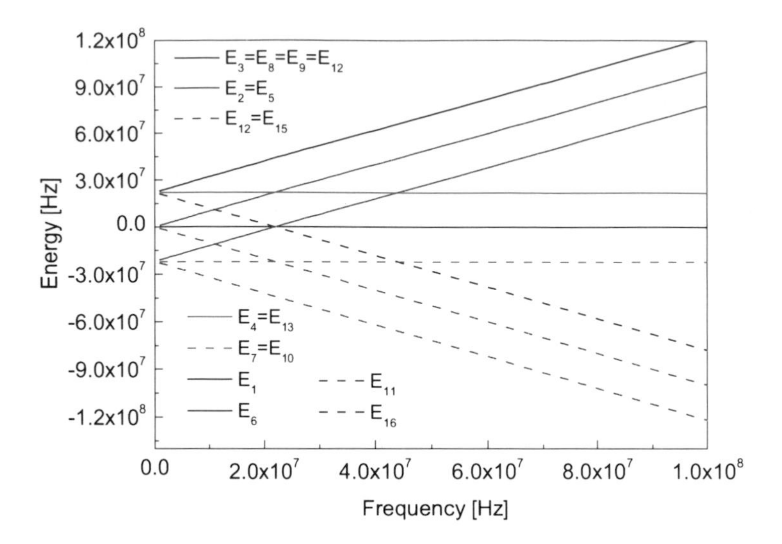

Figure 10.4 Energy level structure of a system consisting of two electron- and two ^{14}N spins neglecting the mixing of energy levels due to the electron spin–nitrogen spin hyperfine interactions.

S_{2+} operator:

$$S_{2+} \propto |1\rangle\langle 9| + |2\rangle\langle 10| + |3\rangle\langle 11| + |4\rangle\langle 12|$$
$$|5\rangle\langle 13| + |6\rangle\langle 14| + |7\rangle\langle 15| + |8\rangle\langle 16| \qquad (10.7)$$

In this case the relevant transition frequencies are $\omega_{1,9} = \omega_{2,11} = \omega_{3,11} = \omega_{4,12} = -\omega_S - A/2$ and $\omega_{5,13} = \omega_{6,14} = \omega_{7,15} = \omega_{8,16} = -\omega_S + A/2$. As expected, for both spins the same two transition frequencies have been obtained $\omega_+ = \omega_S + A/2$ and $\omega_- = \omega_S - A/2$. This implies that the spectrum consists of two lines separated by the frequency A. In analogy to Eq. 6.10 one can now write:

$$L_S(\omega) \propto \frac{T_+}{1 + (\Delta\omega - A/2)^2 T_+^2} + \frac{T_-}{(1 + (\Delta\omega + A/2)^2 T_-^2)} \qquad (10.8)$$

where $\Delta\omega = \omega - \omega_S$. In Eq. 10.8 we have assumed that some of the electron spin relaxation rates are equal, that is, $R_{1,3,1,3} = R_{5,7,5,7} = R_{9,11,9,11} = R_{13,15,13,15} = T_+^{-1}$ and $R_{2,4,2,4} = R_{6,8,6,8} = R_{2,4,2,4} = R_{6,8,6,8} = R_{10,12,10,12} = R_{14,16,14,16} = T_-^{-1}$ (this will be confirmed soon in Eqs. 10.17 and 10.18).

As many of the transition frequencies are equal, one should carefully consider possible couplings between them. Taking into account the possibility of joint evolution of the spin coherences, the lineshape function should be calculated as the matrix product:

$$L_S(\omega) \propto \mathrm{Re}\left\{[S_+]^+ \cdot [M]^{-1} \cdot [S_+]\right\} \tag{10.9}$$

where the representation of the $S_{1+} \equiv S_+$ operator (we have already shown that for S_{1+} and S_{2+} one gets the same results) is very simple $[S_+]^+ = [1, \ldots, 1]$, as all terms in Eq. 10.6 have the same weight factor. The matrix $[M]$ (of the dimension 8×8) is given as:

$$[M] = \begin{bmatrix} \Lambda_{1,3} & R_{1,3,5,7} & R_{1,3,9,11} & R_{1,3,13,15} & 0 & 0 & 0 & 0 \\ R_{1,3,5,7} & \Lambda_{5,7} & R_{5,7,9,11} & R_{5,7,13,15} & 0 & 0 & 0 & 0 \\ R_{1,3,5,7} & R_{5,7,9,11} & \Lambda_{9,11} & R_{9,11,13,15} & 0 & 0 & 0 & 0 \\ R_{1,3,13,15} & R_{5,7,13,15} & R_{9,11,13,15} & \Lambda_{13.15} & 0 & 0 & 0 & 0 \\ 0 & 0 & 0 & 0 & \Lambda_{2,4} & R_{2,4,6,8} & R_{2,4,10,12} & R_{2,4,14,16} \\ 0 & 0 & 0 & 0 & R_{2,4,6,8} & \Lambda_{6,8} & R_{6,8,10,12} & R_{6,8,14,16} \\ 0 & 0 & 0 & 0 & R_{2,4,10,12} & R_{6,8,10,12} & \Lambda_{10,12} & R_{10,12,14,16} \\ 0 & 0 & 0 & 0 & R_{2,4,14,16} & R_{6,8,14,15} & R_{10,12,14,16} & \Lambda_{14,16} \end{bmatrix} \tag{10.10}$$

where $\Lambda_{\alpha\beta} = i\left(\omega + \omega_{\alpha\beta}\right) + R_{\alpha\beta\alpha\beta}$.

To proceed further we have to set up the Hamiltonian, $H'(t)$, which is responsible for the electron spin relaxation. Although it has been anticipated that the S_1–S_2 dipole–dipole coupling contributes to the electronic relaxation, let us for a while limit ourselves to the anisotropic parts of the hyperfine interactions:

$$H'(P_1, S_1, P_2, S_2)(t) = H^{(\mathrm{L})}_{A,\mathrm{aniso}}(P_1, S_1)(t) + H^{(\mathrm{L})}_{A,\mathrm{aniso}}(P_2, S_2)(t) \tag{10.11}$$

The form of the Hamiltonian of anisotropic hyperfine coupling has been given in Chapter 6 (Eq. 6.14); let us remind it for convenience:

$$H^{(\mathrm{L})}_{A,\mathrm{aniso}}(P, S)(t) = \sum_{m=-2}^{2} (-1)^m G^{2(\mathrm{L})}_{-m} T^2_m(P, S) \tag{10.12}$$

where $G_m^{2(\mathrm{L})}$ is given by Eq. 6.15 (the angle Ω denotes the orientation of the P–S axis with respect to the direction of the external magnetic field):

$$G_m^{2(\mathrm{L})} = D_{0,-m}^2(\Omega)\sqrt{\frac{2}{3}}\left[A_{zz} - \frac{1}{2}\left(A_{xx} + A_{yy}\right)\right]$$

$$+ \left(D_{-2,-m}^2(\Omega) + D_{2,-m}^2(\Omega)\right)\frac{1}{2}\left(A_{xx} - A_{yy}\right) \quad (10.13)$$

As in the present case, we consider two pairs of electron spin–nitrogen spin systems, we shall use the notations, that is, for the P_1–S_2 pair, $G_m^{2(\mathrm{L})} \equiv G_m^{(1)}$ and for the P_2–S_2 pair, $G_m^{2(\mathrm{L})} \equiv G_m^{(2)}$.

The matrix representation of $H_{A,\mathrm{aniso}}^{(\mathrm{L})}(P, S)$ (in the Zeeman basis) for a single pair of spins, P–S, is given by Eq. 6.16. In the present case the matrix representation of the Hamiltonian $H'(t)$ consists of four blocks (I′, II′, III′, and IV′), in analogy to the representation of the static Hamiltonian, H_0, and some other non-zero elements. The blocks have the following form:

$$[\mathrm{I}'] = [\mathrm{IV}'] =$$

$$\begin{bmatrix} \frac{1}{2\sqrt{6}}G_0^{(1)} + \frac{1}{2\sqrt{6}}G_0^{(2)} & -\frac{1}{4}G_{-1}^{(1)} & -\frac{1}{4}G_{-1}^{(1)} & \frac{1}{2}G_{-2}^{(1)} \\ \frac{1}{4}G_1^{(1)} & -\frac{1}{2\sqrt{6}}G_0^{(1)} + \frac{1}{2\sqrt{6}}G_0^{(2)} & -\frac{1}{2\sqrt{6}}G_0^{(1)} & \frac{1}{4}G_{-1}^{(1)} \\ \frac{1}{4}G_1^{(1)} & -\frac{1}{2\sqrt{6}}G_0^{(1)} & -\frac{1}{2\sqrt{6}}G_0^{(1)} + \frac{1}{2\sqrt{6}}G_0^{(2)} & \frac{1}{4}G_{-1}^{(1)} \\ \frac{1}{2}G_2^{(1)} & -\frac{1}{4}G_1^{(1)} & -\frac{1}{4}G_1^{(1)} & \frac{1}{2\sqrt{6}}G_0^{(t)} + \frac{1}{2\sqrt{6}}G_0^{(2)} \end{bmatrix}$$

$$(10.14a)$$

$$[\mathrm{II}'] = [\mathrm{III}'] =$$

$$\begin{bmatrix} \frac{1}{2\sqrt{6}}G_0^{(1)} - \frac{1}{2\sqrt{6}}G_0^{(2)} & -\frac{1}{4}G_{-1}^{(1)} & -\frac{1}{4}G_{-1}^{(1)} & \frac{1}{2}G_{-2}^{(1)} \\ \frac{1}{4}G_1^{(1)} & -\frac{1}{2\sqrt{6}}G_0^{(1)} - \frac{1}{2\sqrt{6}}G_0^{(2)} & -\frac{1}{2\sqrt{6}}G_0^{(1)} & \frac{1}{4}G_{-1}^{(1)} \\ \frac{1}{4}G_1^{(1)} & -\frac{1}{2\sqrt{6}}G_0^{(1)} & -\frac{1}{2\sqrt{6}}G_0^{(1)} - \frac{1}{2\sqrt{6}}G_0^{(2)} & \frac{1}{4}G_{-1}^{(1)} \\ \frac{1}{2}G_2^{(1)} & -\frac{1}{4}G_1^{(1)} & -\frac{1}{4}G_1^{(1)} & \frac{1}{2\sqrt{6}}G_0^{(t)} - \frac{1}{2\sqrt{6}}G_0^{(2)} \end{bmatrix}(t)$$

$$(10.14b)$$

The other non-zero elements yield:

$$\begin{aligned}\langle 1|\,H'\,|5\rangle &= \langle 1|\,H'\,|9\rangle = \langle 2|\,H'\,|6\rangle = \langle 2H'10\rangle = \langle 3|\,H'\,|7\rangle \\ &= \langle 3|\,H'\,|11\rangle = \langle 4|\,H'\,|8\rangle = \langle 4|\,H'\,|12\rangle = \langle 4|\,H'\,|12\rangle \\ &= -\langle 5|\,H'\,|13\rangle = -\langle 6|\,H'\,|14\rangle = -\langle 7|\,H'\,|15\rangle \\ &= -\langle 8|\,H'\,|16\rangle = -\langle 9|\,H'\,|13\rangle = -\langle 10|\,H'\,|14\rangle \\ &= -\langle 11|\,H'\,|15\rangle = -\langle 12|\,H'\,|16\rangle = -\frac{1}{4}G_{-1}^{(2)} \quad (10.15a)\end{aligned}$$

$$\langle 1|\,H'\,|13\rangle = \langle 2|\,H'\,|14\rangle = \langle 3|\,H'\,|15\rangle = \langle 4|\,H'\,|16\rangle = -\frac{1}{2}G_{-2}^{(2)} \quad (10.15b)$$

$$\langle 5|\,H'\,|9\rangle = \langle 6|\,H'\,|10\rangle = \langle 7|\,H'\,|11\rangle = \langle 8|\,H'\,|12\rangle = -\frac{1}{2\sqrt{6}}G_{0}^{(2)} \quad (10.15c)$$

The representation of the Hamiltonian $H'(t)$ is shown in Fig. 10.5. The form of the matrix brings us to the subject of cross-correlation effects. The question which one should ask is whether $G_m^1(t) = G_m^{(2)}(t)$. The quantities $G_m^{(1)}(t)$ and $G_m^{(2)}(t)$ depend on the orientation of the P_1–S_1 and P_2–S_2 axes with respect to the direction of the external magnetic field at time t. As we do not consider internal dynamics of the molecule, we treat it as a rigid body undergoing an overall tumbling, the angles $\Omega_1(t)$ and

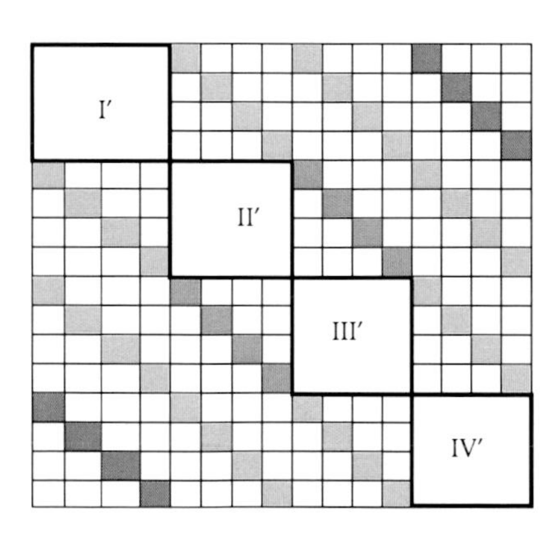

Figure 10.5 Structure of the Hamiltonian matrix of $H'(P_1, P_2, S_1, S_2)$, Eq. 10.11. Elements containing $G_{|m|=1}^{(2)}$ (green), $G_0^{(2)}$ elements (pink), and $G_{|m|=2}^{(2)}$ elements (blue).

$\Omega_2(t)$ (describing the orientations of the P_1–S_1 and P_2–S_2 axis, respectively) are fully correlated. They do not need to be equal, but knowing one of the orientations one can determine the second one. Without losing the generality, one can set $\Omega_1(t) = \Omega_2(t)$ (i.e., $G_m^{(1)}(t) = G_m^{(2)}(t)$).

Let us begin with calculating the first set of the electron spin relaxation rates $R_{1,3,1,3}$, $R_{1,7,1,7}$, $R_{9,11,9,11}$, and $R_{13,15,13,15}$, applying the formula of Eq. 3.19 of Chapter 3. We use the notation $R_{\alpha,\beta,\alpha,\beta}$, instead of the previously used $R_{\alpha\alpha\beta\beta}$, to avoid confusion in the numbering of spin states. The calculations might seem somewhat cumbersome, but they are manageable as many elements of the Hamiltonian matrix, $H'(t)$ are zero. For instance, for the relaxation rate $R_{1,3,1,3}$ one gets:

$$R_{1,3,1,3} = \frac{1}{16}J(\omega_{1,2}) + \frac{1}{8}J(\omega_{1,3}) + \frac{1}{4}J(\omega_{1,4}) + \frac{1}{16}J(\omega_{1,5}) + \frac{1}{16}J(\omega_{1,9}) + \frac{1}{4}J(\omega_{1,13}) + \frac{1}{24}J(\omega_{2,3}) + \frac{1}{16}J(\omega_{3,4}) + \frac{1}{16}J(\omega_{3,7}) + \frac{1}{16}J(\omega_{3,11}) + \frac{1}{4}J(\omega_{3,15}) \tag{10.16}$$

which, after substituting the transition frequencies simplifies to:

$$R_{1,3,1,3} = \frac{1}{4}J\left(\frac{A}{2}\right) + \frac{1}{4}J\left(\omega_S + \frac{A}{2}\right) + \frac{19}{24}J(\omega_S) \tag{10.17}$$

Following this procedure one can check that indeed $R_{1,3,1,3} = R_{1,7,1,7} = R_{9,11,9,11} = R_{13,15,13,15} = T_+^{-1}$. Analogously, for the second term in Eq. 10.8, one obtains:

$$R_{2,4,2,4} = \frac{1}{4}J\left(\frac{A}{2}\right) + \frac{1}{4}J\left(\omega_S - \frac{A}{2}\right) + \frac{19}{24}J(\omega_S) \tag{10.18}$$

and $R_{2,4,2,4} = R_{6,8,6,8} = R_{10,12,10,12} = R_{14,16,14,16} = T_-^{-1}$. Next, one should calculate the further electron spin relaxation rates entering

Eq. 10.10; they are given as:

$$R_{1,3,5,7} = R_{5,7,13,15} = R_{9,11,13,15} = R_{2,4,6,8} = R_{6,8,14,16} = R_{10,12,14,16} = \frac{1}{8}J\left(\frac{A}{2}\right) \quad (10.19a)$$

$$R_{1,3,9,11} = \frac{1}{8}J\left(\omega_S + \frac{A}{2}\right),\ R_{2,4,10,12} = \frac{1}{8}J\left(\omega_S - \frac{A}{2}\right) \quad (10.19b)$$

$$R_{5,7,9,11} = R_{6,8,10,12} = \frac{1}{12}J(\omega_S) \quad (10.19c)$$

$$R_{1,3,13,15} = R_{2,4,14,16} = \frac{1}{4}J(\omega_S) \quad (10.19d)$$

Thus, Eq. 10.19 clearly shows that the description of the lineshape function given by Eq. 10.8 is an oversimplification and one should use the matrix product of Eq. 10.9 for this purpose.

In the next step it is interesting to explore the influence of the S_1–S_2 dipole–dipole coupling on the ESR lineshape. The dipolar interaction between the two electron spins is described by the Hamiltonian $H_{DD}(S_1, S_2)$. It has been defined in Eq. 2.12 and Eq. 2.13, but now the expressions have to be adjusted to the geometry of our system:

$$H_{DD}^{(L)}(S_1, S_2)(t) = a_{DD} \sum_{m=-2}^{2} (-1)^m F_{-m}^{2(L)}(t)\, T_m^2(S_1, S_2) \quad (10.20)$$

with $F_{-m}^{2(L)}(t) = D_{0,m}^2(\Omega_{DDL}(t))$, where $\Omega_{DDL}(t)$ describes the orientation of the S_1–S_2 dipole–dipole axis with respect to the laboratory frame (the direction of the external magnetic field); $\Omega_{DDL}(t)$ is also modulated by the molecular tumbling. One can hardly expect that $\Omega_{DDL}(t) = \Omega_1(t) = \Omega_2(t)$; this would mean that all nitrogen and electron spins form one axis. Nevertheless, the assumption simplifies the calculations, not affecting the quantum-mechanical picture of the system.

The perturbation Hamiltonian, $H'(P_1, P_2, S_1, S_2)(t)$, now consists of three terms:

$$H'(P_1, P_2, S_1, S_2)(t) = H_{A,aniso}^{(L)}(P_1, S_1)(t) + H_{A,aniso}^{(L)}(P_2, S_2)(t) + H_{DD}^{(L)}(S_1, S_2)(t) \quad (10.21)$$

The dipolar Hamiltonian contributes to the blocks I, II, III, and IV given by Eq. 10.14a,b (simplified by setting $G_m^{(1)} = G_m^{(2)} = G_m$), which yield (it has been set, $F_m^{2(\mathrm{L})} = F_m$):

$$[\mathrm{I}'] \doteq \begin{bmatrix} \frac{1}{\sqrt{6}}G_0 + \frac{1}{2\sqrt{6}a_{\mathrm{DD}}F_0} & -\frac{1}{4}G_{-1} & -\frac{1}{4}G_{-1} - \frac{1}{4}a_{\mathrm{DD}}F & \frac{1}{2}G_{-2} \\ \frac{1}{4}G_1 & -\frac{1}{2\sqrt{6}}a_{\mathrm{DD}}F_0 & -\frac{1}{2\sqrt{6}}G_0 & \frac{1}{4}G_{-1} + \frac{1}{4}a_{\mathrm{DD}}F_{-1} \\ \frac{1}{4}G_1 + \frac{1}{4}a_{\mathrm{DD}}F_1 & -\frac{1}{2\sqrt{6}}G_0 & -\frac{1}{2\sqrt{6}}a_{\mathrm{DD}}F_0 & \frac{1}{4}G_{-1} \\ \frac{1}{2}G_2 & -\frac{1}{4}G_1 - \frac{1}{4}a_{\mathrm{DD}}F_1 & -\frac{1}{4}G_1 & \frac{1}{\sqrt{6}}G_0 + \frac{1}{2\sqrt{6}}a_{\mathrm{DD}}F_1 \end{bmatrix}(t)$$

(10.22a)

$$[\mathrm{II}'] = \begin{bmatrix} \frac{1}{2\sqrt{6}}a_{\mathrm{DD}}F_0 & -\frac{1}{4}G_{-1} & -\frac{1}{4}G_{-1} - \frac{1}{4}a_{\mathrm{DD}}F & \frac{1}{2}G_{-2} \\ \frac{1}{4}G_1 & -\frac{1}{\sqrt{6}}G_0 + \frac{1}{2\sqrt{6}}a_{\mathrm{DD}}F_0 & -\frac{1}{2\sqrt{6}}G_0 & \frac{1}{4}G_{-1} + \frac{1}{4}a_{\mathrm{DD}}F_{-1} \\ \frac{1}{4}G_1 + \frac{1}{4}a_{\mathrm{DD}}F_1 & -\frac{1}{2\sqrt{6}}G_0 & -\frac{1}{\sqrt{6}}G_0 - \frac{1}{2\sqrt{6}}a_{\mathrm{DD}}F_0 & \frac{1}{4}G_{-1} \\ \frac{1}{2}G_2 & -\frac{1}{4}G_1 - \frac{1}{4}a_{\mathrm{DD}}F_{-} & -\frac{1}{4}G_1 & \frac{1}{2\sqrt{6}}a_{\mathrm{DD}}F_0 \end{bmatrix}(t)$$

(10.22b)

$$[\mathrm{III}'] = \begin{bmatrix} -\frac{1}{2\sqrt{6}}a_{\mathrm{DD}}F_0 & -\frac{1}{4}G_{-1} & -\frac{1}{4}G_{-1} + \frac{1}{4}a_{\mathrm{DD}}F & \frac{1}{2}G_{-2} \\ \frac{1}{4}G_1 & -\frac{1}{\sqrt{6}}G_0 - \frac{1}{2\sqrt{6}}a_{\mathrm{DD}}F_0 & -\frac{1}{2\sqrt{6}}G_0 & \frac{1}{4}G_{-1} - \frac{1}{4}a_{\mathrm{DD}}F_{-1} \\ \frac{1}{4}G_1 - \frac{1}{4}a_{\mathrm{DD}}F_1 & -\frac{1}{2\sqrt{6}}G_0 & -\frac{1}{\sqrt{6}}G_0 - \frac{1}{2\sqrt{6}}a_{\mathrm{DD}}F_0 & \frac{1}{4}G_{-1} \\ \frac{1}{2}G_2 & -\frac{1}{4}G_1 + \frac{1}{4}a_{\mathrm{DD}}F_{-} & -\frac{1}{4}G_1 & -\frac{1}{2\sqrt{6}}a_{\mathrm{DD}}F_0 \end{bmatrix}(t)$$

(10.22c)

$$[\mathrm{IV}'] = \begin{bmatrix} \frac{1}{\sqrt{6}}G_0 + \frac{1}{2\sqrt{6}}a_{\mathrm{DD}}F_0 & -\frac{1}{4}G_{-1} & -\frac{1}{4}G_{-1} + \frac{1}{4}a_{\mathrm{DD}}F & \frac{1}{2}G_{-2} \\ \frac{1}{4}G_1 & -\frac{1}{2\sqrt{6}}a_{\mathrm{DD}}F_0 & -\frac{1}{2\sqrt{6}}G_0 & \frac{1}{4}G_{-1} - \frac{1}{4}a_{\mathrm{DD}}F_{-1} \\ \frac{1}{4}G_1 - \frac{1}{4}a_{\mathrm{DD}}F_1 & -\frac{1}{2\sqrt{6}}G_0 & -\frac{1}{2\sqrt{6}}a_{\mathrm{DD}}F_0 & \frac{1}{4}G_{-1} \\ \frac{1}{2}G_2 & -\frac{1}{4}G_1 + \frac{1}{4}a_{\mathrm{DD}}F_1 & -\frac{1}{4}G_1 & \frac{1}{\sqrt{6}}G_0 + \frac{1}{2\sqrt{6}}a_{\mathrm{DD}}F_0 \end{bmatrix}(t) \quad (10.22\mathrm{d})$$

The S_1–S_2 dipole–dipole interaction also contributes to other Hamiltonian matrix elements. We shall not present here the full calculations as we believe that after the examples given, the procedure is clear. For instance, the dipole–dipole contribution to the relaxation rate $R_{1,1,2,2}$ is $R^{\mathrm{DD}}_{1,1,2,2} = \frac{1}{8}a^2_{\mathrm{DD}}J(0)$. Also the electronic relaxation rates contain terms originating from interference effects (Section 5.3 of Chapter 5) between the anisotropy of the hyperfine coupling and the dipole–dipole coupling as both interactions are modulated by the rotational dynamics. This subject shall be discussed in detail in Section 10.2 below.

In Fig. 6.9, an illustrative ESR spectrum for di-*tert* butyl is shown; the broad line indicates fast electron spin relaxation originating from the strong dipole–dipole coupling between the electron spins. In Fig. 10.6, examples of ^{1}H spin–lattice relaxation dispersion data for propylene carbonate solution of di-*tert* butyl are presented.

It is interesting to compare the shapes of the relaxation dispersion with those obtained for solutions of radicals containing only one paramagnetic center shown (Fig. 7.6).

10.2 Interference Effects for Nitroxide Radicals

In Section 10.1, we have discussed the influence of the dipole–dipole coupling between the electron spins, S_1–S_2, on ESR lineshape of a system containing two electron spin–nuclear (^{15}N) spin pairs, P–S (a biradical system). It has been explained that the presence of the

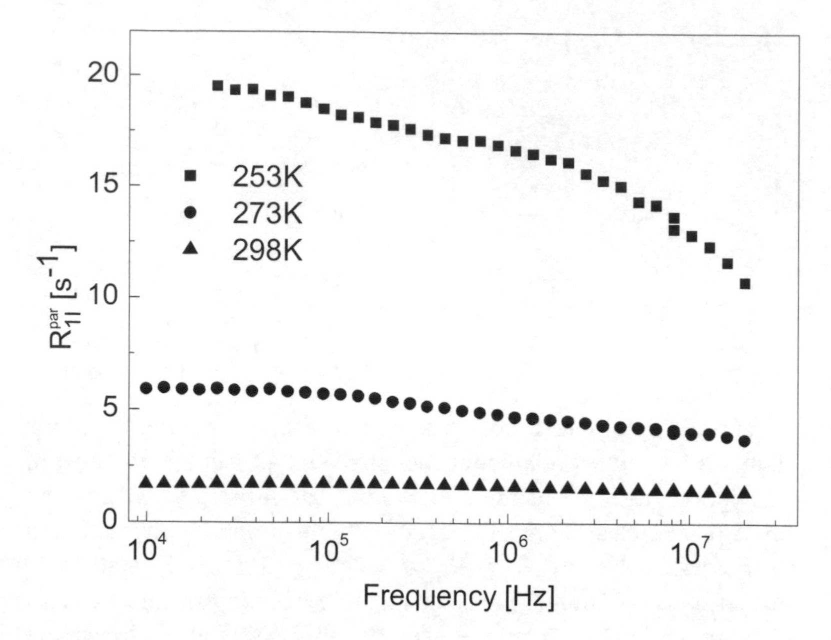

Figure 10.6 ^1H spin–lattice relaxation for propylene carbonate solution of di-tert butyl matrix (10 mMol concentration).

dipole–dipole interaction leads not only to relaxation rates entirely associated with this coupling, but also to additional relaxation terms caused by interference between the dipole–dipole coupling and the hyperfine interactions. Interference effects are also present for simpler systems, for instance, a ^{14}N radical (another example of interference effects has been described in Section 5.3) [1, 2, 3–5].

As for ^{14}N, $P = 1$, the nitrogen nucleus experiences a quadrupolar interaction, $H_Q(P)$. The quadrupolar coupling is also modulated by the rotational motion of the paramagnetic molecule (for instance the anisotropic part of the P–S hyperfine coupling, $H_{A,\mathrm{aniso}}^{(L)}(P, S)$). In consequence, the interactions interfere, creating additional terms contributing to the electronic relaxation. To understand the influence of the ^{14}N quadrupolar coupling on the electron spin relaxation one has to extend Eq. 6.31 by including the quadrupole interaction to the relaxation Hamiltonian. For this purpose it is useful to remind the labeling of the Zeeman states $|m_S, m_P\rangle$: $|1\rangle = |1/2, 1\rangle$, $|2\rangle = |1/2, 0\rangle$, $|3\rangle = |1/2, -1\rangle$, $|4\rangle = |-1/2, 1\rangle$,

$|5\rangle = |-1/2, 0\rangle$, and $|6\rangle = |-1/2, -1\rangle$ (Chapter 6) and the form of the quadrupolar Hamiltonian, $H_{\mathrm{Q}}^{(\mathrm{L})}(P)$ (Eq. 4.9, 4.10):

$$H_{\mathrm{Q}}^{(\mathrm{L})}(P)(t) = \frac{1}{2}\sqrt{\frac{3}{2}}\frac{a_{\mathrm{Q}}}{P(2P-1)}\sum_{m=-2}^{2}(-1)^m V_{-m}^{2(\mathrm{L})} T_m^2(P) \tag{10.23a}$$

with

$$V_{-m}^{2(\mathrm{L})}(t) = D_{0,m}^2(\Omega_{\mathrm{PL}}(t)) + \frac{\eta}{\sqrt{6}}\left(D_{2,m}^2(\Omega_{\mathrm{PL}}(t))\, D_{-2,m}^2(\Omega_{\mathrm{PL}}(t))\right) \tag{10.23b}$$

where $\Omega_{\mathrm{PL}}(t)$ describes the orientation of the principal axis system of the electric field gradient (P) with respect to the laboratory axis (L). The contribution of the quadrupole interaction leads to the following representation of the perturbing Hamiltonian, $H'(t)$:

The notation in Eq. 10.24 has been simplified: $G_m^{2(\mathrm{L})} \equiv G_m$ and $V_m^{2(\mathrm{L})} \equiv V_m$. Equation 6.27 gives the relationship between the Zeeman basis $\{|i\rangle\}$ and the eigenbasis $\{|\psi_\alpha\rangle\}$ of the $I-S$ spin system, $|\psi_1\rangle = |1\rangle$, $|\psi_2\rangle = a|2\rangle + b|4\rangle$, $|\psi_3\rangle = c|2\rangle + d|4\rangle$, $|\psi_4\rangle = e|3\rangle + f|5\rangle$, $|\psi_5\rangle = g|3\rangle + h|5\rangle$, and $|\psi_6\rangle = |6\rangle$. Using this relationship the representation of the Hamiltonian $H_{\mathrm{A,aniso}}^{(\mathrm{L})}(P, S)$ in the eigenbasis $\{|\psi_\alpha\rangle\}$ has been given by Eq. 6.41. In the same way one can set up the representation of the quadrupolar Hamiltonian, $H_{\mathrm{Q}}^{(\mathrm{L})}(P)$ in the $\{|\psi_\alpha\rangle\}$ basis:

$$\left\lfloor H_{\mathrm{Q}}^{(\mathrm{L})}(P)\right\rfloor(t) =$$

$$\begin{bmatrix}
\frac{1}{2}\sqrt{\frac{3}{2}}V_0 & a\frac{\sqrt{3}}{2}V_{-1} & c\frac{\sqrt{3}}{2}V_{-1} & e\sqrt{\frac{3}{2}}V_{-1} & g\sqrt{\frac{3}{2}}V_{-2} & 0 \\
-a\frac{\sqrt{3}}{2}V_{-1} & b^2\frac{1}{2}\sqrt{\frac{3}{2}}V_0 & bd\frac{1}{2}\sqrt{\frac{3}{2}}V_0 & (ae+bf)\frac{\sqrt{3}}{2}V_{-1} & (ag+bh)\frac{\sqrt{3}}{2}V_{-1} & b\sqrt{\frac{3}{2}}V_{-2} \\
-c\frac{\sqrt{3}}{2}V_1 & bd\frac{1}{2}\sqrt{\frac{3}{2}}V_0 & d^2\frac{1}{2}\sqrt{\frac{3}{2}}V_0 & (ce+df)\frac{\sqrt{3}}{2}V_{-1} & (cg+dh)\frac{\sqrt{3}}{2}V_{-1} & d\sqrt{\frac{3}{2}}V_{-2} \\
e\sqrt{\frac{3}{2}}V_2 & -(ae+bf)\frac{\sqrt{3}}{2}V_1 & -(ce+df)\frac{\sqrt{3}}{2}V_1 & e^2\frac{1}{2}\sqrt{\frac{3}{2}}V_0 & eg\frac{1}{2}\sqrt{\frac{3}{2}}V_0 & f\frac{\sqrt{3}}{2}V_{-1} \\
g\sqrt{\frac{3}{2}}V_2 & -(ag+bh)\frac{\sqrt{3}}{2}V_{-1} & -(cg+dh)\frac{\sqrt{3}}{2}V_1 & eg\frac{1}{2}\sqrt{\frac{3}{2}}V_0 & g^2\frac{1}{2}\sqrt{\frac{3}{2}}V_0 & h\frac{\sqrt{3}}{2}V_{-1} \\
0 & b\sqrt{\frac{3}{2}}V_2 & d\sqrt{\frac{3}{2}}V_2 & -f\frac{\sqrt{3}}{2}V_1 & -h\frac{\sqrt{3}}{2}V_{-1} & \frac{1}{2}\sqrt{\frac{3}{2}}V_0
\end{bmatrix} \tag{10.25}$$

$$[H'](t) = \left[H^{(L)}_{A,\text{aniso}}(P,S) + H^{(L)}_{Q}(P) \right](t)$$

$$\begin{bmatrix} \frac{1}{\sqrt{6}}G_0 + \frac{1}{2}\sqrt{\frac{3}{2}}a_Q V_0 & \frac{1}{2\sqrt{2}}G_{-1} + \frac{\sqrt{3}}{2}a_Q V_{-1} & \sqrt{\frac{3}{2}}a_Q V_{-2} & \frac{1}{2}G_{-1} & \frac{1}{2}G_{-2} & 0 \\ -\frac{1}{2\sqrt{2}}G_1 - \frac{\sqrt{3}}{2}a_Q V_1 & 0 & \frac{1}{2\sqrt{2}}G_{-1} + \frac{\sqrt{3}}{2}a_Q V_{-1} & -\frac{1}{2\sqrt{3}}G_0 & 0 & \frac{1}{\sqrt{2}}G_{-2} \\ \sqrt{\frac{3}{2}}a_Q V_2 & -\frac{1}{2\sqrt{2}}G_1 - \frac{\sqrt{3}}{2}a_Q V_1 & -\frac{1}{\sqrt{6}}G_0 + \frac{1}{2}\sqrt{\frac{3}{2}}a_Q V_0 & 0 & -\frac{1}{2\sqrt{3}}G_0 & -\frac{1}{2}G_{-1} \\ -\frac{1}{2}G_1 & -\frac{1}{2\sqrt{3}}G_0 & 0 & -\frac{1}{\sqrt{6}}G_0 + \frac{1}{2}\sqrt{\frac{3}{2}}a_Q V_0 & -\frac{1}{2\sqrt{2}}G_{-1} + \frac{\sqrt{3}}{2}a_Q V_{-1} & \sqrt{\frac{3}{2}}a_Q V_{-2} \\ \frac{1}{\sqrt{2}}G_2 & 0 & -\frac{1}{2\sqrt{3}}G_0 & \frac{1}{2\sqrt{2}}G_1 - \frac{\sqrt{3}}{2}a_Q V_1 & 0 & -\frac{1}{2\sqrt{2}}G_{-1} + \frac{\sqrt{3}}{2}a_Q V_{-1} \\ 0 & \frac{1}{\sqrt{2}}G_2 & \frac{1}{2}G_1 & \sqrt{\frac{3}{2}}a_Q V_2 & \frac{1}{2\sqrt{2}}G_1 - \frac{\sqrt{3}}{2}a_Q V_1 & \frac{1}{\sqrt{6}}G_0 + \frac{1}{2}\sqrt{\frac{3}{2}}a_Q V_0 \end{bmatrix} \tag{10.24}$$

Equations 6.41a-c give expressions for some of the electron spin relaxation rates, $R_{\alpha\beta\alpha\beta}$, resulting from the anisotropic part of the hyperfine coupling; below we shall denote them as $R^{A}_{\alpha\beta\alpha\beta}$. The total electron spin relaxation rates, $R^{\text{tot}}_{\alpha\beta\alpha\beta}$, are given as a sum of three terms:

$$R^{\text{tot}}_{\alpha\beta\alpha\beta} = R^{A}_{\alpha\beta\alpha\beta} + R^{Q}_{\alpha\beta\alpha\beta} + R^{\text{A-Q}}_{\alpha\beta\alpha\beta} \qquad (10.26)$$

where $R^{\text{Q}}_{\alpha\beta\alpha\beta}$ denotes the contribution to the relaxation rates originating from the quadrupole interaction, while $R^{\text{A-Q}}_{\alpha\beta\alpha\beta}$ describes relaxation terms resulting from interference effects between the anisotropic part of the P–S hyperfine coupling and the quadrupole interaction. The relaxation rates can be obtained using the Redfield relaxation formula of Eq. 3.19; for instance for the quadrupolar relaxation rates one obtains:

$$\begin{aligned} R^{\text{Q}}_{1212} = {} & \frac{3}{8}\left(1-b^2\right)^2 J^{\text{Q}}(0) + \frac{3}{2}a^2 J^{\text{Q}}(\omega_{12}) + \frac{3}{4}c^2 J^{\text{Q}}(\omega_{13}) \\ & + \frac{3}{2}e^2 J^{\text{Q}}(\omega_{14}) + \frac{3}{2}g^2 J^{\text{Q}}(\omega_{15}) + \frac{3}{8}b^2 d^2 J^{\text{Q}}(\omega_{23}) \\ & + \frac{3}{4}(ae-bf)^2 J^{\text{Q}}(\omega_{24}) + \frac{3}{4}(ag-bh)^2 J^{\text{Q}}(\omega_{25}) \\ & + \frac{3}{2}b^2 J^{\text{Q}}(\omega_{26}) \end{aligned} \qquad (10.27)$$

For isotropic molecular tumbling, the quadrupolar spectral densities, $J^{\text{Q}}(\omega)$ are defined as:

$$J^{\text{Q}}(\omega) = \frac{1}{5}a_{\text{Q}}^2\left(1+\frac{\eta^3}{3}\right)\frac{\tau_{\text{rot}}}{1+\omega^2\tau_{\text{rot}}^2} \qquad (10.28)$$

The method of calculating relaxation rates originating from interference between different spin interactions has been described in Section 5.3. Special attention should be given here Eq. 5.35, which describes the correlation between different spin interactions when they are modulated by the same motional process. As an example, explicit expressions for some of the relaxation rates obtained from the general relaxation formula of Eq. 3.19 are provided:

$$
\begin{aligned}
R_{1212}^{\text{A-Q}} &= \frac{1}{2\sqrt{2}}\left(1-b^2\right)\left[1+b\left(1+a\sqrt{2}\right)\right]J^{\text{A-Q}}(0) \\
&+\frac{\sqrt{3}}{2}a\left(a+b\sqrt{2}\right)J^{\text{A-Q}}(\omega_{12}) \\
&+\frac{\sqrt{3}}{8}c\left(c+d\sqrt{2}\right)^2 J^{\text{A-Q}}(\omega_{13})+\frac{\sqrt{3}}{4}efJ^{\text{A-Q}}(\omega_{14}) \\
&+\frac{\sqrt{3}}{4}ghJ^{\text{A-Q}}(\omega_{15})-\frac{1}{4\sqrt{2}}bd\left(ad+bc+bd\sqrt{2}\right)J^{\text{A-Q}}(\omega_{23}) \\
&+\frac{1}{4}\sqrt{\frac{3}{2}}(ae+bf)(ae-bf)J^{\text{A-Q}}(\omega_{24}) \\
&+\frac{1}{4}\sqrt{\frac{3}{2}}(ag+bh)(ag-bh)J^{\text{A-Q}}(\omega_{25})+\frac{\sqrt{3}}{2}abJ^{\text{A-Q}}(\omega_{26})
\end{aligned}
\tag{10.29a}
$$

$$
\begin{aligned}
R_{1313}^{\text{A-Q}} &= \frac{1}{2\sqrt{2}}\left(1-d^2\right)\left[1+d\left(d+c\sqrt{2}\right)\right]J^{\text{A-Q}}(0) \\
&+\frac{\sqrt{3}}{8}a\left(a+b\sqrt{2}\right)J^{\text{A-Q}}(\omega_{12})+ \\
&\frac{\sqrt{3}}{4}c\left(c+d\sqrt{2}\right)^2 J^{\text{A-Q}}(\omega_{13})+\frac{\sqrt{3}}{4}efJ^{\text{A-Q}}(\omega_{14}) \\
&+\frac{\sqrt{3}}{4}ghJ^{\text{A-Q}}(\omega_{15})-\frac{1}{4\sqrt{2}}bd\left(ad+bc+bd\sqrt{2}\right)J^{\text{A-Q}}(\omega_{23}) \\
&+\frac{1}{4}\sqrt{\frac{3}{2}}(ce+df)(ce-df)J^{\text{A-Q}}(\omega_{34}) \\
&+\frac{1}{4}\sqrt{\frac{3}{2}}(cg+dh)(cg-dh)J^{\text{A-Q}}(\omega_{35})+\sqrt{\frac{3}{2}}cdJ^{\text{A-Q}}(\omega_{36})
\end{aligned}
\tag{10.29b}
$$

where the spectral density $J^{\text{A-Q}}(\omega)$ has the form:

$$
\begin{aligned}
J^{\text{A-Q}}(\omega) &= \frac{1}{10}a_Q\left\{\left[A_{zz}-\frac{1}{2}\left(A_{xx}+A_{yy}\right)\right]\right. \\
&\left.+\frac{\eta}{\sqrt{6}}\left(A_{xx}-A_{yy}\right)\right\}\frac{\tau_{\text{rot}}}{1+\omega^2\tau_{\text{rot}}^2}
\end{aligned}
\tag{10.30}
$$

The expression for the spectral density, $J^{\text{A-Q}}(\omega)$ has been obtained combining Eq. 10.13 and Eq. 10.23b under the assumption explained below.

10.3 Spin Interactions and Molecular Geometry

In Sections 10.1 and 10.2, a very simple molecular geometry has been considered. It has been assumed that the principal axis system of the hyperfine, electron spin–nitrogen spin interaction (its asymmetric part) coincides with the principal axis system of the electric field gradient at the position of nitrogen. It has also been assumed that the principal axes of the P_1–S_1 and P_2–S_2 hyperfine couplings coincide with the S_1–S_2 dipole–dipole axis. Such simplifying assumptions help to focus on the quantum-mechanical concept of the calculations. Nevertheless, one should be prepared to take into account "real" molecular geometries.

In the case of several spins and interactions one has to define a molecule-fixed frame which is treated as a "reference" frame. For the system discussed in Section 10.1, it is natural to choose the S_1–S_2 dipole–dipole axis as the molecular (reference) frame (M). Its orientation with respect to the laboratory frame (the direction of the external magnetic field) is described by the angle, $\Omega_{\mathrm{DDL}}(t) \equiv \Omega_{\mathrm{ML}}(t)$; the S_1–S_2 dipole–dipole Hamiltonian is given (in the laboratory frame) by Eq. 10.20. Then, to express the P_1–S_1(P_2–S_2) hyperfine interaction in the laboratory frame one has to perform a two-step transformation [2, 4, 6]. Firstly, the interaction should be transformed to the molecular frame, yielding:

$$H_{\mathrm{A,aniso}}^{(\mathrm{M})}(P, S) = \sum_{m=-2}^{2} (-1)^m G_{-m}^{2(\mathrm{M})} T_m^2(P, S) \tag{10.31}$$

with (see Eq. 10.12)

$$G_{-m}^{2(\mathrm{M})} = D_{0,-m}^2(\Omega_{\mathrm{AM}}) \sqrt{\frac{2}{3}} \left[A_{zz} - \frac{1}{2}(A_{xx} + A_{yy})\right] + \left(D_{-2,-m}^2(\Omega_{\mathrm{AM}}) + D_{2,-m}^2(\Omega_{\mathrm{AM}})\right) \frac{1}{2}(A_{xx} - A_{yy}) \tag{10.32}$$

where Ω_{AM} denotes the angle between the principal axis system of the P_1–S_1(P_2–S_2) anisotropic hyperfine coupling and the molecular frame (the Hamiltonian $H_{\mathrm{A,aniso}}^{(\mathrm{M})}(P, S)$ is time independent as the angle Ω_{AM} is determined only by the molecular geometry). Secondly, the Hamiltonian $H_{\mathrm{A,aniso}}^{(\mathrm{M})}(P, S)$ has to be transformed to the laboratory frame by the angle $\Omega_{\mathrm{ML}}(t)$. Thus, eventually one gets:

$$H^{(\mathrm{L})}_{\mathrm{A,aniso}}(P,S)(t) = \sum_{m=-2}^{2}(-1)^m \left(\sum_{k=-2}^{2} G_k^{2(\mathrm{M})} D^2_{k-m}(\Omega_{\mathrm{ML}}(t))\right) T^2_m(P,S) \tag{10.33}$$

Analogously, one should proceed with the quadrupolar interaction of ^{14}N. Performing the two step-transformation, from the principal axis system of the electric field gradient tensor (P) to the molecular frame (M) (via the time-independent angle, Ω_{PM}) and then, from the (M) frame to the (L) frame (via the $\Omega_{\mathrm{ML}}(t)$ angle), one obtains:

$$H^{(\mathrm{L})}_{\mathrm{Q}}(P)(t) = \frac{1}{2}\sqrt{\frac{3}{2}}\frac{a_Q}{P(2P-1)}\sum_{m=-2}^{2}(-1)^m \times \left(\sum_{k=-2}^{2} V_k^{2(\mathrm{M})} D^2_{k-m}(\Omega_{\mathrm{ML}}(t))\right) T^2_m(P) \tag{10.34}$$

where (see Eq. 10.23b):

$$V_m^{2(\mathrm{M})} = D^2_{0,m}(\Omega_{\mathrm{PM}}) + \frac{\eta}{\sqrt{6}}\left(D^2_{2,m}(\Omega_{\mathrm{PM}}(t))\, D^2_{-2,m}(\Omega_{\mathrm{PM}})\right) \tag{10.35}$$

In consequence, the time fluctuations of all interactions are described by the angle $\Omega_{\mathrm{ML}}(t)$, while the molecular geometry is encoded to the Hamiltonian forms by the angles Ω_{AM} and Ω_{PM}.

References

1. Goldman, M. (1984). Interference effects in the relaxation of a pair of unlike spin-1/2 nuclei. *J. Magn. Reson.*, **60**, pp. 437–452.
2. Kowalewski, J. and Mäler, L. (2006). *Nuclear Spin Relaxation in Liquids: Theory, Experiments and Applications* (New York: Taylor & Francis).
3. Kumar, A., Grace, R. C. R. and Madhu, P. K. (2000). Cross-correlations in NMR. *Prog. Nucl. Magn. Reson. Spectr.*, **37**, pp. 191–319.
4. Nilsson, T. and Kowalewski, J. (2000). Slow-motion theory of nuclear spin relaxation in paramagnetic low-symmetry complexes: A generalization to high electron spin. *J. Magn. Reson.*, **146**, pp. 345–358.
5. Werbelow, L. G. (1978). The use of interference terms to separate dipolar and chemical shift anisotropy contributions to nuclear spin relaxation. *J. Magn. Reson.*, **71**, pp.151–153.
6. Kruk, D. (2007). *Theory of Evolution and Relaxation of Multi-Spin Systems* (Bury St Edmunds: Arima).

Chapter 11

Dynamic Nuclear Polarization

Dynamic nuclear polarization (DNP) is a phenomenon caused by quantum-mechanical properties of a spin system. Under certain conditions electron spin polarization can be transferred to (taken over by) neighboring nuclear spins leading to a considerable increase of nuclear spin polarization. In this chapter a theoretical background of the DNP phenomenon is presented.

11.1 Principles of Dynamic Nuclear Polarization (DNP)

Consider an electron spin S coupled by dipole–dipole and scalar interactions to a nuclear spin I. The Hamiltonian, $H_0(I, S)$, for the I–S spin system contains the following terms:

$$H_0(I, S) = H_Z(I) + H_Z(S) + H_{A,sc}(I, S) + H_{DD}(I, S) \qquad (11.1)$$

where $H_Z(I)$ and $H_Z(S)$ are Zeeman couplings for the nuclear and electron spins, respectively. The Hamiltonian of the scalar interaction has been defined by Eq. 6.1, which in the present case yields:

$$H_{A,sc}(I, S) = A I_Z S_Z + \frac{1}{2} A (I_+, S_- + I_-, S_+) \qquad (11.2a)$$

Understanding Spin Dynamics
Danuta Kruk

ISBN 978-981-4463-49-2 (Hardcover), 978-981-4463-50-8 (eBook)
www.panstanford.com

$$[H_0(I,S)](\Omega) =$$

$$\begin{bmatrix} \frac{1}{2\sqrt{6}} a_{\mathrm{DD}} D^2_{0,0}(\Omega) + \frac{A}{4} + \frac{1}{2}\omega_{\mathrm{I}} + \frac{1}{2}\omega_{\mathrm{S}} & -\frac{1}{4} a_{\mathrm{DD}} D^2_{0,-1}(\Omega) & -\frac{1}{4} a_{\mathrm{DD}} D^2_{0,-1}(\Omega) & -\frac{1}{2} a_{\mathrm{DD}} D^2_{0,-2}(\Omega) \\ \frac{1}{4} a_{\mathrm{DD}} D^2_{0,1}(\Omega) & -\frac{1}{2\sqrt{6}} a_{\mathrm{DD}} D^2_{0,1}(\Omega) - \frac{A}{4} + \frac{1}{2}\omega_{\mathrm{I}} - \frac{1}{2}\omega_{\mathrm{S}} & -\frac{1}{2\sqrt{6}} a_{\mathrm{DD}} D^2_{0,0}(\Omega) + \frac{A}{2} & \frac{1}{4} a_{\mathrm{DD}} D^2_{0,-1}(\Omega) \\ \frac{1}{4} a_{\mathrm{DD}} D^2_{0,1}(\Omega) & -\frac{1}{2\sqrt{6}} a_{\mathrm{DD}} D^2_{0,0}(\Omega) + \frac{A}{2} & -\frac{1}{2\sqrt{6}} a_{\mathrm{DD}} D^2_{0,0}(\Omega) - \frac{A}{4} - \frac{1}{2}\omega_{\mathrm{I}} + \frac{1}{2}\omega_{\mathrm{S}} & \frac{1}{4} a_{\mathrm{DD}} D^2_{0,-1}(\Omega) \\ \frac{1}{2} a_{\mathrm{DD}} D^2_{0,2}(\Omega) & \frac{1}{4} a_{\mathrm{DD}} D^2_{0,1}(\Omega) & \frac{1}{4} a_{\mathrm{DD}} D^2_{0,1}(\Omega) & \frac{1}{2\sqrt{6}} a_{\mathrm{DD}} D^2_{0,0}(\Omega) + \frac{A}{4} - \frac{1}{2}\omega_{\mathrm{I}} - \frac{1}{2}\omega_{\mathrm{S}} \end{bmatrix} \tag{11.2b}$$

where A is the scalar coupling constant. The form of the dipole-dipole Hamiltonian in the laboratory frame has been given by Eq. 4.51 (in the present chapter we shall omit the index (L) for simplicity); it depends on the orientation of the molecule with respect to the external magnetic field, $\Omega = (0, \beta_{DDL}, \gamma_{DDL}) = (0, \theta, \phi)$. One should note here that we consider a time-independent dipole–dipole interaction which provides (besides the scalar interaction) a polarization transfer pathway between the electron and nuclear spins. The representation of the Hamiltonian $H_0(I, S)$ in the basis $\{|m_I, m_S\rangle\}$ (using the labeling $|1\rangle = |1/2, 1/2\rangle, |2\rangle = |1/2, -1/2\rangle, |3\rangle = |-1/2, 1/2\rangle$, and $|4\rangle = |-1/2, -1/2\rangle$) is given as shown on adjacent page as Eq. 11.2b (see Eq. 4.42 for the dipolar part).

It is important to notice that for $\theta = 0$ only the states $|2\rangle$ and $|3\rangle$ are coupled. As explained in the previous chapters, to get the eigenvalues (energy levels) and eigenvectors of the I–S spin system, the Hamiltonian matrix has to be diagonalized. This leads to four energy levels E_α, $\alpha = 1, \ldots, 4$ and four associated eigenfunctions $|\psi_\alpha\rangle$ given as linear combinations of the Zeeman functions $|n\rangle = |m_I, m_S\rangle$, $|n\rangle = 1, \ldots, 4$ $|\psi_\alpha\rangle = \sum_{n=1}^{4} a_{\alpha n} |n\rangle$. As explained in Chapter 3, the populations of the individual energy levels are represented by the density matrix elements, $\rho_{\alpha\alpha} \equiv |\psi_\alpha\rangle \langle\psi_\alpha|$. They change in time following the set of equations (see Eq. 3.17):

$$\frac{d\rho_{11}}{dt} = R_{1111}\rho_{11} + R_{1122}\rho_{22} + R_{1133}\rho_{33} + R_{1144}\rho_{44} \tag{11.3a}$$

$$\frac{d\rho_{22}}{dt} = R_{2211}\rho_{11} + R_{2222}\rho_{22} + R_{2233}\rho_{33} + R_{2244}\rho_{44} \tag{11.3b}$$

$$\frac{d\rho_{33}}{dt} = R_{3311}\rho_{11} + R_{3322}\rho_{22} + R_{3333}\rho_{33} + R_{3344}\rho_{44} \tag{11.3c}$$

$$\frac{d\rho_{44}}{dt} = R_{4411}\rho_{11} + R_{4422}\rho_{22} + R_{4433}\rho_{33} + R_{4444}\rho_{44} \tag{11.3d}$$

The coefficients, $R_{\alpha\alpha\beta\beta}$ (relaxation rates) describe transition probabilities between the energy levels. A general solution of this set of equations has the form:

$$\rho_{\alpha\alpha}(t) = \sum_{i=1}^{4} C_{\alpha i}\xi_{\alpha i} \exp(\lambda_i t) \tag{11.4}$$

where λ_i are eigenvalues (they are real) of the relaxation matrix containing the elements $R_{\alpha\alpha\beta\beta}$, while $\xi_{\alpha i}$ represents eigenvectors associated with the λ_i eigenvalues. The set of integration constants, $C_{\alpha i}$ can be obtained from the initial populations of the energy levels. The populations are determined by the Boltzman distribution:

$$\rho_{\alpha\alpha}(0) = \frac{\exp(-E_\alpha/k_B T)}{\sum_{\alpha=1}^{4} \exp(-E_\alpha/k_B T)} \tag{11.5}$$

where the energy values are expressed in thermal energy units. In this way, one can calculate how the populations of the individual energy levels depend on time.

To evaluate a quantity referred to as DNP factor (f_{DNP}) one has to come back to the Zeeman representation. Since the matrix containing the coefficients, $a_{\alpha n}$ is Hermitian, one can easily obtain the inverse relationship, $|n\rangle = \sum_{\alpha=1}^{4} a_{n\alpha} |\psi_\alpha\rangle$. The nuclear spin magnetization is proportional do the quantity $\langle I_Z\rangle = |1\rangle\langle 1| + |2\rangle\langle 2| - |3\rangle\langle 3| - |4\rangle\langle 4| \equiv \rho_{11}^z + \rho_{22}^z - \rho_{33}^z - \rho_{44}^z$. The density matrix elements ρ_{nn}^z refer to the Zeeman basis and therefore, are denoted explicitly by the index z. They are related to the density matrix elements $\rho_{\alpha\alpha}$ by:

$$\rho_{nn}^z(t) = \sum_{\alpha=1}^{4} a_{n\alpha} a_{n\alpha}^* \rho_{\alpha\alpha}(t) \tag{11.6}$$

and thus, one obtains for the nuclear spin magnetization:

$$\langle I_z\rangle(t) = \sum_{\alpha=1}^{4} \left(a_{1\alpha}a_{1\alpha}^* + a_{2\alpha}a_{2\alpha}^* - a_{3\alpha}a_{3\alpha}^* - a_{4\alpha}a_{4\alpha}^*\right) \rho_{\alpha\alpha}(t) \tag{11.7}$$

The DNP factor, f_{DNP}, is defined as a ratio between the nuclear magnetization at time t and the initial nuclear magnetization. Thus, it can be calculated from the expression:

$$f_{\mathrm{DNP}} = \frac{\langle I_z\rangle(t)}{\langle I_z\rangle(0)} = \frac{\sum_{\alpha=1}^{4} \left(a_{1\alpha}a_{1\alpha}^* + a_{2\alpha}a_{2\alpha}^* - a_{3\alpha}a_{3\alpha}^* - a_{4\alpha}a_{4\alpha}^*\right) \rho_{\alpha\alpha}(t)}{\sum_{\alpha=1}^{4} \left(a_{1\alpha}a_{1\alpha}^* + a_{2\alpha}a_{2\alpha}^* - a_{3\alpha}a_{3\alpha}^* - a_{4\alpha}a_{4\alpha}^*\right) \rho_{\alpha\alpha}(0)} \tag{11.8}$$

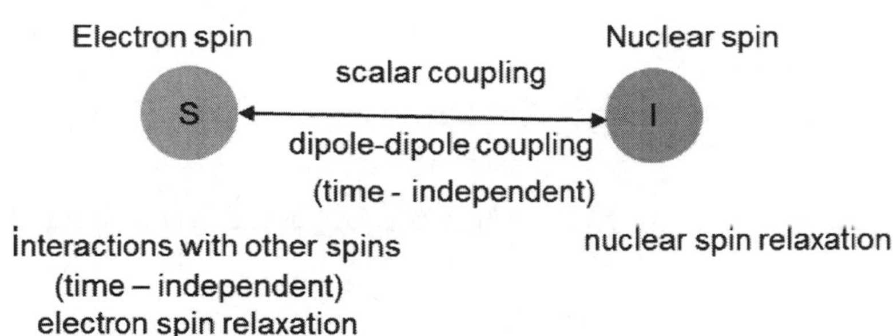

Figure 11.1 Schematic view of a spin system including spin interactions and relaxation processes relevant for DNP effects.

The crucial point of the derivations is to determine the coefficients, $R_{\alpha\alpha\beta\beta}$. There are several factors determining the $R_{\alpha\alpha\beta\beta}$ quantities. In general, they can be classified into two groups, that is, time-independent interactions of the electron spin with other spins affecting the electron spin energy levels and relaxation (electron and nuclear spin) processes. A schematic picture of a spin system in which DNP effects can be observed is shown in Fig. 11.1.

Before we shall discuss the role of other spin interactions and relaxation processes, one should realize that the coefficients $R_{\alpha\alpha\beta\beta}$ are also affected by magnetic field, $B_1(t)$ applied in spin resonance experiments (Chapter 1). This field generates a Hamiltonian, $H'(t)$, which can be expressed as:

$$H'(t) = \gamma_S B_1(t)\left[S_z \cos\varsigma + \frac{1}{2}(S_+ + S_-)\sin\varsigma\right] \quad (11.9)$$

The angle ς gives the relative orientation of the magnetic fields $\vec{B}_0$ and $\vec{B}_1$; typically $\varsigma = \pi/2$. The "nuclear spin" part of this Hamiltonian can be neglected. It is convenient to rewrite the Hamiltonian, $H'(t)$ in the form:

$$H'(t) = \omega_{1,S}\left[S_Z \cos\varsigma + \frac{1}{2}(S_+ + S_-)\sin\varsigma\right] f(t) = H' f(t) \quad (11.10)$$

where $\omega_{1,S} = \gamma_S B_1$; B_1 is the amplitude of the oscillating magnetic field, that is, $B_1(t) = B_1 f(t)$. The function $f(t)$ is a periodic function of a frequency, ω. The Hamiltonian $H'(t)$ expressed in the Zeeman basis $\{|n\rangle = |m_I, m_S\rangle\}$ yields:

$$H'(t) = \frac{1}{2}\gamma_S B_1(t) \begin{bmatrix} \cos\varsigma & \sin\varsigma & 0 & 0 \\ \sin\varsigma & -\cos\varsigma & 0 & 0 \\ 0 & 0 & \cos\varsigma & \sin\varsigma \\ 0 & 0 & \sin\varsigma & -\cos\varsigma \end{bmatrix} \tag{11.11}$$

The coefficients $R_{\alpha\alpha\beta\beta}$ are directly related to the elements of the Hamiltonian $H'(t)$ in the eigenbasis $\{|\psi_\alpha\rangle\}$, which can be obtained using the relationship between the set of vectors $\{|n\rangle\}$ and $\{|\psi_\alpha\rangle\}$:

$$\left\langle \psi_\alpha \middle| H'(t) \middle| \psi_\beta \right\rangle = f(t) \sum_{n,n'=1}^{4} \alpha_{\alpha n}\alpha^*_{\beta n'} \left\langle n \middle| H' \middle| n' \right\rangle \tag{11.12}$$

The coefficients $R_{\alpha\alpha\beta\beta}$, for $\alpha \neq \beta$ are then given as:

$$R_{\alpha\alpha\beta\beta} \propto \left[\sum_{n,n'=1}^{4} \alpha_{\alpha n}\alpha^*_{\beta n'} \left\langle n \middle| H' \middle| n' \right\rangle \right]^2 L_S(\omega_{\alpha\beta}, \omega) \tag{11.13}$$

The diagonal elements $R_{\alpha\alpha\beta\beta}$ are given as $R_{\alpha\alpha\beta\beta} = -\sum_{\beta\neq\alpha} R_{\alpha\alpha\beta\beta}$. The electron spin resonance (ESR) lineshape function, $L_S(\omega_{\alpha\beta}, \omega)$ is important for the efficiency of the DNP process. The amplitude of the ESR spectrum at a frequency ω gives the fraction of electron spins for which the transition frequency is ω. Therefore, the function $L_S(\omega_{\alpha\beta}, \omega)$ acts as a weigh factor for the transition probabilities $R_{\alpha\alpha\beta\beta}$. The spectrum has to be normalized, that is, the amplitude of the ESR spectrum at a given frequency should be divided by the integral (the surface under the lineshape curve). As we have seen in the previous chapters, the ESR lineshape function is determined by many parameters and it has to be calculated starting from the electron spin Hamiltonian appropriate for the considered case. As the first step one can consider the simplest case, that is, a Lorentzian ESR lineshape:

$$L_S(\omega_0, \omega, \tau_S) \propto \frac{1}{1 + (\omega_0 - \omega)^2 \tau_S^2} \tag{11.14}$$

where τ_S is treated as an effective electron spin relaxation time. If the electron spin energy levels would be fully determined by the Zeeman

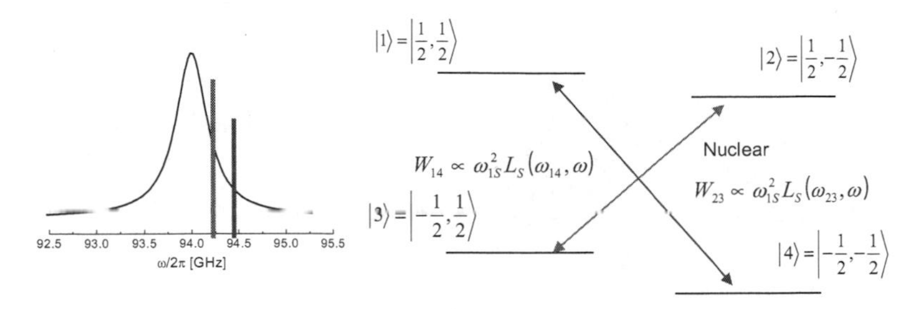

Figure 11.2 The probabilities of spin transitions are related to the ESR lineshape.

interaction, one could set $\omega_0 = \omega_S$ (electron spin Larmor frequency). However, as there are other interactions affecting the electron spin energy level structure (in this case the I–S dipole–dipole and scalar couplings), the "central" transition frequency $\omega_0 = \omega_S$ should be replaced by transition frequencies between the energy levels, $\omega_{\alpha\beta}$. The role of the ESR lineshape function is schematically shown in Fig. 11. 2.

Examples of dependencies of the DNP factor on the frequency ω are shown in Fig. 11.3 for an electron spin–proton spin system. A Lorenzian ESR lineshape has been assumed. The distance between the electron and proton has been varied. One can see that at some frequencies, the DNP factor reaches local maxima; at these frequencies the proton magnetization is significantly larger (the proton signal is strongly enhanced) as the nuclear spin profits from the electron spin polarization due to coupled nuclear spin–electron spin transitions (the enhancement factor cannot exceed the ratio γ_S/γ_I).

Although the presented description captures the essential features of the DNP mechanism it has some deficiencies. Besides the fact that the ESR lineshape is described in a highly simplified way, the electron spin relaxation (which determines the ESR line width) is not taken into account when the transition probabilities $R_{\alpha\alpha\beta\beta}$ are determined. One can imagine that due to fast relaxation processes the contribution to $R_{\alpha\alpha\beta\beta}$ originating from the magnetic field, $B_1(t)$ can become negligible.

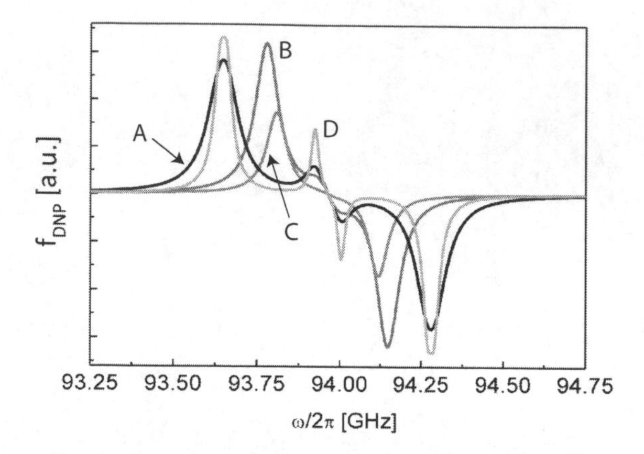

Figure 11.3 DNP factor versus frequency for an electron spin–proton spin system, the I–S scalar interaction has been set to zero, $r = 150$ pm, $\tau_S = 3\times10^{-8}$ s (A), $r = 200$ pm, $\tau_S = 3\times10^{-8}$ s (B), $r = 250$ pm, $\tau_S = 3\times10^{-8}$ s (C), $r = 250$ pm, $\tau_S = 3\times10^{-7}$ s (D).

11.2 DNP and ESR Spectrum

As anticipated in the previous section, calculating the ESR lineshape one should take into account the electron spin–nuclear spin interactions. Thus, the lineshape function should be calculated from Eq. 4.23 which now takes the form:

$$L_S(\omega) = \mathrm{Re}\left\{[S_+]^+ \cdot [M - i\omega_I 1]^{-1} \cdot [S_+]\right\} \tag{11.15}$$

The Liouville basis $\{|\psi_\alpha\rangle\langle\psi_\beta|\}$ in which the matrices of Eq. 11.14 should be represented, is constructed from the eigenfunctions $\{|\psi_\alpha\rangle\}$ of the Hamiltonian $H_0(I, S)$ (Eq. 11.1). The dimension of the Liouville basis is 16. One should distinguish here a 4×4 population block of the matrix $[M - i\omega 1]$ and 8×8 coherence block ($\alpha \neq \beta$), which is diagonal (Chapter 3). For the ESR spectrum, only the diagonal part is of importance. The diagonal elements $[M - i\omega 1]_{\alpha\beta}$ are given as $[M - i\omega 1]_{\alpha\beta} = \tau_S^{-1} + i(\omega_{\alpha\beta} - \omega)$. The representation of the S_+ operator in the basis $\{|\psi_\alpha\rangle\langle\psi_\beta|\}$ can be obtained from the relationship $|\psi_\alpha\rangle = \sum_{n=1}^{4} a_{\alpha n}|n\rangle$, knowing that $S_+ = |1\rangle\langle 2| + |3\rangle\langle 4|$. We still apply here the phenomenological description of the electronic relaxation, introducing just an effective

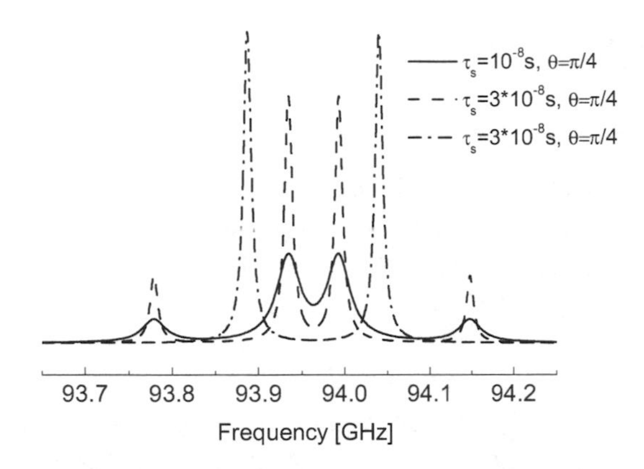

Figure 11.4 ESR spectra for an electron spin–proton spin system; the scalar interaction has been set to zero, $r = 200$ pm .

relaxation time, τ_S. In Fig. 11.4, some examples of ESR spectra for different values of τ_S are shown.

The electron (and nuclear) spin relaxation determines the relaxation rates, $R_{\alpha\alpha\beta\beta}$ in Eq. 11.3. The relaxation rates (as well as $R_{\alpha\beta\alpha\beta}$) should be calculating from the Redfield relaxation theory (Eq. 3.19) knowing the spin interactions which are responsible for the relaxation process and using models of the motion which leads to stochastic fluctuations of these interactions. In most cases this is a rather difficult task; therefore, we shall continue with the phenomenological treatment of relaxation.

Coming back for a while to the Zeeman basis, $\{|n\rangle = |m_I, m_S\rangle\}$, one can distinguish four relaxation rates $R^Z_{1122} = R^Z_{3344} = R_{1Q,S}$ (single-quantum electron spin transitions), $R^Z_{1133} = R^Z_{2244} = R_{1Q,I}$ (single-quantum nuclear spin transitions), $R^Z_{1133} = R_{0Q}$ (zero-quantum transition), and $R^Z_{2244} = R_{2Q}$ (double-quantum transition); $R^z_{nnnn} = -\sum_{n \neq n'} R^z_{n'n'n'n'}$. The relaxation rates can be used to calculate the coefficients $R_{\alpha\beta\alpha\beta}$ (in the basis $\{|\psi_\alpha\rangle\langle\psi_\beta|\}$) which are given as:

$$R_{\alpha\alpha\beta\beta} = \sum_{n,n'=1}^{4} a_{\alpha n} a^*_{\beta n'} R^z_{nnn'n'} \tag{11.16}$$

As the ESR lineshape as well as the DNP factor depend on the orientation of the I–S dipole–dipole coupling, for solid polycrystalline systems averaging over the molecular orientation (powder averaging) should be performed.

11.3 Systems of Many Spins

Discussing the effects of DNP we have considered so far a "model system" consisting of one nuclear spin $I = 1/2$ and one electron spin $S = 1/2$. Real molecular systems are more complex. When, for instance, nitroxide radicals are used as the source of electronic polarization, the spin system which should be considered consists of (at least) two nuclear spins: $I = 1/2$ (for instance ^{1}H), $P = 1/2$ or $P = 1$ (^{15}N or ^{14}N, respectively) and an electron spin $S = 1/2$. For $P = 1/2$, the eigenbasis $\{|\psi_\alpha\rangle\}$ consists of eight functions given as linear combinations of the Zeeman functions $\{|n\rangle = |m_\mathrm{I}, m_\mathrm{P}, m_\mathrm{S}\rangle\}$ (let us use the labeling

$$
\begin{aligned}
&|1\rangle = |1/2, 1/2, 1/2\rangle, \quad |2\rangle = |1/2, 1/2, -1/2\rangle, \\
&|3\rangle = |-1/2, 1/2, 1/2\rangle, |4\rangle = |-1/2, 1/2, -1/2\rangle, \\
&|5\rangle = |1/2, -1/2, 1/2\rangle, |6\rangle = |1/2, -1/2, -1/2\rangle, \\
&|7\rangle = |-1/2, -1/2, 1/2\rangle,
\end{aligned}
$$

and

$$
|8\rangle = |-1/2, -1/2, -1/2\rangle).
$$

The total spin Hamiltonian, $H_0(I, P, S)$ contains, in principle, several terms, but the most important ones are:

$$
H_0(I, P, S) = H_\mathrm{Z}(I) + H_\mathrm{Z}(S) + H_\mathrm{DD}(I, S) + H_\mathrm{A,sc}(P, S) \quad (11.17)
$$

where the Hamiltonian of the scalar, electron spin–nitrogen spin interaction is defined by Eq. 6.1. The general procedure of calculating the DNP factor, f_DNP, has been described in Section 11.1. As nitroxide radicals are often used as a source of polarization, it is useful to provide the matrix representation of the Hamiltonian $H_0(I, P, S)$. The matrix $[H_0(I, P, S)]$ has a block structure shown in Fig. 11.5, where the blocks are given as:

$$[\mathrm{I}] = [\mathrm{II}] = \begin{bmatrix} \frac{1}{2}\omega_I + \frac{1}{2}\omega_S + \frac{A}{4} + \frac{1}{2\sqrt{6}} a_{DD} D^2_{0,0}(\Omega) & -\frac{1}{4} a_{DD} D^2_{0,-1}(\Omega) & \frac{1}{2} a_{DD} D^2_{0,-2}(\Omega) & \frac{1}{2} a_{DD} D^2_{0,-2}(\Omega) \\ \frac{1}{4} a_{DD} D^2_{0,1}(\Omega) & \frac{1}{2}\omega_I - \frac{1}{2}\omega_S - \frac{A}{4} - \frac{1}{2\sqrt{6}} a_{DD} D^2_{0,0}(\Omega) & \frac{A}{4} - \frac{1}{2\sqrt{6}} a_{DD} D^2_{0,0}(\Omega) & \frac{1}{4} a_{DD} D^2_{0,-1}(\Omega) \\ \frac{1}{2} a_{DD} D^2_{0,2}(\Omega) & \frac{A}{4} - \frac{1}{2\sqrt{6}} a_{DD} D^2_{0,0}(\Omega) & -\frac{1}{2}\omega_I + \frac{1}{2}\omega_S - \frac{A}{4} - \frac{1}{2\sqrt{6}} a_{DD} D^2_{0,0}(\Omega) & \frac{1}{4} a_{DD} D^2_{0,-1}(\Omega) \\ \frac{1}{2} a_{DD} D^2_{0,-2}(\Omega) & -\frac{1}{4} a_{DD} D^2_{0,-1}(\Omega) & -\frac{1}{4} a_{DD} D^2_{0,1}(\Omega) & -\frac{1}{2}\omega_I - \frac{1}{2}\omega_S + \frac{A}{4} + \frac{1}{2\sqrt{6}} a_{DD} D^2_{0,0}(\Omega) \end{bmatrix} \tag{11.18}$$

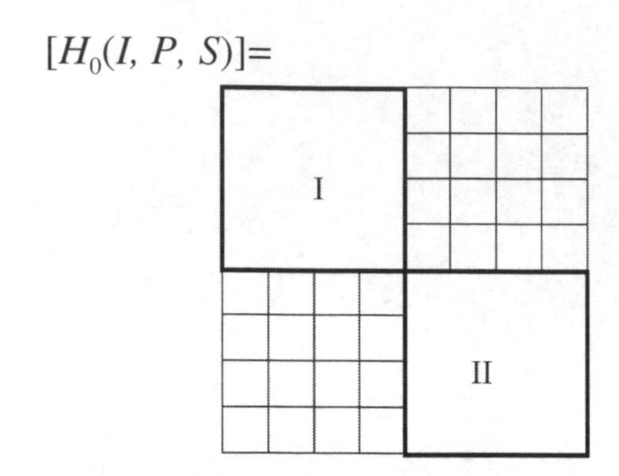

Figure 11.5 Structure of the Hamiltonian matrix, $[H_0\,(I,\ P,\ S)]$ of Eq. 11.17.

Taking into account that in the present case the representation of the I spin magnetization is given as

$$\langle I_Z\rangle = \rho_{11}^Z + \rho_{22}^Z - \rho_{33}^Z - \rho_{44}^Z + \rho_{55}^Z + \rho_{66}^Z - \rho_{77}^Z - \rho_8^Z,$$

one can set up an expression for f_{DNP} in full analogy to Eq. 11.8, while the ESR lineshape should be calculated from Eq. 11.15. As we know from Chapter 6, a relevant contribution to the relaxation rates, $R_{\alpha\alpha\beta\beta}$, stems from the anisotropic part of the electron spin–nitrogen spin hyperfine coupling.

Another example which is worth considering is the case of two coupled electron spins, $S_1 = S_2 = 1/2$ and one nuclear (proton) spin, $I = 1/2$. Then the Hamiltonian, $H_0\,(S,\ S_1,\ S_2)$ contains the following terms (scalar couplings are neglected):

$$H_0\,(S,\ S_1,\ S_2) = H_Z\,(I) + H_Z\,(S_1) + H_Z\,(S_2) + H_{DD}\,(I,\ S_1) + H_{DD}\,(I,\ S_2) + H_{DD}\,(S_1,\ S_2) \quad (11.19)$$

The eigenbasis $\{|\psi_\alpha\rangle\}$ also consists of eight functions, but now they are given as linear combinations of the Zeeman functions, $\left\{|n\rangle = \left|m_{\mathrm{I}},\ m_{\mathrm{S}_1},\ m_{\mathrm{S}_2}\right\rangle\right\}$. It is convenient to label the functions as follows:

$$|1\rangle = |1/2,\ 1/2,\ 1/2\rangle\,, \quad |2\rangle = |-1/2,\ 1/2,\ 1/2\rangle\,,$$
$$|3\rangle = |1/2,\ -1/2,\ 1/2\rangle\,, |4\rangle = |-1/2,\ -1/2,\ 1/2\rangle\,,$$
$$|5\rangle = |1/2,\ 1/2,\ -1/2\rangle\,, |6\rangle = |-1/2,\ 1/2,\ -1/2\rangle\,,$$
$$|7\rangle = |1/2,\ -1/2,\ -1/2\rangle\,,$$

and

$$|8\rangle = |-1/2,\ -1/2,\ -1/2\rangle\,.$$

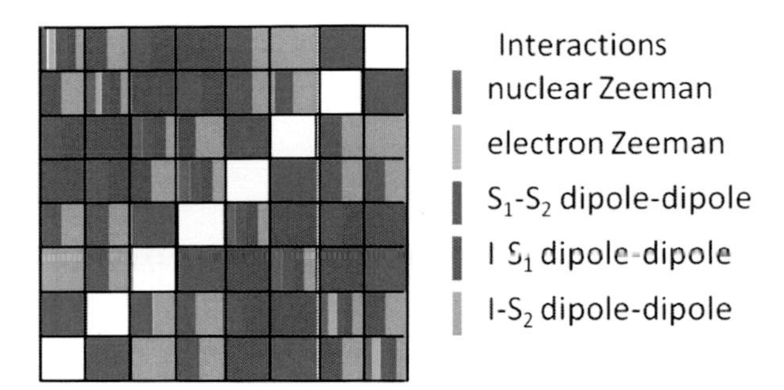

Figure 11.6 Structure of the Hamiltonian matrix $[H_0\,(S,\,S_1,\,S_2)]$ of Eq. 11.19. Different colors show the contributions of the various spin interactions to the individual elements of the Hamiltonian matrix.

The Hamiltonian matrix $[H_0\,(S,\,S_1,\,S_2)]$ in the Zeeman basis $\{|n\rangle = |m_{\mathrm{I}},\, m_{\mathrm{S}_1},\, m_{\mathrm{S}_2}\rangle\}$ has the structure shown in Fig. 11.6.

Assuming that the Larmor frequencies of both electron spins are identical, the Hamiltonian matrix elements yield:

$$[H_0]\,(1,\,1) = \frac{1}{2}\omega_{\mathrm{I}} + \omega_{\mathrm{S}} + \frac{1}{2\sqrt{6}}F_0^{(1)} + \frac{1}{2\sqrt{6}}F_0^{(2)} + \frac{1}{2\sqrt{6}}F_0^{(e)} \tag{11.20a}$$

$$[H_0]\,(1,\,2) = -\,[H_0]\,(7,\,8) = -\frac{1}{4}F_{-1}^{(1)} - \frac{1}{4}F_{-1}^{(2)} \tag{11.20b}$$

$$[H_0]\,(1,\,3) = -\,[H_0]\,(6,\,8) = -\frac{1}{4}F_{-1}^{(1)} - \frac{1}{4}F_{-1}^{(e)} \tag{11.20c}$$

$$[H_0]\,(1,\,4) = -\,[H_0]\,(5,\,8) = -\frac{1}{2}F_{-2}^{(1)} \tag{11.20d}$$

$$[H_0]\,(1,\,5) = -\frac{1}{4}F_{-1}^{(2)} - \frac{1}{4}F_{-1}^{(e)} \tag{11.20e}$$

$$[H_0]\,(1,\,6) = [H_0]\,(3,\,8) = \frac{1}{2}F_{-2}^{(2)} \tag{11.20f}$$

$$[H_0]\,(1,\,7) = -\,[H_0]\,(2,\,8) = \frac{1}{2}F_{-2}^{(e)} \tag{11.20g}$$

$$[H_0](2,2) = \frac{1}{2}\omega_\mathrm{I} + \omega_\mathrm{S} - \frac{1}{2\sqrt{6}}F_0^{(1)} - \frac{1}{2\sqrt{6}}F_0^{(2)} + \frac{1}{2\sqrt{6}}F_0^{(e)} \tag{11.20h}$$

$$[H_0](2,3) = -[H_0](6,7) = \frac{1}{2\sqrt{6}}F_0^{(1)} \tag{11.20i}$$

$$[H_0](2,4) = [H_0](5,7) = \frac{1}{4}F_{-1}^{(1)} - \frac{1}{4}F_{-1}^{(e)} \tag{11.20j}$$

$$[H_0](2,5) = -[H_0](4,7) = -\frac{1}{2\sqrt{6}}F_0^{(2)} \tag{11.20k}$$

$$[H_0](2,6) = \frac{1}{4}F_{-1}^{(2)} - \frac{1}{4}F_{-1}^{(e)} \tag{11.20l}$$

$$[H_0](3,3) = \frac{1}{2}\omega_\mathrm{I} - \frac{1}{2\sqrt{6}}F_0^{(1)} + \frac{1}{2\sqrt{6}}F_0^{(2)} - \frac{1}{2\sqrt{6}}F_0^{(e)} \tag{11.20m}$$

$$[H_0](3,4) = [H_0](5,6) = \frac{1}{4}F_{-1}^{(1)} - \frac{1}{4}F_{-1}^{(2)} \tag{11.20n}$$

$$[H_0](3,5) = [H_0](4,6) = \frac{1}{2\sqrt{6}}F_0^{(e)} \tag{11.20o}$$

$$[H_0](3,7) = -\frac{1}{4}F_{-1}^{(2)} + \frac{1}{4}F_{-1}^{(e)} \tag{11.20p}$$

$$[H_0](4,4) = \frac{1}{2}\omega_\mathrm{I} + \frac{1}{2\sqrt{6}}F_0^{(1)} + \frac{1}{2\sqrt{6}}F_0^{(2)} - \frac{1}{2\sqrt{6}}F_0^{(e)} \tag{11.20q}$$

$$[H_0](4,8) = \frac{1}{4}F_{-1}^{(2)} + \frac{1}{4}F_{-1}^{(e)} \tag{11.20r}$$

$$[H_0](5,5) = \frac{1}{2}\omega_\mathrm{I} + \frac{1}{2\sqrt{6}}F_0^{(1)} - \frac{1}{2\sqrt{6}}F_0^{(2)} - \frac{1}{2\sqrt{6}}F_0^{(e)} \tag{11.20s}$$

$$[H_0](6,6) = -\frac{1}{2}\omega_\mathrm{I} - \frac{1}{2\sqrt{6}}F_0^{(1)} + \frac{1}{2\sqrt{6}}F_0^{(2)} - \frac{1}{2\sqrt{6}}F_0^{(e)} \tag{11.20t}$$

$$[H_0](7,7) = \frac{1}{2}\omega_\mathrm{I} - \omega_\mathrm{S} - \frac{1}{2\sqrt{6}}F_0^{(1)} - \frac{1}{2\sqrt{6}}F_0^{(2)} + \frac{1}{2\sqrt{6}}F_0^{(e)} \tag{11.20u}$$

$$[H_0](8,8) = \frac{1}{2}\omega_I - \omega_S + \frac{1}{2\sqrt{6}}F_0^{(1)} + \frac{1}{2\sqrt{6}}F_0^{(2)} + \frac{1}{2\sqrt{6}}F_0^{(e)} \tag{11.20v}$$

The elements not listed above are zero. The functions, $F_m^{(e)}$, $F_m^{(1)}$, $F_m^{(2)}$ correspond to the S_1–S_2, I–S_1, and I–S_2 dipole-dipole interactions, respectively. Choosing the S_1–S_2 dipole–dipole axis as the molecular axis (M), the laboratory forms of the dipole-dipole interactions yield:

$$\begin{aligned} H_{\mathrm{DD}}^{(\mathrm{L})}(S_1, S_2) &= a_{\mathrm{DD}}^{S_1-S_2} \sum_{m=-2}^{2} (-1)^m F_m^{(e)} T_m^2 (S_1, S_2) \\ &= a_{\mathrm{DD}}^{S_1-S_2} \sum_{m=-2}^{2} (-1)^m D_{0-m}^2 (\Omega_{\mathrm{ML}}) T_m^2 (S_1, S_2) \end{aligned} \tag{11.21}$$

$$\begin{aligned} H_{\mathrm{DD}}^{(\mathrm{L})}(I, S_2) &= a_{\mathrm{DD}}^{\mathrm{I}-S_1} \sum_{m=-2}^{2} (-1)^m F_m^{(1)} T_m^2 (I, S_1) \\ &= a_{\mathrm{DD}}^{\mathrm{I}-S_1} \sum_{m=-2}^{2} (-1)^m \left(\sum_{k=-2}^{2} D_{0\mathrm{k}}^2 (\Omega_{\mathrm{I}-S_1,\mathrm{M}}) D_{k-m}^2 (\Omega_{\mathrm{ML}}(t)) \right) \\ &\quad T_m^2 (I, S_1) \end{aligned} \tag{11.22}$$

$$\begin{aligned} H_{\mathrm{DD}}^{(\mathrm{L})}(I, S_2) &= a_{\mathrm{DD}}^{\mathrm{I}-S_2} \sum_{m=-2}^{2} (-1)^m F_m^{(2)} T_m^2 (I, S_2) \\ &= a_{\mathrm{DD}}^{I-S_2} \sum_{m=-2}^{2} (-1)^m \left(\sum_{k=-2}^{2} D_{0\mathrm{k}}^2 (\Omega_{\mathrm{I}-S_2,\mathrm{M}}) D_{k-m}^2 (\Omega_{\mathrm{ML}}(t)) \right) \\ &\quad T_m^2 (I, S_2) \end{aligned} \tag{11.23}$$

where the angles $\Omega_{\mathrm{I}-S_1}$ and $\Omega_{\mathrm{I}-S_2}$ describe the orientation of the I–S_1 and I–S_2 dipole-dipole axes with respect to the S_1–S_2 axis. Using this representation of the Hamiltonian $[H_0(I, S_1, S_2)]$ and taking into account that now the $\langle I_Z \rangle$ quantity is given as:

$$\langle I_Z \rangle \propto |1\rangle\langle 1| + |3\rangle\langle 3| + |5\rangle\langle 5| + |7\rangle\langle 7| - |2\rangle\langle 2| - |4\rangle\langle 4| - |6\rangle\langle 6| - |8\rangle\langle 8| \tag{11.24}$$

one can calculate the DNP factor for a spin system containing two electron spins $S_1 = S_2 = 1/2$ and one nuclear spin $I = 1/2$ following the procedure described in Section 11.1.

This approach can be straightforwardly extended to a larger spin system, containing, for instance two nuclear spins, $I_1 = 1/2$ and $I_2 = 1/2$ (one can think here about ^{1}H and ^{15}N nuclei) and two electron spins, $S_1 = S_2 = 1/2$. The total Hamiltonian, $H_0(I_1, I_2, S_1, S_2)$ is then given as:

$$\begin{aligned} H_0(I_1, I_2, S_1, S_2) = {} & H_Z(I_1) + H_Z(I_2) + H_Z(S_1) + H_Z(S_2) \\ & + H_{A,sc}(I_1, I_2) + H_{DD}(I_1, S_1) + H_{DD}(I_1, S_2) \\ & + H_{DD}(I_1, S_1) + H_{DD}(I_1, S_2) + H_{DD}(I_1, S_2) \end{aligned} \tag{11.25}$$

where $H_{A,sc}(I_1, I_2)$ denotes a scalar interaction between the nuclear spins (in the case of ^{1}H and ^{15}N nuclei, the scalar interaction is very relevant, see Chapter 6).

For another approaches to the DNP phenomenon, the reader is referred to [1–5].

References

1. Hu, K.-N., Yu, H., Swager, T. M. and Griffin, R. G. (2004). Dynamic nuclear polarization with biradicals. *J. Am. Chem. Soc.*, **126**, pp. 10844–10845.
2. Maly, T., Debelouchina, G. T., Bajaj, S. V., Hu, K.-N., Joo, Ch.-G., Mak-Jurkauskas, M. L., Sirigiri, J. R., van der Wel, P. C. A., Herzfeld, J., Temkin, R. J. and Griffin, R. G. (2008). Dynamic nuclear polarization at high magnetic fields. *J. Chem. Phys.*, **128**, pp. 052211–052230.
3. Petta, J. R., Taylor, J. M., Johnson, A. C., Yacoby, A., Lukin, M. D., Marcus, C. M., Hanson M. P. and Gossard, A. C. (2008). Dynamic nuclear polarization with single electron spins. *Phys. Rev. Lett.*, **100**, pp. 067601–067605.
4. Prisner, T. F. (2012). Dynamic Nuclear Polarization. In *NMR of Biomolecules: Towards Mechanistic Systems*. (Wiley Online Library), DOI: 10.1002/9783527644506.ch25.
5. Shimon, D., Hovav, J., Feintuch A., Goldfarb, D. and Vega, S. (2012). Dynamic nuclear polarization in the solid state: A transition between the cross effect and the solid effect. *Phys. Chem. Chem. Phys.*, **14**, pp. 5729–5743.

Chapter 12

Anisotropic and Internal Dynamics

In previous chapters, isotropic molecular tumbling has been discussed and the molecules have been treated as rigid spherical objects. These assumptions are very useful, but in many cases they lead to oversimplifications. In this chapter, the issue of anisotropic rotation described by a diffusion tensor is discussed. Moreover, a combined effect of internal dynamics and overall rotation of the entire molecule is considered.

12.1 Anisotropic Rotation

As already explained in the previous chapters, the dipole–dipole Hamiltonian $H_{\mathrm{DD}}(I, S)(t)$ takes the form:

$$H_{\mathrm{DD}}(I, S)(t) = a_{\mathrm{DD}} \sum_{m=-2}^{2} (-1)^m D^2_{0,-m}(\Omega(t))\, T^2_m(I, S) \qquad (12.1)$$

The Wigner rotation matrices, $D^2_{0,-m}(\Omega(t))$, describe the orientation of the I–S dipole–dipole axis (i.e., the molecular orientation as the dipolar axis is fixed in the molecule) with respect to an arbitrarily chosen direction (for instance, the direction of the

Understanding Spin Dynamics
Danuta Kruk

ISBN 978-981-4463-49-2 (Hardcover), 978-981-4463-50-8 (eBook)
www.panstanford.com

external magnetic field). When the molecule undergoes isotropic tumbling the rotational correlation function, $C^{(2)}_{m,\text{rot}}(t) = C_{\text{rot}}(t) = \left\langle D^2_{0,-m}(t)\, D^{*}_{0,-m}(0)\right\rangle$ is single exponential with the rank-two rotational correlation time, $\tau_{\text{rot}} \equiv \tau^{(2)}_{\text{rot}}$ (Eq. 2.35). As the correlation function is m-independent, the corresponding (Lorentzian) spectral density (Eq. 2.36) also does not dependent of m. The corresponding rank-one rotational correlation time, $\tau^{(1)}_{\text{rot}}$ is associated with the rank-one correlation function (dielectric spectroscopy):

$$C^{(1)}_{m,\text{rot}}(t) = \left\langle D^1_{0,-m}(t)\, D^{1^*}_{0,-m}(0)\right\rangle = \frac{1}{3}\exp\left(-\frac{t}{\tau^{(1)}_{\text{rot}}}\right) \tag{12.2}$$

where the factor 1/3 stems from $1/(2l+1)$ for $l = 1$. The exponential form of the correlation functions stems from the isotropic diffusion equation (Eq. 2.32) containing instead of diffusion tensor a single diffusion coefficient, and the form of the Hamiltonian (Eq. 12.1) in the case of $C_{\text{rot}}(t')$.

If the interacting spins, I and S belong to a molecule which undergoes anisotropic tumbling, the form of the rotational correlation function (spectral density) becomes considerably more complex. The first source of complexity is the generalized diffusion equation including the rotational diffusion tensor (Eq. 2.47). The second reason is the appropriate form of the interaction Hamiltonian ($H_{\text{DD}}(I, S)(t)$ in this case). A two-step transformation has to be applied to the Hamiltonian; an example has been described in Section 10.3. Let us begin with the molecular (M) frame representation of the dipole–dipole Hamiltonian (Eq. 2.12), which we invoke here for convenience:

$$H^{(\text{M})}_{\text{DD}}(I, S) = a_{\text{DD}} \sum_{m=-2}^{2} (-1)^m F^{2(\text{M})}_{-m} T^2_m(I, S) \tag{12.3}$$

where $F^{2(\text{M})}_0 = 1$ and $F^{2(\text{M})}_{\pm 1} = F^{2(\text{M})}_{\pm 2} = 0$. For isotropic molecular tumbling, this interaction has directly been transformed to the laboratory (L) frame. For anisotropic rotation, the Hamiltonian has to be, in the first step, transformed to principal axis system of the diffusion tensor (principal diffusion frame) (P_{D}), according to the general transformation rule:

$$H_{\mathrm{DD}}^{(\mathrm{P_D})}(I,S) = a_{\mathrm{DD}} \sum_{m=-2}^{2} (-1)^m \left(\sum_{k=-2}^{2} F_k^{2(\mathrm{M})} D_{k-m}^2 \left(\Omega_{\mathrm{MP_D}} \right) \right) T_m^2 (I,S)$$

$$= a_{\mathrm{DD}} \sum_{m=-2}^{2} (-1)^m D_{0-m}^2 \left(\Omega_{\mathrm{MP_D}} \right) T_m^2 (I,S) \qquad (12.4)$$

where the set of Euler angles $\Omega_{\mathrm{MP_D}} \equiv (\alpha_{\mathrm{D}}, \beta_{\mathrm{D}}, \gamma_{\mathrm{D}})$ describes the orientation of the dipole-dipole axis relative to the principal diffusion frame; the representation of the dipolar Hamiltonian in the (P_{D}) frame, $H_{\mathrm{DD}}^{(P_{\mathrm{D}})}(I,S)$ is still time independent (the angle $\Omega_{\mathrm{MP_D}}$ is determined by the molecular geometry). Then, in the second step, the Hamiltonian is transformed from the (P_{D}) frame to the laboratory (L) frame, yielding:

$$H_{\mathrm{DD}}^{(\mathrm{L})}(I,S)(t) = a_{\mathrm{DD}} \sum_{m=-2}^{2} (-1)^m \times \left(\sum_{k=-2}^{2} D_{0k}^2 \left(\Omega_{\mathrm{MP_D}} \right) D_{k-m}^2 \left(\Omega_{\mathrm{P_D L}}(t) \right) \right) T_m^2 (I,S) \qquad (12.5)$$

where the angle $\Omega_{\mathrm{P_D L}}(t)$ between the (P_{D}) and laboratory (L) frames is modulated by the anisotropic rotation.

Taking into account the form of the dipolar Hamiltonian (Eq. 12.5), one sees that now the rotational correlation function is defined as:

$$C_{m.\mathrm{rot}}^{(2)}(t) = \left\langle \left[\sum_{k=-2}^{2} D_{0k}^2 \left(\Omega_{\mathrm{MP_D}} \right) D_{k-m}^2 \left(\Omega_{\mathrm{P_D L}}(0) \right) \right] \times \left[\sum_{k=-2}^{2} D_{0k'}^{2^*} \left(\Omega_{\mathrm{MP_D}} \right) D_{k'-m}^{2^*} \left(\Omega_{\mathrm{P_D L}}(t) \right) \right] \right\rangle \qquad (12.6)$$

This expression can be factorized as follows:

$$C_{m.\mathrm{rot}}^{(2)}(t) = \sum_{k=-2}^{2} \sum_{k'=-2}^{2} D_{0k}^2 \left(\Omega_{\mathrm{MP_D}} \right) D_{0k'}^{2^*} \left(\Omega_{\mathrm{MP_D}} \right) \times \left\langle D_{k-m}^2 \left(\Omega_{\mathrm{P_D L}}(0) \right) D_{k'-m}^{2^*} \left(\Omega_{\mathrm{P_D L}}(t) \right) \right\rangle \qquad (12.7)$$

The explicit form of the resulting correlation function:

$$C_{m,\mathrm{rot}}^{(2)}(t) = \left\langle D_{k-m}^2 \left(\Omega_{\mathrm{P_D L}}(0) \right) D_{k'-m}^{2^*} \left(\Omega_{\mathrm{P_D L}}(t) \right) \right\rangle \qquad (12.8)$$

can directly be obtained from the definition of Eq. 2.30 (in analogy to Eq. 2.31). The conditional probability density, $P(\Omega, \Omega_0, t)$, being a solution of Eq. 2.47 (neglecting the translational terms) has now the form:

$$P(\Omega, \Omega_0, t) = \sum_{l,p,n,n'} \frac{2l+1}{4\pi} \left(\xi^l_{p,n,n'}\right)^2 D^l_{nn'}(\Omega)\, D^{1*}_{nn'}(\Omega_0) \exp\left(-\lambda_{l,p} t\right) \tag{12.9}$$

where $\lambda_{l,p}$ are eigenvalues of the anisotropic rotational diffusion operator (Eq. 2.47), while the functions $\Lambda_{I,m} = \frac{2l+1}{4\pi} \sum_{nn'} \left(\xi^l_{p,n,n'}\right)^2 D^l_{nn'}(\Omega)\, D^{l*}_{nn'}(\Omega_0)$ are the corresponding eigenfunctions. This leads to the following form of the correlation function [1, 6–8]:

$$C^{(2)}_{m.\mathrm{rot}}(t) = C^{(2)}_{m.\mathrm{rot}}(t) \sum_{k=-2}^{2} \sum_{k'=-2}^{2} D^2_{0k}(\Omega_{\mathrm{MP_D}})\, D^{2*}_{0k'}(\Omega_{\mathrm{MP_D}}) \times \left\{ \sum_{p=2}^{2} \zeta_{kp}\zeta_{k'p} \exp\left(-\lambda_{-2,p} t\right) \right\} \tag{12.10}$$

In consequence, the corresponding spectral density $J^{(2)}_{m.\mathrm{rot}}(\omega)$ is given as a sum of Lorentzian functions:

$$J^{(2)}_{\mathrm{rot}}(\omega) = \sum_{k=-2}^{2} \sum_{k'=-2}^{2} D^2_{0k}(\Omega_{\mathrm{MP_D}})\, D^{2*}_{0k'}(\Omega_{\mathrm{MP_D}}) \times \left\{ \sum_{p=2}^{2} \varsigma_{kp}\varsigma_{k'p} \frac{\left(\lambda_{2,p}\right)^{-1}}{1 + \left(\omega \left(\lambda_{2,p}\right)^{-1}\right)^2} \right\} \tag{12.11}$$

The eigenvalues, $\lambda_{2,p}$, and the elements of the corresponding eigenfunctions, ς_{pk}, are defined as follows [1, 6–8]:

$$\lambda_{2,0} = 6D + 6\sqrt{D^2 - \tilde{D}^2}, \quad \varsigma_{00} = \sin(\chi/2), \; \varsigma_{02} = \varsigma_{0-2} = \cos(\chi/2)/\sqrt{2} \tag{12.12a}$$

$$\lambda_{2,1} = 4D_{xx} + D_{yy} + D_{zz}, \quad \varsigma_{11} = \varsigma_{1-1} = 1/\sqrt{2} \tag{12.12b}$$

$$\lambda_{2,-1} = D_{xx} + 4D_{yy} + D_{zz}, \quad \varsigma_{-1-1} = -\varsigma_{-11} = -1/\sqrt{2} \tag{12.12c}$$

$$\lambda_{2,2} = 6D - 6\sqrt{D^2 - \tilde{D}^2},$$
$$\varsigma_{20} = \cos(\chi/2), \quad \varsigma_{2-2} = \varsigma_{22} = \sin(\chi/2)/\sqrt{2} \qquad (12.12\text{d})$$

$$\lambda_{2,-2} = D_{xx} + D_{yy} + 4D_{zz}, \quad \varsigma_{-2-2} = \varsigma_{-22} = -1/\sqrt{2} \qquad (12.12\text{e})$$

where $D = \frac{1}{3}(D_{xx} + 4D_{yy} + D_{zz})$, $\tilde{D}^2 = \frac{1}{3}(D_{xx}D_{yy} + D_{yy}D_{zz} + D_{zz}D_{xx})$, and $\chi = \tan^{-1}\left[\frac{\sqrt{3(D_{xx}-D_{yy})^2}}{2D_{zz}-D_{xx}-D_{yy}}\right]$; D_{xx}, D_{yy}, and D_{zz} are Cartesian components of the anisotropic diffusion tensor. The spectral density of Eq. 12.11 can considerably be simplified for axially symmetric diffusion tensor and axially symmetric interactions (like the dipole–dipole coupling) [1, 6–8]:

$$J^{(2)}_{\text{rot}}(\omega) = \left(\frac{3\cos^2\beta - 1}{2}\right)^2 \frac{6D_\perp}{(6D_\perp)^2 + \omega^2}$$
$$+3\cos^2\beta\sin^2\beta \frac{5D_\perp + D_\parallel}{(5D_\perp + D_\parallel)^2 + \omega^2}$$
$$+\frac{3}{4}\sin^4\beta \frac{2D_\perp + 4D_\parallel}{(2D_\perp + 4D_\parallel)^2 + \omega^2} \qquad (12.13)$$

where $\beta = \beta_{\text{MP}_\text{D}}$ is the angle between the dipole–dipole axis and the diffusion axis. The spectral density can be written in terms of correlation times defined as [1, 6–8]:

$$\left(\tau^{(2)}_{\text{m}}\right)^{-1} = 6D_\perp + m^2\left(D_\parallel - D_\perp\right) \qquad (12.14)$$

It yields:

$$J^{(2)}_{\text{rot}}(\omega) = \left(\frac{3\cos^2\beta - 1}{2}\right)^2 \frac{\tau^{(2)}_0}{1 + \left(\omega\tau^{(2)}_0\right)^2}$$
$$+3\cos^2\beta\sin^2\beta \frac{\tau^{(2)}_1}{1 + \left(\omega\tau^{(2)}_1\right)^2} + \frac{3}{4}\sin^4\beta \frac{\tau^{(2)}_2}{1 + \left(\omega\tau^{(2)}_2\right)^2} \qquad (12.15)$$

When the dipole–dipole and the diffusion tensor axes coincide ($\beta = 0$), Eq. 12.15 reduces to the well-known form of Eq. 2.36, as $\tau^{(2)}_{\text{rot}} = \tau_{\text{rot}}$.

In Chapter 8 the issue of quadrupolar lineshape depending on the model of motion has been discussed. The theory presented in

Chapter 8 is valid for arbitrary motional conditions, also beyond the perturbation range. This has been achieved by constructing a basis set (in principle infinite) including spin- and dynamical-variables (for instance rotational states) characterizing the system (Eq. 8.2). Then, the Liouville operator, L_{rot}, describing rotational diffusion of the molecule ($L_{\text{rot}} = i D_{\text{rot}} \nabla^2_{\Omega}$, where $D_{\text{rot}} = 1/6\tau_{\text{rot}}$) has been represented in the generalized basis using Wigner–Eckart theorem. For isotropic rotation the representation is given by Eq. 8.14. This equation can be generalized to the case of anisotropic rotation, for $\beta = 0$:

$$\left(L'K'M' \middle| \left(\Sigma'\sigma' \, | L_{\text{rot}} || LKM \right) | \Sigma\sigma \right)$$
$$= \delta_{LL'}\delta_{KK'}\delta_{MM'}\delta_{\Sigma\Sigma'}\delta_{\sigma\sigma'} i \left[D_{\perp} L(L+1) + \left(D_{\parallel} - D_{\perp} \right) M^2 \right] \quad (12.16)$$

12.2 Internal Dynamics

So far it has been assumed that molecules do not exhibit internal dynamics, that is, they behave like a rigid body. Real molecules include various "molecular units", like for instance methyl groups. The motion of such groups is a result of their own fast dynamics mediated by the overall molecular tumbling. Internal dynamics can also lead to changes in the inter-spin distance as the molecule can change its shape.

Let us begin with the simpler case of internal and overall rotation (in such a case one does not need to bother about the fluctuations of the inter-spin distance due to the internal motion). As usually one has to begin with determining a molecular (molecule fixed) axis (M). Then, the I–S dipole–dipole Hamiltonian, $H^{(\text{L})}_{\text{DD}}(I, S)(t)$, has a form which results from the two-step transformation; from the dipolar frame (D) (determined by the I–S dipole–dipole axis) to the (M) frame and then from the (M) frame to the laboratory (L) frame:

$$H^{(\text{L})}_{\text{DD}}(I, S)(t) = a_{\text{DD}} \sum_{m=-2}^{2} (-1)^m$$
$$\times \left(\sum_{k=-2}^{2} D^2_{0k}\left(\Omega_{\text{DM}}(t)\right) D^2_{k-m}\left(\Omega_{\text{ML}}(t)\right) \right) T^2_m(I, S) \quad (12.17)$$

Then, the correlation function, $C^{(2)}_{m,\text{tot}}(t)$, representing the combined effect of the internal dynamics and the overall relation, is given as:

$$C^{(2)}_{m,\text{tot}}(t) = \left\langle \left[\sum_{k=-2}^{2} D^{2}_{0k}(\Omega_{\text{DM}}(0))\, D^{2}_{k-m}(\Omega_{\text{ML}}(0)) \right] \times \left[\sum_{k'=-2}^{2} D^{2*}_{0k'}(\Omega_{\text{DM}}(t))\, D^{2*}_{k'-m}(\Omega_{\text{ML}}(t)) \right] \right\rangle \quad (12.18)$$

where the angle $\Omega_{\text{DM}}(t)$ is modulated by the internal (local) dynamics, why the angle $\Omega_{\text{ML}}(t)$ fluctuates in time due to the overall (global) motion. Assuming that, the internal and global dynamics are stochastically uncorrelated due to time scale separation (the internal dynamics is usually much faster) or due to different physical origins. This implies that the correlation function can be factorized into a product of two correlation functions describing the local and the global dynamics, respectively:

$$C^{(2)}_{m,\text{tot}}(t) = C_{\text{local}}(t)\, C_{m,\text{global}} = \left\langle \left[\sum_{k,k'=-2}^{2} D^{2}_{0k}(\Omega_{\text{DM}}(0))\, D^{2*}_{0k}(\Omega_{\text{DM}}(t)) \right] \right\rangle \times \left\langle \left[\sum_{k,k'=-2}^{2} D^{2}_{k-m}(\Omega_{\text{ML}}(0))\, D^{2*}_{k'-m}(\Omega_{\text{ML}}(t)) \right] \right\rangle \quad (12.19)$$

When the local rotation is isotropic, the function of Eq. 12.19 can be further simplified as only terms with $k = k'$ contribute to the expression. This also applies to some cases of restricted diffusion. When the local dynamics is restricted (due to, for instance, geometrical constrains), the correlation function $C_{\text{local}}(t)$ does not decay to zero, but to a value denoted as S^2 [5, 6]:

$$C_{\text{local}}(t) = \left(1 - S^2\right) \exp\left(-t/\tau_{\text{local}}\right) + S^2 = \left(1 - S^2\right) \tilde{c}(t) + S^2 \quad (12.20)$$

In Eq. 12.20, it has been assumed that the local correlation function decays in a single exponential way, with a correlation time, τ_{loc}. The parameter, S^2 is referred to as "order parameter"; for $S^2 = 0$ the function decays to zero. Nevertheless, the correlation function $\tilde{c}(t)$ can take more complex form, for instance, stretched exponential

(Eq. 2.50 Chapter 2). The local dynamics can also be modeled as, for instance, anisotropic tumbling. In such a case the correlation function $\tilde{c}(t)$ takes the form [5–8]:

$$\tilde{c}(t) = \left(\frac{3\cos^2\beta - 1}{2}\right)^2 \exp\left(-\frac{\tau_0^{(2)}}{t}\right) + 3\cos^2\beta\sin^2\beta\exp\left(-\frac{\tau_1^{(2)}}{t}\right) + \frac{3}{4}\sin^4\beta\left(-\frac{\tau_2^{(2)}}{t}\right) \quad (12.21)$$

As far as restricted diffusion is concerned, it is of interest to consider diffusion on a cone; a situation when the dipole–dipole I - S axis moves on a cone determined by a semi-angle $\tilde{\beta}$. It has been shown that the spectral density of Eq. 12.21 also describes the diffusion on a cone for $\beta = \tilde{\beta}$ [10]. Independently of the form of the local correlation function, $\tilde{c}(t)$, the total correlation function, $C_{m,\mathrm{tot}}^{(2)}(t)$ is given as [5, 6]:

$$\begin{aligned} C_{m,\mathrm{tot}}^{(2)}(t) &= \left\lfloor\left(1 - S^2\right)\tilde{c}(t) + S^2\right\rfloor C_{m,\mathrm{global}}(t) \\ &= \left(1 - S^2\right)\tilde{c}(t)\,C_{m,\mathrm{global}}(t) + S^2 C_{m,\mathrm{global}}(t) \end{aligned} \quad (12.22)$$

If the global correlation function, $C_{m,\mathrm{tot}}^{(2)}(t)$ is also single exponential with a correlation time, τ_{global}: $C_{\mathrm{m,global}}^{(2)}(t) = C_{\mathrm{global}}^{(2)}(t) = \exp\left(-t/\tau_{\mathrm{global}}\right)$, the total correlation function, $C_{m,\mathrm{tot}}^{(2)}(t) = C_{\mathrm{tot}}^{(2)}(t)$ yields:

$$C_{\mathrm{tot}}^{(2)}(t) = \left(1 - S^2\right)\exp\left(-t/\tau_{\mathrm{eff}}\right) + S^2\exp\left(-t/\tau_{\mathrm{global}}\right) \quad (12.23)$$

where $\tau_{\mathrm{eff}}^{-1} = \tau_{\mathrm{local}}^{-1} + \tau_{\mathrm{global}}^{-1}$ Eq. 12.23 is referred to as "model-free" Lipari–Szabo approach [10, 11]. As in most cases the condition $\tau_{\mathrm{loc}} \ll \tau_{\mathrm{global}}$ is fulfilled, Eq. 12.23 simplifies to:

$$C_{\mathrm{tot}}^{(2)}(t) = \left(1 - S^2\right)\exp\left(-t/\tau_{\mathrm{local}}\right) + S^2\exp\left(-t/\tau_{\mathrm{global}}\right) \quad (12.24)$$

which implies the following form of the corresponding spectral density:

$$J_{\mathrm{rot}}^{(2)}(\omega) = \left(1 - S^2\right)\frac{\tau_{\mathrm{local}}}{1 + \left(\omega\tau_{\mathrm{local}}\right)} + S^2\frac{\tau_{\mathrm{global}}}{1 + \left(\omega\tau_{\mathrm{global}}\right)^2} \quad (12.25)$$

One should realize here that the global correlation function, $C_{m,\mathrm{global}}^{(2)}(t)$, can also describe a complex motion, like anisotropic rotation.

This theory has also its general counterpart, valid for arbitrary motional conditions. In the presence of internal dynamics the basis set $|O_i) = |LKM) \otimes |\Sigma\sigma)$ (Eq. 8.2) has to be extended by introducing degrees of freedom for the internal dynamics, $|\tilde{L}\tilde{K}\tilde{M})$ which implies that [2–4]:

$$|O_i) = |\tilde{L}\tilde{K}\tilde{M}) \otimes |LKM) \otimes |\Sigma\sigma) \qquad (12.26)$$

where $|\tilde{L}\tilde{K}\tilde{M}) = |\tilde{L}\tilde{K}\tilde{M}\rangle\langle\tilde{L}\tilde{K}\tilde{M}| = \sqrt{\frac{2L+1}{8\pi^2}} D^{\tilde{L}}_{\tilde{K}\tilde{M}}(\Omega)$ (in analogy to Eq. 8.3). This implies that the basis becomes considerably larger. Let us assume that both the internal as well as the overall dynamics can be modeled as isotropic rotation represented by the Liouville operators $L_{\text{rot,local}} = i D_{\text{rot,local}} \nabla^2_{\Omega_{\text{DM}}}$ and $L_{\text{rot,global}} = i D_{\text{rot,global}} \nabla^2_{\Omega_{\text{ML}}}$, respectively. Then, in analogy to Eq. 8.14 one can write the representation of these operators in the basis $|O_i)$

$$\begin{aligned}&(\tilde{L}'\tilde{K}'\tilde{M}'|(L'K'M'|\Sigma'\sigma'|L_{\text{rot,local}}|\tilde{L}\tilde{K}M)|LKM\ |\Sigma\sigma)\\&= \delta_{\tilde{L}\tilde{L}'}\delta_{\tilde{K}\tilde{K}'}\delta_{\tilde{M}\tilde{M}'}\delta_{LL'}\delta_{KK'}\delta_{MM'}\delta_{\Sigma\Sigma}\delta_{\sigma\sigma'} i\left[(1-S^2)D_{\text{rot,local}}\tilde{L}(\tilde{L}+1)+S^2\right]\end{aligned} \qquad (12.27)$$

where $\tau_{\text{rot,local}} = 1/6D_{\text{rot,local}}$, while

$$\begin{aligned}&(\tilde{L}'\tilde{K}'\tilde{M}'|\ (L'K'M'|\ (\Sigma'\sigma'|\ L_{\text{rot,global}}|\tilde{L}\tilde{K}M)\ |LKM)\ |\Sigma\sigma)\\&= \delta_{\tilde{L}\tilde{L}'}\delta_{\tilde{K}\tilde{K}'}\delta_{\tilde{M}\tilde{M}'}\delta_{LL'}\delta_{KK'}\delta_{MM'}\delta_{\Sigma\Sigma'}\delta_{\sigma\sigma'} i\, D_{\text{rot,global}} L(L+1)\end{aligned} \qquad (12.28)$$

where $\tau_{\text{rot,global}} = 1/6D_{\text{rot,global}}$.

As already anticipated, the internal dynamics can also lead to changes in the interspin distance. In such a case the internal correlation function should have a form analogous to Eq. 2.43:

$$C_{\text{local}}(t) = \sum_{k,k'=-2}^{2} \left\langle \frac{D^2_{0,k}(\Omega_{\text{DM}}(t))}{r^3(t)} \frac{D^{2*}_{0,k'}(\Omega_{\text{DM}}(0))}{r^3(0)} \right\rangle \qquad (12.29)$$

The interspin distance can change due to, for instance, local vibrational dynamics, which can be described in terms of normal modes (internal coordinates) of the molecule. The vibrational dynamics can be considered in terms of classical physics by solving diffusion equation for harmonic oscillator potential or in a quantum-mechanical way [2, 3].

References

1. Grant, D. M. and Brown, R. A. (1996) Encyclopedia of nuclear magnetic resonance, relaxation of coupled spins from rotational diffusion (Chichester: Wiley), pp. 4003–4018.
2. Kruk, D. and Kowalewski, J. (2002). Vibrational motion and nuclear spin relaxation in paramagnetic complexes: Hexaaquonickel (II) as an example. *J. Chem. Phys.*, **116**, PP. 4079–4086.
3. Kruk, D., Kowalewski, J. and Westlund, P. O. (2004). Nuclear and electron spin relaxation in paramagnetic complexes in solution: Effects of the quantum nature of molecular vibrations. *J. Chem. Phys.*, **121**, PP. 2215–2220.
4. Kruk, D. (2007). *Theory of Evolution and Relaxation of Multi-Spin Systems* (Bury St Edmunds: Arima).
5. Lipari, G. and Szabo, A. (1982). Model-free approach to the interpretation of nuclear magnetic resonance relaxation in macromolecules 1. Theory and range of validity. *J. Am. Chem. Soc.*, **104**, pp. 4546–4559.
6. Kowalewski, J. and Mäler, L. (2006). *Nuclear Spin Relaxation in Liquids: Theory, Experiments and Applications* (New York: Taylor & Francis).
7. Woessner, D. E. (1962). Nuclear spin relaxation in ellipsoids undergoing rotational motion. *J. Chem. Phys.*, **37**, pp. 647–654.
8. Woessner, D. E. (1996). *Encyclopedia of Nuclear Magnetic Resonance, Brownian Motion and Correlation Times* (Chichester: Wiley), pp. 1068–1084.

Index